冰河技术丛书

深入理解高并发编程

JDK核心技术

冰河（@冰河技术）◎著

电子工业出版社
Publishing House of Electronics Industry
北京·BEIJING

内 容 简 介

本书从实际需求出发，全面细致地介绍了JDK高并发编程的基础知识、核心工具和线程池核心技术。通过阅读和学习本书，读者可以对JDK高并发编程有更加全面、深入、透彻的理解，提高对高并发编程问题的处理能力和项目实战能力，并提高站在更高层面解决高并发编程系统架构问题的能力。

本书适合互联网行业从业人员、高校师生阅读，尤其适合中高级开发人员、架构师、技术经理及技术专家阅读，也适合对高并发编程感兴趣的人员阅读。

未经许可，不得以任何方式复制或抄袭本书之部分或全部内容。
版权所有，侵权必究。

图书在版编目（CIP）数据

深入理解高并发编程：JDK 核心技术 / 冰河著. —北京：电子工业出版社，2023.2
ISBN 978-7-121-44933-8

Ⅰ.①深… Ⅱ.①冰… Ⅲ.①JAVA 语言一程序设计 Ⅳ.①TP312.8

中国国家版本馆 CIP 数据核字（2023）第 016322 号

责任编辑：张　晶
印　　刷：三河市君旺印务有限公司
装　　订：三河市君旺印务有限公司
出版发行：电子工业出版社
　　　　　北京市海淀区万寿路 173 信箱　邮编：100036
开　　本：787×980　1/16　印张：25.25　字数：565.6 千字
版　　次：2023 年 2 月第 1 版
印　　次：2023 年 2 月第 1 次印刷
定　　价：129.00 元

凡所购买电子工业出版社图书有缺损问题，请向购买书店调换。若书店售缺，请与本社发行部联系，联系及邮购电话：(010) 88254888，88258888。
质量投诉请发邮件至 zlts@phei.com.cn，盗版侵权举报请发邮件至 dbqq@phei.com.cn。
本书咨询联系方式：(010) 51260888-819，faq@phei.com.cn。

推荐语

（排名不分先后）

冰河是 CSDN 的专家博主，专注于高并发编程领域，本书是他的高并发系列图书的第 2 本，系统梳理了 JDK 并发编程的工具类和线程池的源码，给出了丰富的实战案例。高并发是现代开发者要面对的主要场景，本书深入分析 JDK 高并发的实践案例，帮助 Java 开发者深入理解并熟练掌握高并发技术，以应对日益复杂的高并发场景。

——CSDN 创始人、总裁　蒋涛

冰河是 CSDN 的资深专家博主，多年来一直在 CSDN 输出高质量技术文章。当今开发者的技术栈和开发模式都在走向云原生，云原生技术的核心是分布式系统。

冰河这次出版的《深入理解高并发编程：JDK 核心技术》是继《深入理解高并发编程：核心原理与案例实战》后的又一本高并发领域佳作。

在高并发实践中，各种与时序相关的 bug 会让工程师感到棘手，这些问题很难重现并进行分析。本书的每个章节都配有 JDK 核心工具类的源码及实战案例，有助于读者解决实际工作中出现的各种问题。

本书使用大量篇幅剖析了 JDK 中线程池的核心源码，在随书源码中给出了完整的线程池的案例程序，我建议读者自己写代码进行探索，并和冰河直接交流。通过认真学习本书，读者可以"知其然并知其所以然"。

祝贺冰河又出版了一本国内这一领域的高质量原创图书，这本书非常值得读。希望冰河的这本书和相应的社区能把中国 JDK 工程师的水平提高一大截。

——CSDN 副总裁、《编程之美》《构建之法》作者　邹欣

CPU 是宝贵的资源，应得到有效利用。据统计，大约有 75%的 CPU 时间用于等待内存访问，使得 CPU 不能被充分利用，因此，在单核上进行并发编程提升 CPU 利用率成为一个难点。现代计算机具备多核（允许线程级并行，启用指令级并行），计算能力非常强悍，而并发编程也是提高多核 CPU 利用率的一个重要技术。无论在单核、多核、NUMA 等硬件条件下，并发编程技术都是一根"硬骨头"，工程师需要通过系统化地学习并不断参与实战锤炼技术。

冰河老师所著的《深入理解高并发编程：JDK 核心技术》基于 Java 语言，全面细致地介绍了 JDK 高并发编程的基础知识、核心工具，并通过大量的实践案例进行演示，是一本具备高实用价值的图书。期待该书能为读者开启 Java 语言领域的高并发编程大门，帮助读者掌握并发编程的精髓。

——腾讯数据库资深研究员、首席架构师、
《数据库查询优化器的艺术：原理解析与 SQL 性能优化》
《数据库事务处理的艺术：事务管理与并发控制》
《分布式数据库原理、架构与实践》作者　李海翔

　　得知冰河的新书《深入理解高并发编程：JDK 核心技术》即将出版，我拿来样章阅读了一下，这本书是继《深入理解高并发编程：核心原理与案例实战》后的又一佳作，全面并且细致地介绍了 JDK 并发编程的相关知识，从线程进程、调度、上下文切换，到 JDK 工具类的内核源码，再到线程池的核心源码，"手把手"地指导读者学习使用 Java 并发编程技术，让读者进行深层次的理解。

——阿里巴巴 JVM 技术专家、CCF 系统软件专委会执行委员　林子熠博士

　　本书系统讲述了 JDK 并发编程相关的类和工具，对 JDK 的并发工具类源码进行解析，并辅以案例，比单纯介绍 API 的书籍丰满许多。千里之行，始于足下，本书特别适合学习和工作时间较短的朋友阅读，有一定工作经验的朋友查阅本书也可以对工作有所助益。

——资深技术专家、公众号"技术琐话"作者、
《深入分布式缓存》《程序员的三门课》联合作者　于君泽

　　早些时候，冰河撰写的《深入理解高并发编程：核心原理与案例实战》一经出版，立刻引起了广泛的关注，此书从微观内核到并发应用，再到业务架构，层层深入地介绍高并发编程技术，读起来非常过瘾。广大读者热情高涨，一直呼吁赶紧出"续集"，这不，《深入理解高并发编程：JDK 核心技术》来了。

　　"续集"由浅入深，从线程进程、调度、上下文切换等概念开始，逐步过渡到 JDK 的各种阻塞非阻塞队列、锁工具、同步异步等工具集，最终深入 JDK 并发工具的内核源码，让读者能够知其所以然地深刻理解线程进程以及各种工具的核心原理、执行流程，以及设计思路与细节。

　　不管你是工程师、架构师、技术经理，又或者是一名对掌握高并发编程、JDK 内核技术有着执着追求的互联网从业人员，《深入理解高并发编程：JDK 核心技术》都值得一看。

——互联网架构专家、公众号"架构师之路"作者　沈剑

　　在当前这个分布式技术与多核系统被广泛应用的时代，掌握多线程和并发编程技术是设计和实现高性能应用程序的必经之道。而多线程和并发编程的复杂性使得全面掌握它们从来不是

一件简单的事儿。冰河的这本《深入理解高并发编程：JDK 核心技术》全面而系统地讲解了 JDK 中提供的多线程技术和并发工具类，从基础的线程到并发集合类和工具类，再到线程池的原理和应用都进行了深入浅出的分析，细节充实、诚意满满，推荐大家阅读学习。

——(kimmking) Apache Dubbo/ShardingSphere PMC、
长亮科技平台技术部副总经理、《高可用可伸缩微服务架构》合著作者　秦金卫

随着业务规模和复杂度的不断上升，即使做了很好的鲁棒性设计，也可能遇到诡异的并发问题，如果没有扎实的并发编程技术，是不容易定位并解决这些问题的，最终可能将其归为偶发问题或机器问题而不了了之。本书详尽地介绍了 Java 并发编程的核心技术，并辅以冰河工作中遇到的实际案例来帮助读者更好地理解和落地。大家可以体系化阅读冰河的并发编程系列图书，有助于更好地理解并发编程。

——《亿级流量网站架构核心技术》作者　张开涛

本书是继《深入理解高并发编程：核心原理与案例实战》后的又一佳作，用平实的语言和大量代码示例系统讲解了 Java 并发编程中的核心技术，侧重 JDK 层面并发工具的内核源码剖析与实战案例，同时深入系统地分析了线程池的核心原理和源码执行流程。这些也是广大 Java 开发需要具备的基本功，值得去花时间深入地研究与学习。

——Seata 开源社区创始人　季敏

近来搬家收拾图书，把非技术类的都当废纸卖了，年近不惑，对什么样的书真正对人的事业成长有帮助有了更深刻的体会。技术类图书专业性强、写作难度较大，一般生产周期较长，没想到冰河的《深入理解高并发编程：核心原理与案例实战》姐妹篇——《深入理解高并发编程：JDK 核心技术》这么快就来了，而且保持了一贯的高水准。

掌握 Java 高并发编程技术是进阶高级工程师的前提条件，对于不在高并发业务线的研发工程师而言，通过读书来获取高并发编程相关经验必不可少。本书聚焦高并发编程技术的底层——JDK 高并发编程的基础知识以及核心工具和线程池核心技术，对于有志于从事高并发编程相关工作的工程师是一个很好的参考。本书和冰河的《深入理解高并发编程：核心原理与案例实战》一起学习效果更佳，强烈推荐大家购买学习。

——杭州任你说智能科技 CTO　李鹏云

并发编程是所有 Java 程序员都必须学习和掌握的最基础且最重要的知识，向所有 Javaer 推荐冰河的新书《深入理解高并发编程：JDK 核心技术》。

——饿了前么技术总监、公众号"军哥手记"作者　程军

并发编程是 Java 进阶路上的难点之一，是大厂程序员的必备技能。本书作者详细地解读了

JDK 高并发编程的各项技术，同时引用了大量工作中的实战案例，对读者系统地学习 Java 并发编程有很好的指导意义，值得一看。

——京东零售架构师　骆俊武

目前市场上少有专业且系统地介绍 JDK 并发编程的图书，冰河的这本书填补了这个空缺，在互联网愈发成熟的今天，并发编程也变得愈发重要，几乎每一家互联网公司都不可避免地用到它。

考察一个 Java 工程师的能力，最重要的技术之一就是并发编程，因为并发编程中的知识点相对复杂，并且在项目中占据最核心的位置，一旦出现问题影响巨大。

冰河这本书的内容由浅入深，有案例、有图解、有源码，可以有效帮助处于不同阶段的 Java 程序员学习了解并发编程，值得推荐！

——公众号"纯洁的微笑"作者　纯洁的微笑

读完样书的前几个章节，我忍不住将此书推荐给技术团队。作者在高并发编程方面的功力相当深厚，从对线程与进程到各种容器的特点及使用方法的讲解，再到对并发工具/锁工具的介绍，以及对线程池的阐述，都能够深入浅出、娓娓道来，我将这本书推荐给每一位想要学习高并发编程的朋友！

——"顿悟山丘"咨询创始人、公众号"技术领导力"作者　黄哲铿/Mr.K

1996 年 5 月以来，随着互联网高速发展，Java 的复杂性也不断增加。本书通过严谨细致的描述和讲解将知识点化繁为简，详细地罗列了 JDK 的常用数据结构和使用方式，同时辅以大量的图文、Demo 代码，是一本不可多得的 handbook。建议初学者在学习时下载书中源码并运行，对照本书进行验证。

——Apache RocketMQ 北京社区联合发起人 && Commiter 李伟

高并发是 Java 开发领域的一个核心问题，Java 开发者在进阶过程中会遇到各种各样的高并发场景。由于高并发编程的特殊性，它不像普通的业务流程编程那么容易理解，甚至一些从事 Java 开发多年的开发者在编写高并发程序时也会犯错。如果您对于并发问题的处理也不那么得心应手，那么我推荐阅读本书。本书涵盖大量实战案例和源码，可以帮助您直观地理解并掌握 JDK 中的各种核心工具及线程池技术，助您轻松实现高性能的 Java 应用。

——公众号"程序猿 DD"维护者、《Spring Cloud 微服务实战》作者　翟永超

前　言

为什么要写这本书

随着计算机与互联网技术的不断发展，CPU 硬件的核心数也在不断提升，并发编程越来越普及，但是并发编程并不像其他业务那样简单明了。在编写并发程序时，往往会出现各种各样的 Bug，这些 Bug 常常以某种"诡异"的形式出现，然后迅速消失，并且在大部分场景下难以复现。所以，高并发编程着实是一项让程序员头疼的技术。在"冰河技术丛书"之"深入理解高并发编程"系列的第 1 部作品——《深入理解高并发编程：核心原理与案例实战》一书中，全面细致地介绍了高并发编程的基础知识、核心原理、实战案例和系统架构等内容，帮助读者从根本上理解并发编程出现各种 Bug 的根源，并从原理与实战层面找到解决问题的方案。

本书是"冰河技术丛书"之"深入理解高并发编程"系列的第 2 部作品，从实际需求出发，全面细致地介绍了 JDK 高并发编程的基础知识、核心工具和线程池核心技术。每个章节根据实际需要配有相关的原理图、流程图和实战案例。在线程池核心技术篇，还提供了完整的手动编写线程池的案例源码。

通过阅读和学习本书，读者可以更加全面、深入、透彻地理解 JDK 高并发编程知识，提高对高并发编程问题的处理能力和项目实战能力，并提高站在更高层面解决高并发编程系统架构问题的能力。

读者对象

- 互联网行业从业人员
- 高校师生
- 中高级开发人员
- 架构师
- 技术经理及技术专家
- 对高并发编程感兴趣的人员

本书特色

1. 系统介绍 JDK 高并发编程的图书

目前，图书市场少有全面细致地介绍有关 JDK 高并发编程的基础知识、核心工具和线程池核心技术的图书。本书从以上三方面入手，全面、细致并且层层递进地介绍了 JDK 高并发编程相关知识。

2. 大量图解和开发案例

为了便于理解，笔者在介绍 JDK 高并发编程的基础知识和核心工具章节中会配有适量的图解和图表，以及对应的实战案例。在线程池核心技术章节中会配有完整的手写线程池案例。读者按照本书的案例学习，并运行案例代码，能够更加深入地理解和掌握相关知识。另外，这些案例代码和图解的 draw.io 原文件会一起收录于随书资料里。读者也可以访问下面的链接，获取完整的实战案例源码和相关的随书资料。

- GitHub：https://github.com/binghe001/mykit-concurrent-jdk。
- Gitee：https://gitee.com/binghe001/mykit-concurrent-jdk。
- GitCode：https://gitcode.net/binghe001/mykit-concurrent-jdk。

3. 技术点与案例结合

对于 JDK 高并发编程的各项技术，书中都配有相关的典型案例，具有很强的实用性，方便读者随时查阅和参考。

4. 具备较高的实用价值

本书中大量的实战案例来源于笔者实际的工作总结，尤其是核心工具篇与线程池核心技术篇涉及的内容，具有非常高的参考与实用价值。

本书内容及知识体系

第一篇　基础篇（第 1~2 章）

本篇简单地介绍了进程与线程的基本概念、线程调度与上下文切换、进程与线程的综合对比、如何查看进程与线程的运行时信息，以及线程和线程组的基本操作。

第二篇　核心工具篇（第 3~13 章）

本篇通过大量源码和案例详细介绍了 JDK 的各种并发工具，涵盖同步集合、并发 List 集合类、并发 Set 集合类、并发 Map 集合类、并发阻塞队列、并发非阻塞队列、并发工具类、锁工具类、无锁原子类、线程工具类和异步编程工具类。几乎每个章节都配有 JDK 核心工具类的源码及实战案例，有助于读者理解。

第三篇　线程池核心技术篇（第 14~16 章）

本篇深入剖析了 JDK 中线程池的核心源码。包括线程池顶层接口和抽象类、线程池正确运行的核心流程、线程池执行任务的核心流程、Worker 线程的核心流程、线程池优雅退出的核心流程、ScheduledThreadPoolExecutor 类与 Timer 类的区别、定时任务线程池的初始化、调度流程和优雅关闭流程等。通过对本篇的学习，读者能够从源码级别深刻理解线程池的核心原理和执行流程。

为了进一步加深读者对线程池的理解，在本篇的随书源码中，会给出完整的手写线程池的案例程序。

如何阅读本书

- 没有使用过 JDK 的核心工具和线程池进行高并发编程，或者对高并发编程技术掌握薄弱的读者，建议从第 1 章开始顺序阅读，并实现书中的每一个案例。
- 有一定多线程和并发编程基础的读者，可根据自身实际情况，有选择性地阅读相关篇章。
- 对本书中涉及的高并发编程案例，读者可以先自行思考其实现方式，再阅读相关内容，以达到事半功倍的学习效果。
- 先阅读一遍书中的高并发编程案例，再阅读各技术对应的 JDK 底层源码，理解会更加深刻。

勘误和支持

由于作者的水平有限，编写时间仓促，书中难免会出现一些错误或者不妥之处，恳请读者批评指正。如果读者对本书有任何建议或者想法，请联系笔者。也可以将勘误信息提交到随书源码的勘误.md 文件中。

- 微信：hacker_binghe。
- 邮箱：1028386804@qq.com。
- 公众号：冰河技术。

如果想学习更多有关高并发编程的知识，可以关注"冰河技术"微信公众号及冰河的个人博客站点阅读相关的技术文章。冰河的个人博客站点如下。

- GitHub 站点：https://binghe001.github.io。
- GitCode 站点：https://binghe.gitcode.host。

致谢

感谢蒋涛（CSDN 创始人、总裁）、邹欣（CSDN 副总裁）、李海翔（腾讯数据库资深研究员、首席架构师）、林子熠（阿里巴巴 JVM 技术专家、CCF 系统软件专委会执行委员）、于君泽（资深技术专家、公众号"技术琐话"作者）、沈剑（互联网架构专家，公众号"架构师之路"作者）、秦金卫（长亮科技平台技术部副总经理、Apache Dubbo/ShardingSphere PMC）、张开涛（《亿级流量网站架构核心技术》作者）、季敏（Seata 开源社区创始人）、李鹏云（杭州任你说智能科技 CTO）、程军（前饿了么技术总监，公众号"军哥手记"作者）、骆俊武（京东零售架构师）、纯洁的微笑（公众号"纯洁的微笑"作者）、黄哲铿/Mr.K（"顿悟山丘"咨询创始人、公众号"技术领导力"作者）、李伟（Apache RocketMQ 北京社区联合发起人 && Commiter）、翟永超（公众号"程序猿DD"维护者、《Spring Cloud 微服务实战》作者）（以上排名不分先后）等行业大佬对本书的大力推荐。

感谢冰河技术社区的兄弟姐妹们，感谢你们长期对社区的支持和贡献。你们的支持是我写作的最大动力。

感谢我的团队和许许多多一起合作、交流过的朋友们，感谢博客、公众号的粉丝，以及在我博客、公众号留言及鼓励我的朋友们。

感谢电子工业出版社博文视点的编辑张晶，在这几个月的时间中始终支持我写作，在你的鼓励和帮助下，我顺利完成全部书稿。

感谢我的家人，他们都在以自己的方式在我写作期间默默地给予我支持与鼓励，并时时刻刻为我灌输着信心和力量！

最后，感谢所有支持、鼓励和帮助过我的人。谨以此书献给我最亲爱的家人，以及众多关注、认可、支持、鼓励和帮助过我的朋友们！

冰 河

目 录

基础篇

第 1 章 进程与线程的基本概述 1

1.1 进程 1
    1.1.1 进程的基本概念及构成 1
    1.1.2 进程的三态模型、五态模型和七态模型 3
1.2 线程 6
    1.2.1 线程的基本概念及构成 6
    1.2.2 线程的生命周期 7
1.3 线程调度与上下文切换 11
    1.3.1 时间片 11
    1.3.2 线程调度 12
    1.3.3 上下文切换 12
1.4 进程与线程的综合对比 14
1.5 查看进程与线程 14
    1.5.1 编写测试程序 15
    1.5.2 查看进程与线程的方式 16

第 2 章 线程的基本操作 23

2.1 创建线程 23
    2.1.1 继承 Thread 类 23
    2.1.2 实现 Runnable 接口 24
    2.1.3 实现 Callable 接口 26
    2.1.4 FutureTask 配合 Thread 28
    2.1.5 使用线程池 28

2.2 线程的基本操作 30
    2.2.1 线程设置 30
    2.2.2 线程操作 32
2.3 线程组 38
    2.3.1 线程关联线程组 39
    2.3.2 线程组自动归属 40
    2.3.3 顶级线程组 42
    2.3.4 向线程组里添加线程组 42
    2.3.5 获取线程组内的对象 42
    2.3.6 批量中断线程组内的线程 43

核心工具篇

第 3 章 同步集合 45

3.1 Vector 同步集合类及案例 45
3.2 Stack 同步集合类及案例 49
3.3 Hashtable 同步集合类及案例 51
3.4 同步包装器及测试 53
3.5 同步集合的缺陷 56

第 4 章 并发 List 集合类 60

4.1 CopyOnWriteArrayList 概述 60
4.2 写时复制技术 61
4.3 初始化 64
4.4 添加数据 65
4.5 读取数据 65
4.6 修改数据 66

4.7 删除数据 69
4.8 遍历数据 70
4.9 使用案例 72

第 5 章 并发 Set 集合类 74

5.1 CopyOnWriteArraySet 集合类 74
    5.1.1 概述 74
    5.1.2 初始化 75
    5.1.3 添加数据 77
    5.1.4 删除数据 80
    5.1.5 遍历数据 82
    5.1.6 使用案例 83
5.2 ConcurrentSkipListSet 集合类 84
    5.2.1 概述 84
    5.2.2 跳表 85
    5.2.3 初始化 86
    5.2.4 添加数据 87
    5.2.5 删除数据 87
    5.2.6 遍历数据 88
    5.2.7 关系运算 88
    5.2.8 有序集合操作 89
    5.2.9 使用案例 90

第 6 章 并发 Map 集合类 93

6.1 ConcurrentHashMap 集合类 93
    6.1.1 概述 94
    6.1.2 结构 94
    6.1.3 成员变量 97
    6.1.4 内部类 99
    6.1.5 构造方法 100
    6.1.6 初始化 101
    6.1.7 扩容 103
    6.1.8 再谈 sizeCtl 成员变量 112
    6.1.9 添加数据 113

    6.1.10 读取数据 115
    6.1.11 修改数据 116
    6.1.12 删除数据 118
    6.1.13 遍历数据 119
    6.1.14 使用案例 119
6.2 ConcurrentSkipListMap 集合类 121
    6.2.1 概述 121
    6.2.2 内部类 122
    6.2.3 初始化 123
    6.2.4 添加数据 124
    6.2.5 读取数据 128
    6.2.6 修改数据 129
    6.2.7 删除数据 130
    6.2.8 遍历数据 131
    6.2.9 使用案例 133

第 7 章 并发阻塞队列 135

7.1 并发阻塞队列简介 135
    7.1.1 概述 135
    7.1.2 类继承关系 136
    7.1.3 常用方法 137
7.2 ArrayBlockingQueue 139
    7.2.1 概述 139
    7.2.2 核心成员变量 140
    7.2.3 初始化 140
    7.2.4 添加数据 141
    7.2.5 删除数据 144
    7.2.6 获取数据 147
7.3 LinkedBlockingQueue 148
    7.3.1 概述 148
    7.3.2 核心成员变量 148
    7.3.3 初始化 148
    7.3.4 添加数据 149
    7.3.5 删除数据 151

目录

7.3.6 获取数据 ... 152
7.4 PriorityBlockingQueue ... 152
    7.4.1 概述 ... 152
    7.4.2 核心成员变量 ... 153
    7.4.3 初始化 ... 153
    7.4.4 添加数据 ... 155
    7.4.5 删除数据 ... 157
    7.4.6 获取数据 ... 158
7.5 DelayQueue ... 158
    7.5.1 概述 ... 158
    7.5.2 核心成员变量 ... 159
    7.5.3 初始化 ... 159
    7.5.4 添加数据 ... 160
    7.5.5 删除数据 ... 160
    7.5.6 获取数据 ... 162
7.6 SynchronousQueue ... 163
    7.6.1 概述 ... 163
    7.6.2 重要常量与内部类 ... 164
    7.6.3 初始化 ... 165
    7.6.4 添加数据 ... 165
    7.6.5 删除数据 ... 165
    7.6.6 不支持的方法 ... 166
7.7 LinkedTransferQueue ... 167
    7.7.1 概述 ... 167
    7.7.2 重要常量与成员变量 ... 168
    7.7.3 重要内部类 ... 169
    7.7.4 初始化 ... 169
    7.7.5 添加数据 ... 169
    7.7.6 删除数据 ... 174
    7.7.7 获取数据 ... 174
    7.7.8 新增方法 ... 175
7.8 LinkedBlockingDeque ... 175
    7.8.1 概述 ... 175
    7.8.2 核心成员变量 ... 176
    7.8.3 重要内部类 ... 177
    7.8.4 初始化 ... 177
    7.8.5 添加数据 ... 178
    7.8.6 删除数据 ... 179
    7.8.7 获取数据 ... 180
7.9 并发阻塞队列案例 ... 181
    7.9.1 生产者与消费者模型 ... 181
    7.9.2 按周期执行的定时任务 ... 184

第 8 章 并发非阻塞队列 ... 188

8.1 并发非阻塞队列简介 ... 188
    8.1.1 概述 ... 188
    8.1.2 类继承关系 ... 189
    8.1.3 常用方法 ... 189
    8.1.4 并发非阻塞队列与并发阻塞队列的区别 ... 193
8.2 ConcurrentLinkedQueue ... 193
    8.2.1 概述 ... 193
    8.2.2 核心成员变量 ... 194
    8.2.3 重要内部类 ... 194
    8.2.4 初始化 ... 195
    8.2.5 添加数据 ... 195
    8.2.6 删除数据 ... 197
    8.2.7 获取数据 ... 198
    8.2.8 性能对比案例 ... 198
8.3 ConcurrentLinkedDeque ... 201
    8.3.1 概述 ... 201
    8.3.2 核心成员变量 ... 201
    8.3.3 重要内部类 ... 202
    8.3.4 初始化 ... 202
    8.3.5 添加数据 ... 203
    8.3.6 删除数据 ... 204
    8.3.7 获取数据 ... 206
    8.3.8 性能对比案例 ... 207

第 9 章 并发工具类 209

9.1 CountDownLatch 工具类 209
9.1.1 概述及重要方法说明 209
9.1.2 使用案例 211
9.2 CyclicBarrier 工具类 213
9.2.1 概述及重要方法说明 213
9.2.2 使用案例 215
9.3 Phaser 工具类 217
9.3.1 概述及重要方法说明 217
9.3.2 使用案例 223
9.4 Semaphore 工具类 225
9.4.1 概述及重要方法说明 225
9.4.2 使用案例 228
9.5 Exchanger 工具类 229
9.5.1 概述及重要方法说明 229
9.5.2 使用案例 231

第 10 章 锁工具类 233

10.1 Lock 接口 233
10.1.1 概述及核心方法 233
10.1.2 使用案例 235
10.2 Condition 接口 236
10.2.1 概述及核心方法 237
10.2.2 使用案例 239
10.3 ReentrantLock 可重入锁 240
10.3.1 概述及核心方法 241
10.3.2 使用案例 243
10.4 ReadWriteLock 读写锁 245
10.4.1 概述及核心方法 245
10.4.2 使用案例 246
10.5 StampedLock 读写锁 248
10.5.1 概述及核心方法 248
10.5.2 StampedLock 使用案例 256
10.6 锁性能对比案例 259

10.6.1 案例需求 259
10.6.2 案例实现 259
10.6.3 案例测试 262

第 11 章 无锁原子类 264

11.1 无锁原子类概述及分类 264
11.2 操作基本类型的原子类 265
11.2.1 概述 265
11.2.2 AtomicInteger 类核心方法
解析 266
11.3 操作引用类型的原子类 269
11.3.1 概述 270
11.3.2 AtomicReference 类核心方法
解析 270
11.4 操作字段类型的原子类 273
11.4.1 概述 273
11.4.2 AtomicReferenceFieldUpdater 类
核心方法解析 273
11.5 操作数组类型的原子类 277
11.5.1 概述 277
11.5.2 AtomicLongArray 类核心方法
解析 277
11.6 累加器类型的原子类 281
11.6.1 概述 281
11.6.2 LongAdder 类核心方法解析 283
11.7 性能对比案例 285
11.7.1 锁与基本类型原子类性能对比
案例 285
11.7.2 锁与引用类型原子类性能对比
案例 287
11.7.3 锁与字段类型原子类性能对比
案例 290
11.7.4 AtomicLong 与 LongAdder 性能
对比案例 292

第 12 章 线程工具类 295

12.1 Thread 类 295
12.1.1 继承关系 295
12.1.2 定义 296
12.1.3 核心代码解析 296
12.2 ThreadLocal 类 306
12.3 Fork/Join 框架 306
12.3.1 概述 306
12.3.2 核心类 307
12.4 线程工具类案例 307
12.4.1 Thread 类线程中断案例 ... 308
12.4.2 Fork/Join 框架分组合并案例 ... 309

第 13 章 异步编程工具类 312

13.1 Callable 接口 312
13.1.1 概述 312
13.1.2 PrivilegedCallable 实现类 ... 313
13.1.3 PrivilegedCallableUsing CurrentClassLoader 实现类 ... 315
13.1.4 RunnableAdapter 实现类 ... 317
13.1.5 TaskCallable 实现类 317
13.2 异步编程接口 319
13.2.1 两种异步模型 319
13.2.2 Future 接口 320
13.2.3 RunnableFuture 接口 321
13.2.4 FutureTask 类 321
13.3 CompletableFuture 类 330
13.3.1 概述 330
13.3.2 初始化 330
13.3.3 串行执行任务 332
13.3.4 并行执行任务 333
13.3.5 AND 聚合任务 334
13.3.6 OR 聚合任务 335
13.3.7 处理结果 336

13.3.8 使用案例 338
13.4 CompletionService 338
13.4.1 概述 338
13.4.2 接口定义 339
13.4.3 ExecutorCompletionService 类的核心实现 339
13.4.4 使用案例 342

线程池核心技术篇

第 14 章 线程池总体结构 343

14.1 线程池简介 343
14.1.1 线程池核心类继承关系 ... 343
14.1.2 线程池的优点 344
14.1.3 Executors 类 345
14.1.4 ThreadPoolExecutor 类 ... 346
14.2 线程池顶层接口和抽象类 ... 347
14.2.1 接口和抽象类总览 347
14.2.2 Executor 接口 348
14.2.3 ExecutorService 接口 348
14.2.4 AbstractExecutorService 抽象类 350
14.2.5 ScheduledExecutorService 接口 356

第 15 章 线程池核心流程 357

15.1 线程池正确运行的核心流程 ... 357
15.1.1 ThreadPoolExecutor 类的重要属性 357
15.1.2 ThreadPoolExecutor 类的重要内部类 359
15.2 线程池执行任务的核心流程 ... 362
15.2.1 核心流程概述 362
15.2.2 execute()方法解析 362

- 15.2.3 addWorker()方法解析 364
- 15.2.4 addWorkerFailed()方法解析 366
- 15.2.5 拒绝策略执行流程 367
- 15.3 Worker 线程的核心流程 367
  - 15.3.1 Worker 线程的核心流程概述 ...367
  - 15.3.2 runWorker()方法解析 368
  - 15.3.3 getTask()方法解析 369
  - 15.3.4 beforeExecute()方法解析 370
  - 15.3.5 afterExecute()方法解析 370
  - 15.3.6 processWorkerExit()方法解析 ...370
  - 15.3.7 tryTerminate()方法解析 371
  - 15.3.8 terminated()方法解析 372
- 15.4 线程池优雅退出的核心流程 373
  - 15.4.1 shutdown()方法解析 373
  - 15.4.2 shutdownNow()方法解析 375
  - 15.4.3 awaitTermination()方法解析 376

第 16 章 定时任务线程池 378

- 16.1 ScheduledThreadPoolExecutor 类与 Timer 类的区别 378
  - 16.1.1 线程实现的区别 378
  - 16.1.2 系统时间的区别 378
  - 16.1.3 处理异常的区别 379
  - 16.1.4 任务编排的区别 379
  - 16.1.5 任务优先级的区别 379
  - 16.1.6 返回结果的区别 379
- 16.2 定时任务线程池的初始化 380
- 16.3 定时任务线程池的调度流程 380
  - 16.3.1 schedule()方法解析 380
  - 16.3.2 decorateTask()方法解析 381
  - 16.3.3 scheduleAtFixedRate()方法解析 381
  - 16.3.4 scheduleWithFixedDelay()方法解析 382
  - 16.3.5 riggerTime()方法解析 383
  - 16.3.6 overflowFree()方法解析 384
  - 16.3.7 delayedExecute()方法解析 385
  - 16.3.8 reExecutePeriodic()方法解析 386
- 16.4 定时任务线程池优雅关闭流程 386

基础篇

第 1 章

进程与线程的基本概述

随着计算机与互联网技术的不断发展，计算机的处理模型已经从最早期的单用户单任务处理模式，发展为当前的多用户多任务的高并发处理模式，这得益于计算机内部支持以多进程与多线程的方式来处理任务。

1.1 进程

从某种程度上来讲，进程其实是内存中正在运行的程序实例，被分配了一定的空间，各个进程在运行的过程中互不干扰。

1.1.1 进程的基本概念及构成

一般来说，一个进程就是内存中正在运行的一个实例，是系统进行资源分配的基本单位。目前，大部分程序可以同时启动并运行多个进程实例，有些程序只能启动并运行一个进程实例。例如，当打开浏览器时，会启动多个进程实例；当运行 Java 程序时，会启动一个对应的 JVM 进程实例。

进程具有动态性、并发性、独立性和异步性。除此之外，读者还需掌握如下几个基本概念。

- 进程 ID：进程的唯一编号。
- 进程互斥：当多个进程操作同一个资源时，同一时刻只允许一个进程使用资源，其他进程在当前进程使用资源时必须等待，直到当前进程释放资源。
- 临界资源：同一时刻只允许一个进程访问的资源。
- 临界区：临界区本质上是一段代码片段，即在进程中访问临界资源的代码片段，需要保证进程互斥地进入各自的临界区。
- 进程同步：多个并发运行的进程按照一定的顺序执行的过程。

- 进程调度：按照一定的规则，从一组准备就绪的进程中选择一个占用 CPU 的资源运行。进程调度可以分为抢占式和非抢占式两种，包括先来先服务调度、端进程调度、时间片轮转算法调度、高优先级优先服务调度等。
- 进程死锁：多个进程之间竞争系统资源形成的一种相互等待的现象，当发生进程死锁时，如果没有其他外力的作用，这些进程会一直相互等待下去。

总之，进程是系统进行资源分配的基本单位。

在通常情况下，一个进程由程序段、数据段和进程控制块 3 部分组成，如图 1-1 所示。

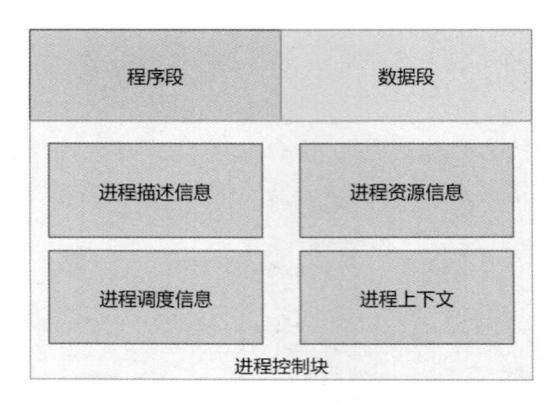

图 1-1　进程的组成

（1）程序段：在一般情况下，程序段又被称为代码段，是进程中的程序指令在内存中的位置，也就是程序指令在内存中的地址，程序段中包含进程需要执行的指令的集合。

（2）数据段：数据段是进程中操作的数据在内存中的地址，包含进程所要处理的数据的集合。

（3）进程控制块：进程控制块中记录了进程当前的运行情况和能够控制进程运行的全部信息，是操作系统中最重要的记录进程运行情况和控制信息的区域，也是进程存在的唯一标志。

进程控制块中包含进程描述信息、进程资源信息、进程调度信息和进程上下文，每个部分的简单介绍如下。

（1）进程描述信息：进程描述信息中主要记录进程的 ID 和名称，同时描述进程的一些状态信息，例如运行态、就绪态、阻塞态等。其中，进程的 ID 代表进程的身份，可以表示唯一的进程。进程的优先级也包含在进程描述信息中，操作系统会根据优先级对进程进行调度，优先调度优先级高的进程。

（2）进程资源信息：进程资源信息中记录着进程在运行过程中需要的系统资源，主要包括系统的内存信息（内存的使用情况和管理内存的数据结构）、I/O 设备信息（I/O 设备的编号和对应的数据结构）、文件引用信息（文件句柄）。

（3）进程调度信息：进程调度信息中主要记录进程在运行过程中的调度信息，包括运行程序的起始地址（程序的第 1 行指令所在的内存地址）以及进程的通信信息。

（4）进程上下文：进程上下文中主要记录进程的运行时环境信息。主要包括进程运行时的各种变量信息、寄存器中的值和运行的环境。当操作系统发生进程切换时，会将当前进程的运行时环境信息保存到进程上下文中，以便在下次切换回当前进程时，能够快速知道当前进程在操作系统发生进程切换之前的状态。

注意：如今大部分操作系统中的进程是能够并发运行的，操作系统中的多个进程可以同时执行。

1.1.2　进程的三态模型、五态模型和七态模型

我们可以通过进程的三态模型、五态模型和七态模型理解进程的生命周期。进程的三态模型将进程的生命周期分为就绪状态、运行状态和等待状态，如图 1-2 所示。

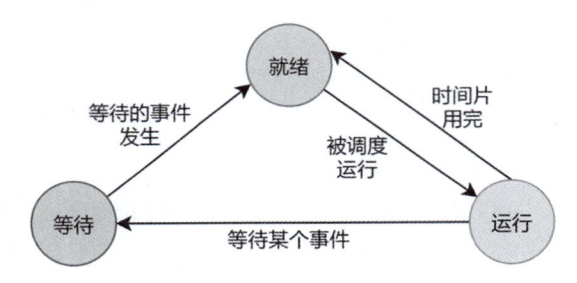

图 1-2　进程的三态模型图

其中，每个状态的简单说明如下。

- 就绪状态：进程准备就绪，只要获取到 CPU 资源就可以运行。
- 运行状态：进程已经获取到运行所需要的资源，且正在运行。
- 等待状态：进程不具备运行的条件。如果某个进程由于等待某一事件而暂时停止运行，那么就算把 CPU 资源分配给它，它也无法运行，此时这个进程处于等待或者阻塞的状态。

三态模型中的进程状态之间的转换关系如下。

（1）就绪状态转换成运行状态：处于就绪状态的进程只要获取到 CPU 的资源就会运行，从就绪状态转换成运行状态。

（2）运行状态转换成就绪状态：如果 CPU 时间片用完，处于运行状态的进程就会释放 CPU 的资源，从运行状态转换成就绪状态。

（3）运行状态转换成等待状态：某个处于运行状态的进程如果需要等待某个事件发生（例

如 I/O 完成），这个进程就会暂时停止运行，从运行状态转换成等待状态。

（4）等待状态转换成就绪状态：如果等待的事件已经发生（例如等待的 I/O 事件已经完成），进程就会从等待状态转换成就绪状态。

注意：进程不可能从就绪状态直接转换成等待状态，也不可能直接从等待状态转换成运行状态。

进程的五态模型比三态模型多了两个状态，分别是新建状态和终止状态，如图 1-3 所示。

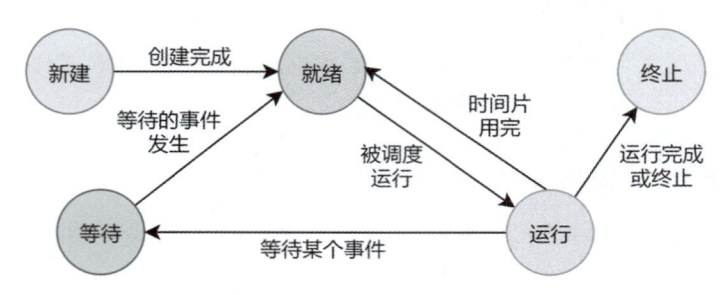

图 1-3 进程的五态模型图

- 新建状态：进程刚刚被创建，还没有进入就绪队列，此时的进程处于新建状态。
- 终止状态：进程运行完成或由于某种原因被终止后进入的最终状态。

五态模型中进程状态之间的转换关系如下。

（1）新建状态转换成就绪状态：进程刚刚被创建时，还没有进入就绪队列。此时，需要为新创建的进程分配资源，并将其放入就绪队列，进程从新建状态转换成就绪状态。

（2）运行状态转换成终止状态：进程正常运行结束、被操作系统终止、出现某种错误或异常终止，就会从运行状态转换成终止状态。处于终止状态的进程不会再次被调度。

注意：在进程的五态模型中，只有处于运行状态的进程才能直接转换成终止状态，处于其他状态的进程不能直接转换成终止状态。

这里只简单介绍了新建状态和终止状态，以及与其对应的状态转换关系。关于运行、就绪、等待三个状态，以及三个状态之间的转换关系，读者可以参考 1.1.2 节的内容。

进程的七态模型比五态模型多了两个状态，分别是挂起就绪状态和挂起等待状态，如图 1-4 所示。

- 挂起就绪状态：进程已经获取到运行所需的资源，但是此时的进程不在内存中，而在外存中，当前进程只有被加载到内存中才能够被操作系统调用执行。
- 挂起等待状态：进程正在等待某个事件，并且此时的进程不在内存中，而在外存中。

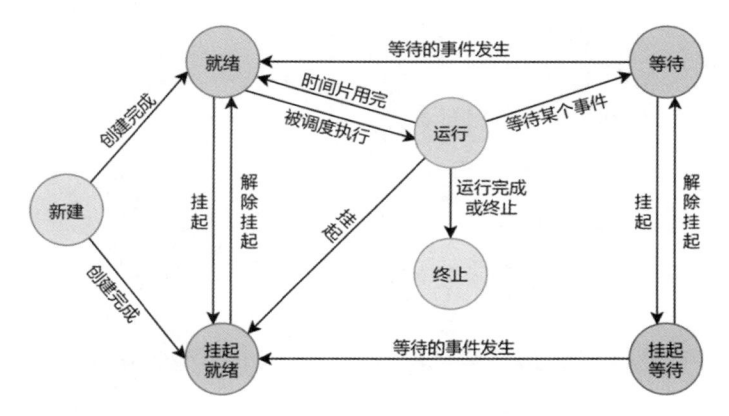

图 1-4　进程的七态模型图

七态模型中进程状态之间的转换关系如下。

（1）新建状态转换成挂起就绪状态：受系统当前的资源状况（例如内存不足等），以及系统的性能等影响，被创建的进程被分配到外存，进程就会从新建状态转换成挂起就绪状态。

（2）运行状态转换成挂起就绪状态：当某个优先级较高、等待的事件已经发生的处于挂起等待状态的进程需要抢占 CPU 资源时，由于内存空间不足，可能导致正在运行的优先级较低的进程从运行状态转换成挂起就绪状态。除此之外，一个处于运行状态的进程也可以将自己挂起。

（3）就绪状态转换成挂起就绪状态：如果当前系统资源短缺（例如内存不足），或者系统的运行性能较差，操作系统可以将就绪状态的进程分配到外存中，进程就会从就绪状态转换成挂起就绪状态。

（4）挂起就绪状态转换成就绪状态：如果当前系统资源比较充足（例如内存充足），系统的运行性能良好，那么操作系统可能将处于挂起就绪状态的进程加载进内存，进程就会从挂起就绪状态转换成就绪状态。

（5）等待状态转换成挂起等待状态：如果当前系统资源短缺（例如内存不足），或者系统的运行性能较差，操作系统可以将处于等待状态的进程分配到外存，进程就会从等待状态转换成挂起等待状态。

（6）挂起等待状态转换成等待状态：如果当前系统资源比较充足（例如内存充足），系统的运行性能良好，那么操作系统可能将处于挂起等待状态的进程加载进内存，进程就会从挂起等待状态转换成等待状态。

（7）挂起等待状态转换成挂起就绪状态：如果处于挂起等待状态的进程等待的某个事件已经发生，则进程会由挂起等待状态转换成挂起就绪状态。

1.2 线程

线程是比进程更小的执行单元，是进程中的一个执行流程，一个进程中可以包含一个或多个线程，每个线程可以执行一个任务。

1.2.1 线程的基本概念及构成

线程是比进程的粒度更小并且能够独立运行的单元，是 CPU 调度的基本单位。线程基本上不会直接拥有系统资源，只在运行时拥有一些系统资源，例如程序计数器、寄存器和栈等。线程运行于进程中，一个进程内的所有线程可以共享进程中的所有资源。

注意： 线程的其他基本概念可以参考《深入理解高并发编程：核心原理与案例实战》一书。

线程的总体结构包括线程的描述信息、栈内存和程序计数器，如图 1-5 所示。

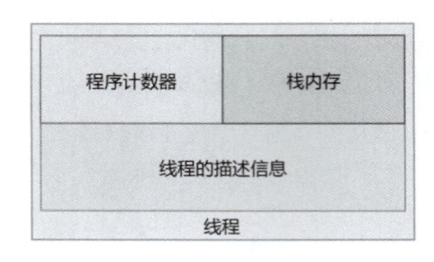

图 1-5 线程的总体结构

（1）线程的描述信息：记录当前线程的基本信息，这些信息能够唯一标识一个线程，并记录线程的状态等信息。

（2）程序计数器：记录当前线程下一条指令的内存地址。

（3）栈内存：每个线程私有的存储局部变量的内存区域，不会在多个线程之间共享，会为运行的方法分配栈帧。

其中，线程的描述信息包括线程 ID、线程名称、线程的优先级、线程状态、是否是守护线程、线程上下文等。

（1）线程 ID：线程的唯一标识。

（2）线程名称：以 Java 为例，在创建线程时，可以手动为其指定一个名称，否则 JVM 会根据一定的规则为线程指定一个默认的名称。建议在创建线程时根据业务情况手动为线程指定名称。

（3）线程的优先级：以 Java 为例，在创建线程时，可以设置线程的优先级。在一般情况下，操作系统会优先调度优先级比较高的线程。

（4）线程状态：标识线程在运行过程中的状态信息。以 Java 为例，线程在运行的过程中可能经历初始化状态、可运行状态、等待状态、超时等待状态、阻塞状态、终止状态。

（5）是否是守护线程：线程大体上可以分为用户线程（也叫非守护线程）和守护线程，守护线程是为用户线程提供服务的一种特殊的线程。在 Java 进程中，不管有没有守护线程在运行，只要有一个非守护线程在运行，Java 进程就不会退出。相反，只要所有的非守护线程运行结束，不管有没有守护线程在运行，Java 进程都会退出。

（6）线程上下文：线程上下文的作用与进程上下文类似。线程上下文中主要记录了线程的运行时环境信息。当发生线程切换时，会将当前线程的运行时环境信息保存到线程上下文中，以便在下次切换回当前线程时快速知道当前线程在切换之前的状态。

1.2.2 线程的生命周期

线程在运行过程中会经历几种状态，而线程在这几种状态之间的转换流程基本上构成了线程的生命周期。

线程的生命周期总体上包括 5 种状态，分别为初始状态、可运行状态、运行状态、休眠状态和终止状态，如图 1-6 所示。

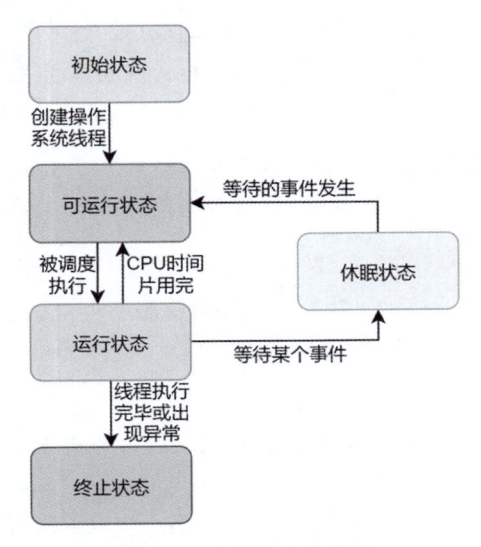

图 1-6　线程的生命周期

（1）初始状态：初始状态比较特殊，属于编程语言层面特有的状态，处于初始状态的线程只是在编程语言层面被创建了，在操作系统层面并没有被真正创建。

（2）可运行状态：线程在操作系统层面被真正创建，并且可以获取 CPU 资源。

（3）运行状态：处于运行状态的线程已经获取到 CPU 资源，正在运行。

（4）休眠状态：线程正在等待某个事件的发生，或者调用了一个阻塞的 API 正处于阻塞状态（例如以阻塞的方式读写文件等），此时处于休眠状态。

（5）终止状态：线程正常运行结束或者出现异常，就会进入终止状态。

线程的通用生命周期中各状态之间的转换关系如下。

（1）初始状态转换成可运行状态：处于初始状态的线程实际上并没有在操作系统层面被真正创建，当在操作层面被真正创建时，线程就会从初始状态转换成可运行状态。

（2）可运行状态转换成运行状态：处于可运行状态的线程获得 CPU 资源后就会转换成运行状态。

（3）运行状态转换成可运行状态：当正在运行的线程用完 CPU 时间片时，就会从运行状态转换成可运行状态。

（4）运行状态转换成休眠状态：如果等待的某个事件发生，或者调用了一个阻塞的 API，处于运行状态的线程就会释放 CPU 的资源，从运行状态转换成休眠状态。

（5）休眠状态转换成可运行状态：如果等待的事件已经发生，或者调用的阻塞 API 已经完成操作，处于休眠状态的线程就会从休眠状态转换成可运行状态。

（6）运行状态转换成终止状态：处于运行状态的线程运行结束，或者出现异常，就会从运行状态转换成终止状态。处于终止状态的线程不会再转换成其他状态，线程的生命周期也就结束了。

注意：在线程的生命周期中，只有处于运行状态的线程可以直接转换成终止状态和休眠状态，处于其他状态的线程不能直接转换成终止状态和休眠状态。处于休眠状态的线程只能直接转换成可运行状态，不能直接转换成其他状态。

Java 线程的生命周期包括初始化状态、可运行状态、等待状态、超时等待状态、阻塞状态和终止状态，其中可运行状态又包括运行状态和就绪状态。Java 中线程的生命周期如图 1-7 所示。

（1）初始化状态：线程在 Java 中被创建，但是还没有调用线程对象的 start()方法，也就是说，还没有创建操作系统层面的线程。

（2）可运行状态：Java 线程生命周期中的可运行状态包括运行状态和就绪状态。

- 运行状态：对应操作系统中的运行状态。
- 就绪状态：对应操作系统中的就绪状态。

（3）等待状态：处于等待状态的线程需要等待其他线程对当前线程进行通知或者中断等操作，从而进入下一个状态。

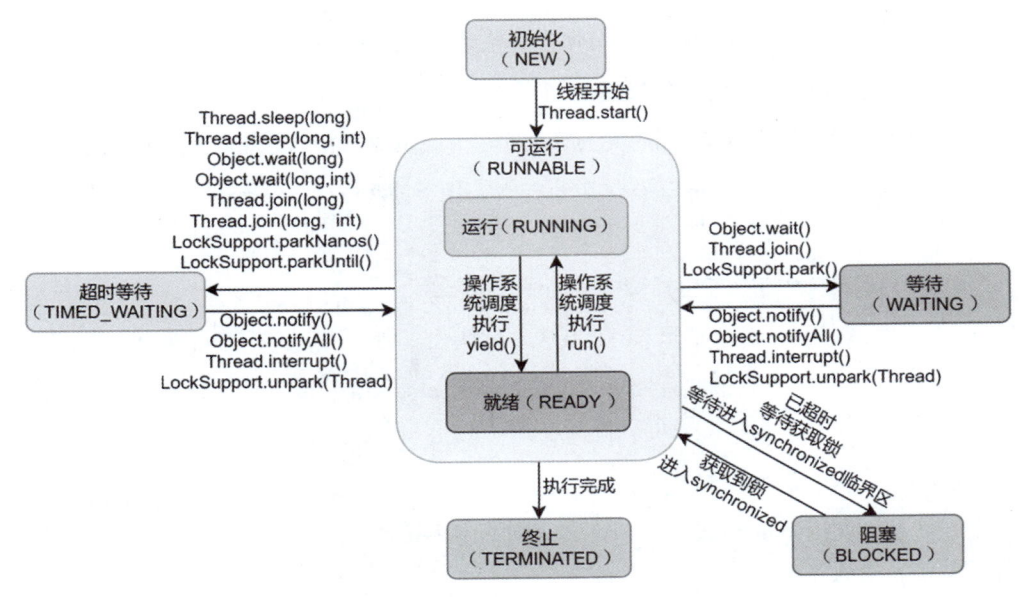

图 1-7　Java 中线程的生命周期

（4）超时等待状态：处于超时等待状态的线程需要等待其他线程在指定时间内对当前线程进行通知或者中断等操作。如果在指定的时间内，其他线程对当前线程进行通知或者中断等操作，则当前线程进入下一个状态。如果超过指定的时间，则当前线程也会进入下一个状态。

（5）阻塞状态：处于阻塞状态的线程需要等待其他线程释放锁，或者等待进入 synchronized 临界区。

（6）终止状态：表示当前线程执行完毕，包括正常执行结束和异常退出。

在 Java 的线程生命周期中，各状态之间转换的场景如下。

1. 初始化状态转换成可运行状态的场景

在 Java 层面，调用线程对象的 start()方法会在操作系统层面创建对应的线程，此时，线程从初始化状态转换成可运行状态。

2. 可运行状态与等待状态互相转换的场景一

（1）线程 a 调用 synchronized(obj)获取到对象锁后，在调用 obj.wait()方法时，会从可运行状态转换成等待状态。

（2）在满足（1）时，线程 b 调用 synchronized(obj)获取到对象锁后，调用 obj.notify()方法、obj.notifyAll()方法、a.interrupt()方法，会有如下两种情况。

- 线程 a 竞争锁成功，由等待状态转换成可运行状态。

- 线程 a 竞争锁失败，由等待状态转换成阻塞状态。

3. 可运行状态与等待状态互相转换的场景二

（1）在线程 a 调用线程 b 的 join()方法时，线程 a 会由可运行状态转换成等待状态。

（2）在满足（1）时，线程 b 运行结束，或者调用了线程 a 的 interrupt()方法，线程 a 会从等待状态转换成可运行状态。

4. 可运行状态与等待状态互相转换的场景三

（1）在线程 a 调用 LockSupport.park()方法时，线程 a 会从可运行状态转换成等待状态。

（2）在满足（1）时，其他线程调用 LockSupport.unpark(a)，或者调用线程 a 的 interrupt()方法，线程 a 会从等待状态转换成可运行状态。

5. 可运行状态与超时等待状态互相转换的场景一

（1）在线程 a 调用 synchronized(obj)获取到对象锁后，调用 obj.wait(long n)方法，线程 a 会从可运行状态转换成超时等待状态。

（2）在满足（1）时，线程 a 的等待时间超过了 n 毫秒，或者在线程 b 调用 synchronized(obj)获取到对象锁后，调用 obj.notify()方法、obj.notifyAll()方法、a.interrupt()方法会有如下两种情况。

- 线程 a 竞争锁成功，由超时等待状态转换成可运行状态。
- 线程 a 竞争锁失败，由超时等待状态转换成阻塞状态。

6. 可运行状态与超时等待状态互相转换的场景二

（1）线程 a 调用 Thread.sleep(long n)方法，从可运行状态转换成超时等待状态。

（2）在满足（1）时，线程 a 的等待时间超过 n 毫秒，从超时等待状态转换成可运行状态。

7. 可运行状态与超时等待状态互相转换的场景三

（1）在线程 a 调用了线程 b 的 join(long n)方法时，从可运行状态转换成超时等待状态。

（2）在满足（1）时，线程 a 的等待时间超过 n 毫秒、线程 b 运行结束，或者调用了线程 a 的 interrupt()方法，线程 a 会从超时等待状态转换成可运行状态。

8. 可运行状态与超时等待状态互相转换的场景四

（1）在线程 a 调用 Locksupport.parkNanos(long nacos)方法或者 LockSupport.parkUntil(long millis)方法时，从可运行状态转换成超时等待状态。

（2）在满足（1）时，其他线程调用 LockSupport.unpark(a)、调用线程 a 的 interrupt()方法，或者线程 a 等待超时，线程 a 会从超时等待状态转换成可运行状态。

9. 可运行状态与阻塞状态互相转换的场景一

（1）线程 a 与线程 b 争抢一个悲观锁，线程 b 争抢成功，则线程 a 会从可运行状态转换成阻塞状态。

（2）在满足（1）时，如果线程 b 释放锁，线程 a 获取锁，线程 a 会从阻塞状态转换成可运行状态。

10. 可运行状态与阻塞状态互相转换的场景二

（1）在线程 a 调用 synchronized(obj) 竞争对象锁时，如果失败，则线程 a 从可运行状态转换成阻塞状态。

（2）在满足（1）时，调用 synchronized(obj) 竞争对象锁成功的线程，执行完同步代码块，就会唤醒所有阻塞在 obj 对象上的线程。这些被唤醒的线程会重新竞争，线程 a 如果竞争成功，则从阻塞状态转换成可运行状态；如果竞争失败，则继续保持阻塞状态。

11. 可运行状态转换成终止状态的场景

线程如果正常执行结束，或者异常退出，就会从可运行状态转换成终止状态。

一个线程转换成终止状态就意味着它已经运行结束，不能再次转换成其他状态。

1.3 线程调度与上下文切换

使用多线程进行并发编程的目的是充分发挥 CPU 并行计算的能力，让程序高效运行。但是，并不是启动的线程越多越好，在使用多线程进行并发编程时，会面临非常多的挑战，例如，线程的调度与上下文切换等问题可能导致并发程序执行效率低下。

1.3.1 时间片

目前主流的 CPU 即使是单核的，也能够支持以多线程的方式来执行程序代码。也就是说，单核 CPU 可以在同一时间内运行多个线程执行程序代码。

CPU 采用时间片轮转策略分配资源。当多个线程在单核 CPU 上执行时，CPU 会为每个线程分配一定的时间片让线程周期性占用 CPU 的资源执行任务，当前线程使用完时间片后就会处于就绪状态，并让出 CPU 资源以便其他线程执行对应的任务。时间片是分配给各个线程的时间，其周期非常短，一般只有几十毫秒，CPU 会不停地切换执行的线程，给人的感觉就像多个线程在同时执行。时间片轮转策略如图 1-8 所示。

当线程 1 占用 CPU 时间片执行任务时，线程 2 处于就绪状态；当线程 2 在占用 CPU 时间片执行任务时，线程 1 处于就绪状态，CPU 会在线程 1 和线程 2 之间不断切换。

时间片轮转策略是单核 CPU 在同一时间段内支持多个线程运行的基础。

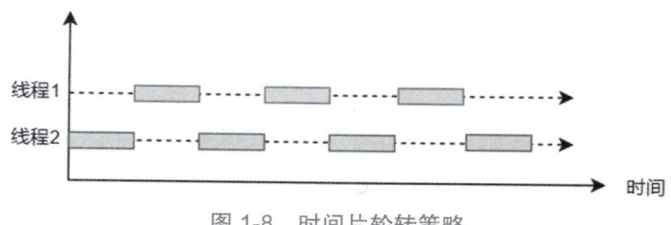

图 1-8　时间片轮转策略

1.3.2　线程调度

目前主流操作系统中主要基于时间片轮转策略进行线程调度。线程的调度模型可以分为分时调度模型和抢占式调度模型。

（1）分时调度模型：分时调度模型会平均分配 CPU 时间片，每个线程占用的 CPU 时间片都是一样的，所有的线程会轮流占用 CPU 时间片。分时调度模型对每个线程来说都是公平的。

（2）抢占式调度模型：按照线程的优先级分配 CPU 时间片，线程的优先级越高，分配到 CPU 时间片的概率越大。如果线程的优先级相同，则会在所有线程中随机选择一个线程占用 CPU 时间片执行任务。Java 中采用的就是抢占式调度模型。

1.3.3　上下文切换

当程序中只有一个主线程时，一般不会被调度出 CPU，也就不会发生线程切换。当程序中可运行的线程数量大于 CPU 的核心数量时，CPU 资源会在不同线程之间来回切换，这个过程会导致线程的上下文切换，也叫作线程切换或任务切换。

在线程上下文切换的过程中，系统会保存当前运行线程的执行上下文，在当前线程被调度出 CPU 后，被调度进 CPU 的线程的执行上下文会被设置为当前上下文。也就是说，在当前线程占用 CPU 资源执行一个时间片后，CPU 会切换到下一个线程执行任务，但是在切换前会保存当前线程的执行状态，以便下次切换回这个线程时加载它的状态。

切换线程的上下文需要一定的开销，这种开销不只包括 JVM 和操作系统的开销。如果在线程切换时，切换到 CPU 中的线程需要的数据不在本地缓存中，就会导致一部分缓存失效，此时线程需要从内存中重新读取数据。

当程序中的线程数量超过 CPU 的核心数量时，除时间片用完会发生线程上下文切换外，如果线程暂停或被其他线程中断，另外一个线程被调度进 CPU 执行任务，则也会发生线程上下文切换。

在 Java 中，当线程等待某个被其他线程占用的锁时，JVM 通常会将等待的线程挂起，并允许这个线程被调度出去。

当一个线程的生命周期状态从可运行状态转换成阻塞状态、等待状态或超时等待状态，或者从阻塞状态、等待状态或超时等待状态转换成可运行状态时，就会发生线程上下文切换。

当一个线程从可运行状态转换成阻塞状态、等待状态或超时等待状态时，操作系统会保存线程的上下文，以便在重新转换成可运行状态时快速恢复线程的状态。当一个线程从阻塞状态、等待状态或超时等待状态转换成可运行状态，并且获取 CPU 时间片执行任务时，操作系统会恢复之前保存的线程上下文。

另外，如果系统中频繁地出现垃圾回收（GC）现象，也可能频繁地导致线程上下文切换。

按照是否由当前线程导致线程上下文切换，可以将线程上下文切换分成自发性上下文切换和非自发性上下文切换。

1. 自发性上下文切换

由当前线程导致的线程上下文切换，在 Java 中，运行中的线程调用如下方法会发生线程上下文切换。

- Object.wait()：线程从可运行状态转换成等待状态。
- Thread.join()：线程从可运行状态转换成等待状态。
- LockSupport.park()：线程从可运行状态转换成等待状态。
- Thread.yield()：线程让出 CPU 资源。
- Thread.sleep(long)：线程从可运行状态转换成超时等待状态。
- Thread.sleep(long, int)：线程从可运行状态转换成超时等待状态
- Object.wait(long)：线程从可运行状态转换成超时等待状态。
- Object.wait(long, int)：线程从可运行状态转换成超时等待状态。
- Thread.join(long)：线程从可运行状态转换成超时等待状态。
- Thread.join(long, int)：线程从可运行状态转换成超时等待状态。
- LockSupport.parkNanos()：线程从可运行状态转换成超时等待状态。
- LockSupport.parkUntil()：线程从可运行状态转换成超时等待状态。

另外，当线程发起了读写文件的操作，或者等待其他线程释放锁时，也会发生自发性线程上下文切换。

2. 非自发性上下文切换

非自发性上下文切换指由于外部因素导致的上下文切换，例如 CPU 时间片用完、线程调度器将更高优先级的线程调度进 CPU 执行，或者 JVM 执行垃圾回收（GC）操作，可能需要暂停执行所有应用线程。

在实际项目中，虽然不能百分之百避免线程上下文切换，但是可以通过如下方案或者措施尽量减少线程上下文切换。

（1）使用 CAS 算法：在某些场景下，使用 CAS 算法代替悲观锁，可以减少线程上下文切换。例如，在 Java 的 java.util.concurrent.atomic 包下使用 CAS 算法实现的原子类，在使用过程中，无须对数据进行加锁，减少了线程上下文切换。

（2）创建的线程尽量少：尽量根据实际业务需要创建线程，避免创建不需要的线程。创建过多的线程会导致大量线程处于等待状态，后续会出现大量的线程上下文切换。

（3）无锁并发编程：在多线程并发的某些场景下可以避免使用锁。例如，将关联性不高的数据分别交给不同的线程处理，避免出现多个线程争抢同一条数据的情况。

（4）使用协程：协程可以在单个线程中实现多个任务的调度与切换，并且大部分在用户态执行任务。使用协程进行并发编程能够有效减少线程上下文切换。

（5）降低 JVM 垃圾回收的频率：通过一定的方式对 JVM 的参数进行调优、降低 JVM 垃圾回收的频率，能够有效减少由于 JVM 垃圾回收导致的应用线程停顿的现象，从而减少线程上下文切换。

1.4　进程与线程的综合对比

进程与线程的综合对比如下。

（1）进程是操作系统分配资源的最小单位，线程是 CPU 调度的最小单位。

（2）线程是运行于进程内的执行单元，一个进程由一个或者多个线程组成。

（3）线程的划分粒度小于进程，线程使得单进程程序具备并发处理任务的可能。

（4）进程之间的关系是相互独立的，但是进程内部的各个线程之间不完全独立，它们共享进程内部的堆内存、方法区内存和系统资源等。

（5）线程比进程更能高效地利用 CPU 资源，更能充分发挥 CPU 的计算性能。

（6）进程之间的通信比较复杂，线程之间的通信相对简单。

（7）线程上下文切换的效率比进程高。

1.5　查看进程与线程

在一般情况下，编译好的程序在服务器上运行，当发生异常时，需要通过某些方式来排查程序的问题，其中，可能涉及查看进程与线程的问题。

1.5.1　编写测试程序

（1）新建名称为 mykit-concurrent-jdk 的 Maven 工程，在 mykit-concurrent-jdk 工程的 pom.xml

文件中添加如下配置。

```xml
<build>
    <plugins>
        <plugin>
            <groupId>org.apache.maven.plugins</groupId>
            <artifactId>maven-compiler-plugin</artifactId>
            <configuration>
                <source>1.8</source>
                <target>1.8</target>
            </configuration>
        </plugin>
    </plugins>
</build>
```

（2）在 mykit-concurrent-chapter01 子工程的 io.binghe.concurrent.chapter01 包下新建 ProcessThreadTest 类，代码如下。

```java
//测试进程与线程
public class ProcessThreadTest {
    public static void main(String[] args){
        new Thread(() -> {
            while (true){
            }
        },"thread-binghe-001").start();
        System.out.println("程序启动成功...");
    }
}
```

（3）为了将 mykit-concurrent-chapter01 子工程打包成 Jar 文件运行，在 mykit-concurrent-chapter01 子工程的 pom.xml 文件中添加如下配置。

```xml
<plugin>
    <groupId>org.apache.maven.plugins</groupId>
    <artifactId>maven-jar-plugin</artifactId>
    <version>2.6</version><!--$NO-MVN-MAN-VER$-->
    <configuration>
        <archive>
            <manifest>
                <addClasspath>true</addClasspath>
                <classpathPrefix>lib/</classpathPrefix>
<mainClass>io.binghe.concurrent.chapter01.ProcessThreadTest</mainClass>
            </manifest>
        </archive>
    </configuration>
</plugin>
```

注意：完整的打包配置参见随书源码（地址见前言）。

（4）将 mykit-concurrent-chapter01 子工程打包成 Jar 文件，以便后续测试使用。

1.5.2 查看进程与线程的方式

在 Windows 操作系统中，可以直接通过任务管理器查看进程与线程数，也可以通过任务管理器杀死进程，这种方式比较简单。

这里简单介绍如何在命令行查看进程与线程。

（1）在 cmd 命令行中输入如下命令启动 Java 程序。

```
java -jar mykit-concurrent-chapter01-1.0.0-SNAPSHOT.jar
程序启动成功...
```

（2）在 cmd 命令行中输入 tasklist | findstr "java"命令，可以查看当前 Windows 操作系统中运行的所有 Java 程序。

```
C:\Users\binghe> tasklist | findstr "java"
java.exe                     25256 Console                     10    147,696 K
java.exe                     24824 Console                     10     24,448 K
```

在当前的 Windows 操作系统中，存在两个运行中的 Java 程序，一个 Java 程序的进程 ID 为 25256，另一个 Java 程序的进程 ID 为 24824。

（3）在 Windows 操作系统中查看指定进程下的线程，可以使用微软提供的 PsList 工具，PsList 工具可以到微软官网下载，使用 PsList 查看进程 ID 为 24824 的 Java 程序中线程的命令如下。

```
E:\binghe\PSTools>pslist.exe -dmx 24824

PsList v1.4 - Process information lister
Copyright (C) 2000-2016 Mark Russinovich
Sysinternals - www.sysinternals.com

Process and thread information for BingHe

Name                    Pid     VM      WS      Priv Priv Pk    Faults   NonP Page
java                  24824 4194303   23656   440148 442968      7822     19  196
 Tid Pri   Cswtch         State        User Time      Kernel Time    Elapsed Time
20500   9     34     Wait:UserReq    0:00:00.031    0:00:00.000     0:22:56.664
 6808   8     91     Wait:UserReq    0:00:00.062    0:00:00.015     0:22:56.454
26656   8 1287187         Running    0:22:35.062    0:00:00.546     0:22:56.353
##########################省略部分输出结果##########################
```

从输出结果中可以看到如下信息。

```
26656   8 1287187         Running    0:22:35.062    0:00:00.546     0:22:56.353
```

ID 为 26656 的线程在运行，其他的线程都在等待，说明这个线程是在 ProcessThreadTest 类中执行死循环的线程。

接下来，就可以对这个线程进行进一步的处理，例如，杀死这个线程。

注意：笔者只是简单列举了使用 tasklist 命令和 PsList 工具查看 Windows 操作系统中运行的 Java 进程与线程的方法，关于 tasklist 命令和 PsList 工具的更多用法，读者可自行查阅相关资料。

另外，微软官方也提供了在 Windows 操作系统中查看进程与线程的可视化工具 ProcessExplorer，有关 ProcessExplorer 的用法，读者可自行查阅相关资料。

在 Linux 操作系统中，也可以通过命令的形式查看运行的 Java 进程与线程，具体步骤如下。

使用如下命令在 Linux 操作系统后台启动 Java 程序。

```
nohup java -jar mykit-concurrent-chapter01-1.0.0-SNAPSHOT.jar >> /dev/null &
```

1. 使用 ps 命令查看 Java 进程

（1）使用 ps 命令查看系统中正在运行的 Java 进程。

```
[root@binghe ~]# ps -ef | grep java
root      2591   2298 98 16:00 pts/0    00:00:14 java -jar
mykit-concurrent-chapter01-1.0.0-SNAPSHOT.jar
root      2622   2298  0 16:01 pts/0    00:00:00 grep --color=auto java
```

此时 Java 程序的进程 ID 为 2591。

（2）使用 ps 命令查看 ID 为 2591 的 Java 进程内的所有线程。

```
[root@binghe ~]# ps -Tf -p 2591
UID        PID   SPID   PPID  C STIME TTY          TIME CMD
root      2591   2591   2298  0 16:00 pts/0    00:00:00 java -jar
mykit-concurrent-chapter01-1.0.0-SNAPSHOT.jar
root      2591   2592   2298  0 16:00 pts/0    00:00:00 java -jar
mykit-concurrent-chapter01-1.0.0-SNAPSHOT.jar
root      2591   2593   2298  0 16:00 pts/0    00:00:00 java -jar
##################省略部分输出结果####################
```

结果中输出了 ID 为 2591 的 Java 进程内的所有线程。

2. 使用 top 命令查看 Java 进程

（1）在命令行输入 top 命令查看正在运行的 Java 进程。

```
[root@binghe ~]# top
PID USER      PR  NI    VIRT    RES    SHR S  %CPU %MEM     TIME+ COMMAND
2591 root     20   0 3469900  40596  11480 S 100.0  1.1   46:47.40 java
 987 root     20   0 1170384  40388  15944 S   0.3  1.0    0:07.63 containerd
   1 root     20   0  193716   6896   4200 S   0.0  0.2    0:01.60 systemd
###############################其他进程省略##############################
```

在当前系统中运行的 Java 进程的 ID 为 2591。

（2）使用 top 命令查看进程 ID 为 2591 的 Java 程序的所有线程。

```
[root@binghe ~]# top -H -p 2591
PID USER      PR  NI    VIRT    RES    SHR S  %CPU %MEM     TIME+ COMMAND
```

```
2606 root       20   0  3469900  40596  11480 R  99.9  1.1   54:01.18 java
2591 root       20   0  3469900  40596  11480 S   0.0  1.1    0:00.02 java
##################省略部分输出结果####################
```

使用 top 命令能够查看到 Java 进程下的所有线程。

注意：关于 ps 命令和 top 命令的其他用法，读者可以自行查阅相关资料。另外，笔者使用的 Linux 操作系统为 CentOS 7 版本。

Java 的 JDK 中自带了 jps 命令和 jstack 命令，可以查看运行的 Java 进程与线程。

（1）在命令行输入 jps 命令查看系统中运行的 Java 进程。

```
[root@binghe ~]# jps
2715 Jps
2591 jar
```

在系统中有一个 ID 为 2591 的 Java 进程。

（2）使用 jstack 命令查看 ID 为 2591 的 Java 进程中的线程状态。

```
[root@binghe ~]# jstack 2591
Full thread dump Java HotSpot(TM) 64-Bit Server VM (25.321-b07 mixed mode):

"Signal Dispatcher" #4 daemon prio=9 os_prio=0 tid=0x00007f3748181000 nid=0xa28 runnable [0x0000000000000000]
   java.lang.Thread.State: RUNNABLE

"Finalizer" #3 daemon prio=8 os_prio=0 tid=0x00007f374814d800 nid=0xa27 in Object.wait() [0x00007f373136f000]
   java.lang.Thread.State: WAITING (on object monitor)
        at java.lang.Object.wait(Native Method)
        - waiting on <0x00000000ec588ee8> (a java.lang.ref.ReferenceQueue$Lock)
        at java.lang.ref.ReferenceQueue.remove(ReferenceQueue.java:144)
        - locked <0x00000000ec588ee8> (a java.lang.ref.ReferenceQueue$Lock)
        at java.lang.ref.ReferenceQueue.remove(ReferenceQueue.java:165)
        at java.lang.ref.Finalizer$FinalizerThread.run(Finalizer.java:216)

"VM Thread" os_prio=0 tid=0x00007f374813f800 nid=0xa25 runnable
"GC task thread#0 (ParallelGC)" os_prio=0 tid=0x00007f374801e800 nid=0xa21 runnable
"VM Periodic Task Thread" os_prio=0 tid=0x00007f37481af000 nid=0xa2d waiting on condition
JNI global references: 309
```

上述输出的信息中省略了部分内容，在 ID 为 2591 的 Java 进程中，既有处于可运行状态的线程，也有处于等待状态的线程。

Java 中自带的 jconsole 命令可以以图形界面的形式查看某个 Java 进程中的线程运行情况，如果使用 JConsole 远程监控 Java 程序，那么需要在命令行启动 Java 程序时，添加一些 jmx 的参数。

使用 JConsole 远程监控 Java 线程的步骤如下。

（1）在命令行启动 Java 程序时，添加 jmx 参数。

```
nohup java -Djava.rmi.server.hostname='192.168.184.102'
-Dcom.sun.management.jmxremote -Dcom.sun.management.jmxremote.port='11111'
-Dcom.sun.management.jmxremote.ssl=false
-Dcom.sun.management.jmxremote.authenticate=false -jar
mykit-concurrent-chapter01-1.0.0-SNAPSHOT.jar >> /dev/null &
```

（2）在本机的 Windows 操作系统中安装好 JDK，配置好 JDK 的系统环境变量后，打开 Windows 操作系统的运行面板，输入 jconsole，如图 1-9 所示。

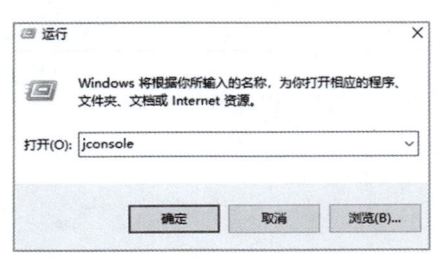

图 1-9　Windows 操作系统的运行面板

（3）单击"确定"按钮后会打开 JConsole 新建连接界面，如图 1-10 所示。

图 1-10　JConsole 新建连接界面

（4）在 JConsole 新建连接界面中选择远程进程，并输入远程连接的 IP 地址和端口号（192.168.184.102:11111），由于在启动程序时，com.sun.management.jmxremote.authenticate 参数

的配置为 false，所以用户名和口令为空即可，如图 1-11 所示。

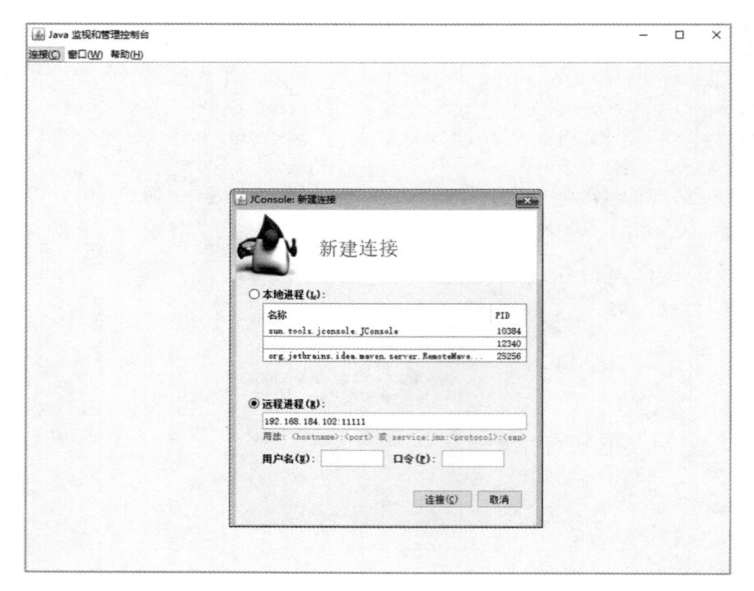

图 1-11　输入远程连接的 IP 地址和端口号

（5）单击"连接"按钮，会提示"安全连接失败"，如图 1-12 所示。

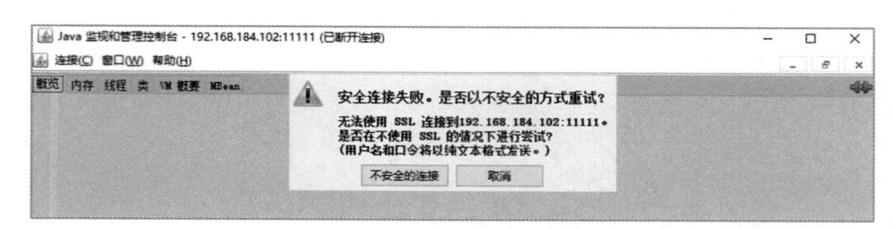

图 1-12　"安全连接失败"提示

这是由于在启动 Java 进程时，com.sun.management.jmxremote.ssl 参数设置为 false。在实际环境下，建议读者将其设置为 true。

（6）单击"不安全的连接"按钮即可进入 JConsole 界面，单击"线程"选项卡，在左下角选择"thread-binghe-001"线程，如图 1-13 所示。

名称为"thread-binghe-001"的线程处于可运行状态。

Arthas 是阿里巴巴集团开源的 Java 应用诊断利器，使用 Arthas 能够非常方便地定位线上系统出现的问题。

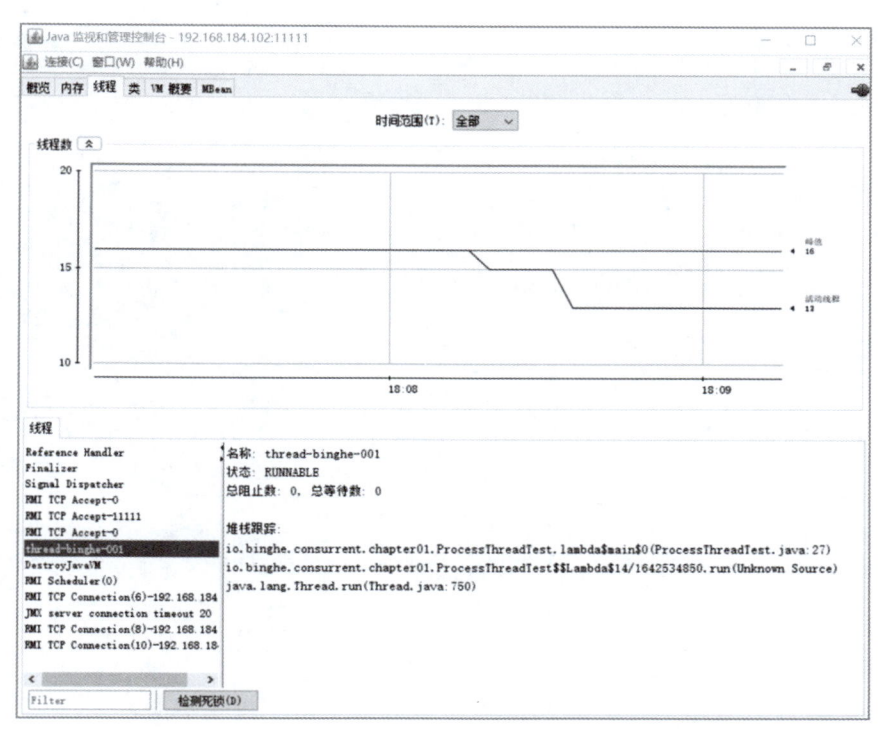

图 1-13　JConsole 界面

使用 Arthas 查看进程和线程的步骤如下。

（1）访问 https://github.com/alibaba/arthas/releases 下载 Arthas。

（2）解压后将 arthas-boot.jar 文件上传到相应的服务器。

（3）使用如下命令启动 Arthas。

```
[root@binghe ~]# java -jar arthas-boot.jar
[INFO] arthas-boot version: 3.6.4
[INFO] Found existing java process, please choose one and input the serial number
of the process, eg : 1. Then hit ENTER.
* [1]: 2925 mykit-concurrent-chapter01-1.0.0-SNAPSHOT.jar
光标处
```

通过 Arthas 检测到系统中存在 ID 为 2925 的 Java 进程，并且 Arthas 的运行光标停留在操作系统的命令行。

（4）在命令行中输入"1"。

```
1
###################省略部分输出信息###############
main_class
pid         2925
```

```
time             2022-08-19 18:25:09
[arthas@2925]$
```

已经正式进入 Arthas 的命令行，并且 Arthas 已经绑定到 ID 为 2925 的 Java 进程。

（5）在 Arthas 的命令行输入 dashboard，输出的部分结果如下。

```
ID      NAME                                            STATE           %CPU
9       thread-binghe-001                               RUNNABLE        100.0
-1      C2 CompilerThread1                              -               0.25
25      Timer-for-arthas-dashboard-50662556-06e4        - RUNNABLE      0.12
-1      C1 CompilerThread2                              -               0.1
-1      VM Periodic Task Thread                         -               0.09
23      arthas-NettyHttpTelnetBootstrap-3-2             RUNNABLE        0.06
-1      C2 CompilerThread0                              -               0.03
-1      VM Thread                                       -               0.01
```

名称为"thread-binghe-001"的线程处于可运行状态，其占用的 CPU 资源为 100%。

注意：在 Arthas 的命令行输入 dashboard 后，我们只截取了部分输出结果，读者可自行验证完整的输出结果。更多有关 Arthas 的用法，读者可参见 Arthas 的官网和其他资料。

另外，本书各章对应的代码地址见前言。

第 2 章

线程的基本操作

线程是 CPU 调度的基本单位，使用多线程编程能够更好地利用 CPU 的资源，充分发挥多核 CPU 的计算性能。在 Java 的多线程编程中，有很多简单易用的 API，使用这些 API 可以方便地创建线程并对线程进行操作。

2.1 创建线程

线程的创建方式主要包括继承 Thread 类、实现 Runnable 接口、实现 Callable 接口、FutureTask 配合 Thread 和使用线程池 5 种。

2.1.1 继承 Thread 类

继承 Thread 类是一种简单常用的创建线程的方式。在使用继承 Thread 类的方式创建线程时，需要创建一个继承 Thread 类，并重写 Thread 类的 run()方法。

（1）创建 ThreadTest 类，并在 ThreadTest 类中创建 MyThreadTask 内部类继承 Thread 类，同时重写 Thread 类的 run()方法，代码如下。

```
//继承 Thread 类创建线程
public class ThreadTest{
    private static class MyThreadTask extends Thread{
        @Override
        public void run() {
            System.out.println("子线程名称===>> " + Thread.currentThread().getName());
        }
    }
}
```

（2）在 main()方法中测试上面的程序，代码如下。

```
public static void main(String[] args){
    Thread thread = new MyThreadTask();
```

```
    thread.start();
    System.out.println("主线程名称===>> " + Thread.currentThread().getName());
}
```

（3）运行 main()方法，输出结果如下。

```
主线程名称===>> main
子线程名称===>> Thread-0
```

从输出结果可以看出，除了主线程，程序还使用继承 Thread 类的方式创建了一个名为 Thread-0 的子线程，并执行了子线程中重写的 run()方法。

注意：使用继承 Thread 类的方式创建线程存在一定的局限性，因为 Java 是单继承的，不支持多继承，在实际项目中创建线程时，不推荐直接使用继承 Thread 类的方式创建线程。

2.1.2 实现 Runnable 接口

在通过实现 Runnable 接口的方式创建线程时，可以将异步执行的任务封装成一个 Runnable 接口的实现类，并将主要的异步任务执行的逻辑封装到 Runnable 接口的 run()方法中，最终将 Runnable 接口的实现类传递给 Thread 类，创建一个线程。

该方式可以细分为通过匿名类创建线程、通过 Lambda 表达式创建线程和通过 Runnable 实现类创建线程 3 种方式。

1. 通过匿名类创建线程

当实现类是一次性类时，可以使用匿名实例的形式创建线程。

（1）创建 RunnableTest 类，并在 RunnableTest 类中创建 createThreadByRunnableAnonymousClass()方法，在 createThreadByRunnableAnonymousClass()方法中将 Runnable 接口的匿名类对象传递给 Thread 类的构造方法，创建 Thread 类的对象，并返回 Thread 类的对象，代码如下。

```
//通过匿名类创建线程
public Thread createThreadByRunnableAnonymousClass(){
    return new Thread(new Runnable() {
        @Override
        public void run() {
            System.out.println("子线程名称===>> " +
                               Thread.currentThread().getName());
        }
    });
}
```

（2）在 main()方法中测试上面的程序，代码如下。

```
public static void main(String[] args) {
    new RunnableTest().createThreadByRunnableAnonymousClass().start();
    System.out.println("主线程名称===>> " + Thread.currentThread().getName());
}
```

(3)运行 main()方法,输出结果如下。

```
主线程名称===>> main
子线程名称===>> Thread-0
```

除了主线程,程序还通过匿名类的方式创建了一个名为 Thread-0 的子线程,并执行了子线程中重写的 run()方法。

2. 通过 Lambda 表达式创建线程

Lambda 表达式是 Java 8 提供的新特性,可以用于创建线程。

(1)创建 createThreadByRunnableLambda()方法,并在 createThreadByRunnableLambda()方法中通过 Lambda 表达式创建一个 Thread 类的对象,并返回 Thread 类的对象,代码如下。

```
//Lambda 表达式创建线程
public Thread createThreadByRunnableLambda(){
    return new Thread(() -> {
        System.out.println("子线程名称===>> " + Thread.currentThread().getName());
    });
}
```

(2)在 main()方法中测试上面的程序,如下所示。

```
public static void main(String[] args) {
    new RunnableTest().createThreadByRunnableLambda().start();
    System.out.println("主线程名称===>> " + Thread.currentThread().getName());
}
```

(3)运行 main()方法,输出结果同上。

注意:Lambda 表达式是 Java 8 提供的新特性,可以关注笔者的微信公众号"冰河技术"阅读有关 Java 8 新特性的专栏文章。

3. 通过 Runnable 实现类创建线程

创建一个 Runnable 实现类,将它传递到 Thread 类的构造方法中,从而创建一个线程。

(1)在 RunnableTest 类中创建一个实现了 Runnable 接口的内部类 MyRunnableTask,并重写 Runnable 接口的 run()方法,代码如下。

```
//通过 Runnable 实现类创建线程
public class RunnableTest{
    private static class MyRunnableTask implements Runnable{
        @Override
        public void run() {
            System.out.println("子线程名称===>> " + Thread.currentThread().getName());
        }
    }
}
```

(2)在 main()方法中测试上面的程序,代码如下。

```
public static void main(String[] args) {
    new Thread(new MyRunnableTask()).start();
    System.out.println("主线程名称===>> " + Thread.currentThread().getName());
}
```

（3）运行 main() 方法，输出结果同上。

2.1.3 实现 Callable 接口

不同于 Runnable 接口，Callable 接口能够返回结果数据，并能在外部捕获 Callable 接口中的 call() 方法抛出的异常。

该方式可以细分为通过匿名类创建线程、通过 Lambda 表达式创建线程和通过 Callable 实现类创建线程 3 种方式。

1. 通过匿名类创建线程

（1）创建 CallableTest 类，在 CallableTest 类中创建 createThreadByCallableAnonymousClass() 方法，并在 createThreadByCallableAnonymousClass() 方法中，将 Callable 接口的匿名类对象传递给 FutureTask 类的构造方法，创建 FutureTask 对象，并返回 FutureTask 对象。代码如下。

```
//通过匿名类创建线程
public FutureTask createThreadByCallableAnonymousClass() {
    return new FutureTask<>(new Callable<String>() {
        @Override
        public String call() throws Exception {
            System.out.println("子线程名称===>> " + Thread.currentThread().getName());
            return Thread.currentThread().getName();
        }
    });
}
```

（2）在 main() 方法中测试上面的程序，代码如下。

```
public static void main(String[] args) throws ExecutionException, InterruptedException {
    FutureTask futureTask = new CallableTest().createThreadByCallableAnonymousClass();
    new Thread(futureTask).start();
    System.out.println("从子线程中获取的数据为===>> " + futureTask.get());
    System.out.println("主线程名称===>> " + Thread.currentThread().getName());
}
```

（3）运行 main() 方法，输出结果如下。

```
子线程名称===>> Thread-0
从子线程中获取的数据为===>> Thread-0
主线程名称===>> main
```

除了主线程，程序还通过匿名类创建了一个名为 Thread-0 的子线程，并执行了子线程中重写的 call()方法。另外，通过 FutureTask 能够获取 Callable 接口中 call()方法的返回结果。

2. 通过 Lambda 表达式创建线程

（1）创建 createThreadByCallableLambda()方法，在 createThreadByCallableLambda()方法中通过 Lambda 表达式将 Callable 接口的对象传递给 FutureTask 类的构造方法，创建 FutureTask 类的对象，并返回 FutureTask 类的对象，代码如下。

```
//通过 Lambda 表达式创建线程
public FutureTask createThreadByCallableLambda() {
    return new FutureTask<>(() -> {
        System.out.println("子线程名称===>> " + Thread.currentThread().getName());
        return Thread.currentThread().getName();
    });
}
```

（2）在 main()方法中测试上面的程序，代码如下。

```
public static void main(String[] args) throws ExecutionException,
InterruptedException {
    FutureTask futureTask = new CallableTest().createThreadByCallableLambda();
    new Thread(futureTask).start();
    System.out.println("从子线程中获取的数据为===>> " + futureTask.get());
    System.out.println("主线程名称===>> " + Thread.currentThread().getName());
}
```

（3）运行 main()方法，输出结果同上。

3. 通过 Callable 实现类创建线程

（1）在 CallableTest 类中创建一个实现了 Callable 接口的内部类 MyCallbleTask，并在 MyCallbleTask 类中重写 Callable 接口的 call()方法，代码如下。

```
//通过 Callable 实现类创建线程
private static class MyCallbleTask implements Callable<String>{
    @Override
    public String call() throws Exception {
        System.out.println("子线程名称===>> " + Thread.currentThread().getName());
        return Thread.currentThread().getName();
    }
}
```

（2）在 main()方法中测试上面的程序，代码如下。

```
public static void main(String[] args) throws ExecutionException,
InterruptedException {
    FutureTask futureTask = new FutureTask(new MyCallbleTask());
    new Thread(futureTask).start();
    System.out.println("从子线程中获取的数据为===>> " + futureTask.get());
    System.out.println("主线程名称===>> " + Thread.currentThread().getName());
}
```

(3)运行 main()方法,输出结果同上。

2.1.4　FutureTask 配合 Thread

FutureTask 类实现了 Runnable 接口和 Callable 接口,所以也可以使用 FutureTask 配合 Thread 的方式创建线程,示例代码如下。

```
public class FutureTaskTest {
    public static void main(String[] args)throws ExecutionException,
InterruptedException{
        FutureTask<String> futureTask = new FutureTask<>(()-> {
            return "使用FutureTask配合Thread的方式创建线程";
        });
        new Thread(futureTask).start();
        System.out.println(futureTask.get());
    }
}
```

直接运行上述代码,输出结果如下。

```
使用FutureTask配合Thread的方式创建线程
```

使用 FutureTask 配合 Thread 的方式创建线程,可以通过 FutureTask 对象获取子线程返回的结果。

2.1.5　使用线程池

通过继承 Thread 类、实现 Runnable 接口、使用 FutureTask 配合 Thread 和实现 Callable 接口创建的线程,在执行完任务后就会被销毁,这些线程资源和线程实例都不能被复用,要知道,在程序处理的过程中,线程的创建与销毁是非常消耗性能的。那如何更好地复用创建的线程呢?答案就是线程池技术。

通过线程池技术创建线程包括通过 Executors 工具类和 ThreadPoolExecutor 类创建线程。

1. 通过 Executors 工具类创建线程

Executors 工具类是 JDK 提供的简化线程池开发工作的工具类,接下来,就以 Executors 工具类的 newFixedThreadPool()方法为例,通过 Executors 工具类创建线程。

(1)创建 ExecutorsTest 类,并在 ExecutorsTest 类中使用 Executors 工具类的 newFixedThreadPool()方法创建 ExecutorService 对象,代码如下。

```
//通过Executors工具类创建线程
public class ExecutorsTest {
    private static ExecutorService threadPool;
    static {
        threadPool = Executors.newFixedThreadPool(3);
    }
}
```

（2）在 main()方法中分别测试 ExecutorService 对象的 submit(Runnable task)、submit(Callable task)和 execute(Runnable command)方法。

- 测试 submit(Runnable task)方法。

main()方法的代码如下。

```java
public static void main(String[] args){
    System.out.println("主线程名称===>> " + Thread.currentThread().getName());
    threadPool.submit(()->{
        System.out.println("子线程名称===>> " + Thread.currentThread().getName());
    });
}
```

输出结果如下。

```
主线程名称===>> main
子线程名称===>> pool-1-thread-1
```

- 测试 submit(Callable task)方法。

main()方法的代码如下。

```java
public static void main(String[] args) throws ExecutionException, InterruptedException{
    System.out.println("主线程名称===>> " + Thread.currentThread().getName());
    Future<String> future = threadPool.submit(() -> {
        System.out.println("子线程名称===>> " + Thread.currentThread().getName());
        return Thread.currentThread().getName();
    });
    System.out.println("从子线程中获取的数据为===>> " + future.get());
}
```

输出结果如下。

```
主线程名称===>> main
子线程名称===>> pool-1-thread-1
从子线程中获取的数据为===>> pool-1-thread-1
```

- 测试 execute(Runnable command)方法。

main()方法的代码如下。

```java
public static void main(String[] args){
    System.out.println("主线程名称===>> " + Thread.currentThread().getName());
    threadPool.execute(() -> {
        System.out.println("子线程名称===>> " + Thread.currentThread().getName());
    });
}
```

输出结果如下。

```
主线程名称===>> main
子线程名称===>> pool-1-thread-1
```

除了主线程，程序还使用线程池创建了名称为 pool-1-thread-1 的线程，并且执行了子线程中重写的 run()方法和 call()方法。同时，线程池中 ExecutorService 对象的 submit()方法存在返回值，execute()方法没有返回值。

2. 通过 ThreadPoolExecutor 类创建线程

《阿里巴巴 Java 开发手册》中不建议直接使用 Executors 工具类创建线程池，因为存在资源耗尽的风险，推荐通过 ThreadPoolExecutor 类创建线程池。这里以 ThreadPoolExecutor 类的如下构造方法为例创建线程。

```
public ThreadPoolExecutor(int corePoolSize,
            int maximumPoolSize,
            long keepAliveTime,
            TimeUnit unit,
            BlockingQueue<Runnable> workQueue)
```

（1）创建 ThreadPoolExecutorTest 类，并在 ThreadPoolExecutorTest 类中创建 ThreadPoolExecutor 类的对象，代码如下。

```
//通过 ThreadPoolExecutor 类创建线程
public class ThreadPoolExecutorTest {
    private static ThreadPoolExecutor threadPool;
    static {
        threadPool = new ThreadPoolExecutor(3, 3, 30,
            TimeUnit.SECONDS, new ArrayBlockingQueue<>(5));
    }
}
```

（2）在 main()方法中分别测试 ThreadPoolExecutor 类对象的 submit(Runnable task)、submit(Callable task)和 execute(Runnable command)方法，与测试 ExecutorService 对象的 submit(Runnable task)、submit(Callable task)和 execute(Runnable command)方法的方式与结果相同。

注意：在 Java 中使用线程池创建线程的更多方式，读者可以参考《深入理解高并发编程：核心原理与案例实战》一书，也可以通过阅读"冰河技术"微信公众号中的相关技术文章自行实现。

2.2　线程的基本操作

Java 的 Thread 类提供了大量操作线程的 API，可以方便地操作 Java 中的线程。

2.2.1　线程设置

Java 提供了两种设置线程名称的方式，一种是在创建 Thread 对象时，通过构造方法传入线程的名称；另一种是在创建 Thread 对象后，通过 Thread 对象的 setName()方法设置线程的名称。

在 Thread 类中，通过构造方法传入线程名称的代码如下。

```
public Thread(String name)
public Thread(ThreadGroup group, String name)
public Thread(Runnable target, String name)
public Thread(ThreadGroup group, Runnable target, String name)
public Thread(ThreadGroup group, Runnable target, String name,long stackSize)
```

在 Thread 类中，通过 setName()方法设置线程名称的方法如下。

```
public final synchronized void setName(String name)
```

（1）通过 Thread 类的构造方法设置线程名称的示例代码如下。

```
//线程的基本操作
public class ThreadHandlerTest {
    public static void main(String[] args){
        Thread thread = new Thread(() -> {
            System.out.println("子线程名称===>> " + Thread.currentThread().getName());
        },"binghe-thread");
        thread.start();
        System.out.println("主线程名称===>> " + Thread.currentThread().getName());
    }
}
```

运行上面的程序，输出结果如下。

```
主线程名称===>> main
子线程名称===>> binghe-thread
```

从输出结果可以看出，通过 Thread 类的构造方法已经成功将子线程的名称设置为 binghe-thread。

（2）通过 Thread 对象的 setName()方法设置线程名称的示例代码如下。

```
public static void main(String[] args){
    Thread thread = new Thread(() -> {
        System.out.println("子线程名称===>> " + Thread.currentThread().getName());
    });
    thread.setName("binghe-thread");
    thread.start();
    System.out.println("主线程名称===>> " + Thread.currentThread().getName());
}
```

输出结果同上。

通过 Thread 类的构造方法为线程指定线程组的方法如下。

```
public Thread(ThreadGroup group, String name)
public Thread(ThreadGroup group, Runnable target, String name)
public Thread(ThreadGroup group, Runnable target, String name,long stackSize)
```

代码示例如下。

```
public static void main(String[] args){
    Thread thread = new Thread(new ThreadGroup("binghe-thread-group"),() -> {
```

```
        System.out.println("子线程名称===>> " + Thread.currentThread().getName());
    });
    thread.start();
    System.out.println("主线程名称===>> " + Thread.currentThread().getName());
    System.out.println("子线程所在的线程组名称为===>>> " +
thread.getThreadGroup().getName());
}
```

运行 main()方法，输出结果如下。

```
主线程名称===>> main
子线程名称===>> Thread-0
子线程所在的线程组名称为===>>> binghe-thread-group
```

从输出结果可以看出，通过 Thread 类的构造方法正确设置了线程所在的线程组。

可以通过 Thread 类的 setPriority()方法设置线程的优先级，setPriority()接收一个范围在 1~10 的 int 类型的数字表示线程的优先级，数字越大优先级越高，最小的优先级为 1，最大的优先级为 10，默认的优先级为 5。操作系统会倾向于优先调度优先级较高的线程。

在 Thread 类中，设置线程优先级的方法如下。

```
public final void setPriority(int newPriority)
```

设置线程优先级的示例程序比较简单，读者可自行实现。

2.2.2 线程操作

1. 启动线程

可以通过 Thread 类的 start()方法启动线程，也可以通过线程池的 submit()方法和 execute()方法启动线程，在 Thread 类中启动线程的方法如下。

```
public synchronized void start()
```

关于启动线程的示例已经在本章中大量使用。

2. 线程休眠

在 Thread 类中，可以通过 sleep()方法实现线程的休眠，sleep()方法如下。

```
public static native void sleep(long millis) throws InterruptedException;
public static void sleep(long millis, int nanos) throws InterruptedException;
```

例如，在子线程中先输出当前时间，再输出当前线程的名称，使用 Thread 类的 sleep()方法让线程休眠 2s，再次输出当前时间，示例代码如下。

```
public static void main(String[] args){
    Thread thread = new Thread(() -> {
        System.out.println("当前时间为===>>> " + new Date());
        System.out.println("子线程名称===>> " + Thread.currentThread().getName());
        try {
            Thread.sleep(2000);
```

```
        } catch (InterruptedException e) {
            e.printStackTrace();
        }
        System.out.println("当前时间为===>>> " + new Date());
    });
    thread.start();
    System.out.println("主线程名称===>> " + Thread.currentThread().getName());
}
```

运行 main()方法，输出结果如下。

```
主线程名称===>> main
当前时间为===>>> Sat Aug 20 17:12:25 CST 2022
子线程名称===>> Thread-0
当前时间为===>>> Sat Aug 20 17:12:27 CST 2022
```

从输出结果可以看出，两次输出的时间间隔 2s，说明使用 Thread.sleep()方法成功让子线程休眠了 2s。

3. 中断线程

在 Thread 类中，可以通过 interrupt()方法实现线程的中断操作，interrupt()方法如下。

```
public void interrupt()
```

interrupt()方法的作用如下。

（1）当线程被 Object.wait()、Thread.join()和 Thread.sleep()方法阻塞时，调用了当前线程的 interrupt()方法，则当前线程会抛出 InterruptedException 异常，并清空线程的中断标记。此时，需要捕获 InterruptedException 异常，并做好异常处理。

（2）如果线程在运行时调用了当前线程的 interrupt()方法，则当前线程的中断标记会被设置为 true。此时，在程序的逻辑中，需要通过 Thread 类的 isInterrupted()方法来检测当前线程是否被中断，如果已经被中断，则退出程序。

接下来，实现线程中断的示例程序。

（1）中断休眠状态的线程示例程序代码如下。

```
public static void main(String[] args) throws InterruptedException {
    Thread thread = new Thread(() -> {
        System.out.println("子线程名称===>> " + Thread.currentThread().getName());
        try {
            Thread.sleep(5000);
        } catch (InterruptedException e) {
            System.out.println("中断休眠中的线程会抛出异常，
                                并清空中断标记，
                                捕获异常后重新设置中断标记");
            Thread.currentThread().interrupt();
        }
    });
```

```
        thread.start();
        //保证子线程已经启动
         Thread.currentThread().sleep(500);
        System.out.println("在主线程中中断子线程");
        //中断子线程
        thread.interrupt();
        System.out.println("主线程名称===>> " + Thread.currentThread().getName());
}
```

运行 main()方法，输出结果如下。

```
子线程名称===>> Thread-0
在主线程中中断子线程
主线程名称===>> main
中断休眠中的线程会抛出异常，并清空中断标记，捕获异常后重新设置中断标记
```

休眠中的子线程在被中断时会抛出 InterruptedException 并清空中断标记，此时需要捕获 InterruptedException 异常，并在捕获异常后的 catch{}代码块中调用当前线程的 interrupt()方法重新设置中断标记，这时子线程才能被中断，退出执行。

（2）中断正在运行的线程的示例代码如下。

```
public static void main(String[] args) throws InterruptedException {
    Thread thread = new Thread(() -> {
        System.out.println("子线程名称===>> " + Thread.currentThread().getName());
        while (!Thread.currentThread().isInterrupted()){
        }
        System.out.println("子线程退出了while循环");
    });
    thread.start();
    //保证子线程已经启动
    Thread.currentThread().sleep(500);
    System.out.println("在主线程中中断子线程");
    //中断子线程
    thread.interrupt();
    System.out.println("主线程名称===>> " + Thread.currentThread().getName());
}
```

在子线程中通过 while (!Thread.currentThread().isInterrupted()){}不断检测当前线程是否被打断，如果当前线程被打断，则退出 while 循环，并输出"子线程退出了 while 循环"。

运行 main()方法，输出结果如下。

```
子线程名称===>> Thread-0
在主线程中中断子线程
主线程名称===>> main
子线程退出了while循环
```

从输出结果可以看出，主线程成功打断了正在运行的子线程，子线程退出了 while 循环。

4. 等待与通知

等待与通知操作可以调用 Object 对象的 wait()、notify()或 notifyAll()方法实现，前提是获取到 synchronized 对象锁。

在 Object 类中，等待与通知方法如下。

```
public final void wait() throws InterruptedException
public final native void wait(long timeout) throws InterruptedException;
public final void wait(long timeout, int nanos) throws InterruptedException
public final native void notify();
public final native void notifyAll();
```

示例程序代码如下。

```
public static void main(String[] args) throws InterruptedException {
    final Object obj = new Object();
    Thread thread = new Thread(() -> {
        System.out.println("子线程名称===>> " + Thread.currentThread().getName());
        System.out.println("子线程等待");
        synchronized (obj){
            try {
                obj.wait();
            } catch (InterruptedException e) {
                e.printStackTrace();
            }
        }
        System.out.println("子线程被唤醒");
    });
    thread.start();
    //保证子线程已经启动
    Thread.currentThread().sleep(500);
    System.out.println("主线程通知子线程");
    synchronized (obj){
        obj.notify();
    }
    System.out.println("主线程名称===>> " + Thread.currentThread().getName());
}
```

在子线程中，首先输出当前线程的信息，然后输出"子线程等待"，随后获取 synchronized 对象锁，获取锁成功后调用锁对象的 wait()方法等待，如果子线程被其他线程唤醒，则输出"子线程被唤醒"。接下来，在主线程获取到 synchronized 对象锁后，调用锁对象的 notify()方法唤醒子线程。

运行 main()方法，输出结果如下。

```
子线程名称===>> Thread-0
子线程等待
主线程通知子线程
主线程名称===>> main
子线程被唤醒
```

注意：wait()、notify()与notifyAll()方法在Object类中，而不在Thread类中。

5. 挂起与执行

Thread类的API提供了一组线程挂起与继续执行的API，对应的方法是suspend()和resume()，调用suspend()方法会使线程挂起，调用resume()方法会使线程继续执行。不过这两种方法已经被标记为废弃，不推荐使用。

在Thread类中，suspend()方法和resume()方法如下。

```
@Deprecated
public final void suspend()
@Deprecated
public final void resume()
```

这里列举一个简单的使用示例。

```
public static void main(String[] args) throws InterruptedException {
    final Object obj = new Object();
    Thread thread = new Thread(() -> {
        System.out.println("子线程名称===>> " + Thread.currentThread().getName());
        synchronized (obj){
            System.out.println("子线程挂起");
            Thread.currentThread().suspend();
        }
        System.out.println("子线程被唤醒");
    });
    thread.start();
    //保证子线程已经启动
    Thread.currentThread().sleep(500);
    System.out.println("主线程通知子线程继续执行");
    thread.resume();
    System.out.println("主线程名称===>> " + Thread.currentThread().getName());
}
```

在子线程中，首先输出当前线程的名称，然后获取synchronized对象锁，获取锁成功后，输出"子线程挂起"，随后调用suspend()方法将子线程挂起，如果有其他线程唤醒子线程继续执行，则输出"子线程被唤醒"。在主线程中，为了让子线程先执行，主线程会先休眠500毫秒，然后输出"主线程通知子线程继续执行"，最后主线程唤醒子线程继续执行。

运行main()方法，输出结果如下。

```
子线程名称===>> Thread-0
子线程挂起
主线程通知子线程继续执行
主线程名称===>> main
子线程被唤醒
```

从输出结果中可以看出，子线程调用suspend()方法被挂起，主线程成功调用子线程的

resume()方法唤醒子线程使其继续执行。

> **注意**：Thread 类中的 suspend()方法和 resume()方法已经被标记为过时，不推荐使用。

6. 等待结束与谦让

Thread 类提供了等待线程结束的 join()方法和先让其他线程执行的 yield()方法，如下所示。

```
public final void join() throws InterruptedException
public final synchronized void join(long millis) throws InterruptedException
public final synchronized void join(long millis, int nanos) throws
InterruptedException
public static native void yield();
```

join()方法会等待线程执行结束，yield()方法是一个静态本地方法，如果在某个线程中调用了 Thread.yield()方法，就会让当前线程让出 CPU 资源。

让出 CPU 资源的线程后续还会和其他线程竞争 CPU 的资源，但不一定能竞争成功。如果某个线程的优先级比较低，或者这个线程执行的是一些低级别的任务，那么可以在适当的时候调用 Thread.yield()方法让出 CPU 资源，以便高优先级的线程尽快执行。

接下来，给出一个 join()方法的使用示例。

```
private static int sum = 0;
public static void main(String[] args) throws InterruptedException {
    Thread thread = new Thread(() -> {
        System.out.println("子线程名称===>> " + Thread.currentThread().getName());
        IntStream.range(0, 1000).forEach((i) -> {sum += 1;});
    });
    thread.start();
    thread.join();
    System.out.println("主线程获取的结果为===>>> " + sum);
    System.out.println("主线程名称===>> " + Thread.currentThread().getName());
}
```

上述程序首先定义了一个 int 类型的静态变量 sum，初始值为 0，接下来在 main()方法中创建一个子线程，输出当前线程的名称，然后循环 1000 次，每次都让 sum 值加 1。在主线程中启动子线程，并调用子线程的 join()方法，等待子线程执行结束后再打印 sum 变量的值。

运行 main()方法，输出结果如下。

```
子线程名称===>> Thread-0
主线程获取的结果为===>>> 1000
主线程名称===>> main
```

从输出结果可以看出，主线程正确获取了子线程中对于 sum 变量的操作结果。

7. 守护线程

使用 Thread 类创建的线程可以分为用户线程和守护线程，用户线程也叫非守护线程，Thread 类提供了守护线程和非守护线程之间的转换方法，如下所示。

```
public final void setDaemon(boolean on)
```

在调用 Thread 类中的 setDaemon()方法时传入 true，会将线程设置为守护线程；传入 false，会将线程设置为非守护线程。使用 Thread 类创建的线程默认为非守护线程。

setDaemon()方法比较简单，读者可自行实现。

8. 终止线程

Thread 类提供了终止线程的方法，如下所示。

```
@Deprecated
public final void stop()
```

示例程序如下。

```
public static void main(String[] args) throws InterruptedException {
    Thread thread = new Thread(() -> {
        System.out.println("子线程名称===>> " + Thread.currentThread().getName());
        while (true){
        }
    });
    System.out.println("启动子线程===>>> " + new Date());
    thread.start();
    //保证子线程已经启动
    Thread.currentThread().sleep(5000);
    System.out.println("强制退出子线程===>>> " + new Date());
    thread.stop();
    System.out.println("主线程名称===>> " + Thread.currentThread().getName());
}
```

在子线程中，首先打印当前线程的名称，然后使用一个 while(true)死循环防止子线程退出，接下来让主线程休眠 5s，调用子线程的 stop()方法强制退出子线程。

运行 main()方法，输出结果如下。

```
启动子线程===>>> Sat Aug 20 20:10:16 CST 2022
子线程名称===>> Thread-0
强制退出子线程===>>> Sat Aug 20 20:10:21 CST 2022
主线程名称===>> main
```

从输出结果可以看出，子线程启动后 5s 就被强制退出了。

注意：stop()方法会强制退出线程，已被标记为过时，不推荐使用。

2.3 线程组

Java 提供的线程组能够更加高效地管理线程，可以将线程归到某一个线程组中，对线程进行分类管理。线程组中可以存在其他的线程组，也可以存在线程对象。

2.3.1 线程关联线程组

线程与线程组之间的关联关系包括一级关联和多级关联，一级关联指一个线程组中只有线程，没有线程组；多级关联指一个线程组中可能有线程，也可能有线程组。通过线程组的方法可以查看线程组中活跃的线程数量。

线程一级关联线程组的示例如下。

```java
public static void main(String[] args) throws InterruptedException {
    ThreadGroup threadGroup = new ThreadGroup("binghe-thread-group");
    Thread thread1 = new Thread(threadGroup, () -> {
        System.out.println("子线程名称===>> " + Thread.currentThread().getName());
        try {
            Thread.currentThread().sleep(1000);
        } catch (InterruptedException e) {
            e.printStackTrace();
        }
    });
    Thread thread2 = new Thread(threadGroup, () -> {
        System.out.println("子线程名称===>> " + Thread.currentThread().getName());
        try {
            Thread.currentThread().sleep(1000);
        } catch (InterruptedException e) {
            e.printStackTrace();
        }
    });
    thread1.start();
    thread2.start();
    //保证子线程已经启动
    Thread.currentThread().sleep(500);
    System.out.println("线程组中活跃的线程数量为===>> " + threadGroup.activeCount());
    System.out.println("主线程名称===>> " + Thread.currentThread().getName());
}
```

在上面的程序代码中，首先创建了一个名为 binghe-thread-group 的线程组，接下来将线程组对象传入 Thread 类的构造方法创建了两个子线程，然后在子线程的 run()方法中输出当前线程的名称，最后将当前线程休眠 1s。在主线程中，分别启动两个子线程，为了保证两个子线程已经启动，将主线程休眠 500 毫秒后输出线程组中活跃线程的数量。

运行 main()方法，输出结果如下。

```
子线程名称===>> Thread-0
子线程名称===>> Thread-1
线程组中活跃的线程数量为===>> 2
主线程名称===>> main
```

线程组中活跃的线程数量为 2，符合预期。

线程多级关联线程组的示例如下。

```java
public static void main(String[] args) throws InterruptedException {
    ThreadGroup threadGroup = new ThreadGroup("binghe-thread-group");
    ThreadGroup subThreadGroup = new
ThreadGroup(threadGroup,"binghe-sub-thread-group");
    Thread thread1 = new Thread(threadGroup, () -> {
        System.out.println("子线程名称===>> " + Thread.currentThread().getName());
        try {
            Thread.currentThread().sleep(1000);
        } catch (InterruptedException e) {
            e.printStackTrace();
        }
    });
    Thread thread2 = new Thread(subThreadGroup, () -> {
        System.out.println("子线程名称===>> " + Thread.currentThread().getName());
        try {
            Thread.currentThread().sleep(1000);
        } catch (InterruptedException e) {
            e.printStackTrace();
        }
    });
    thread1.start();
    thread2.start();
    //保证子线程已经启动
    Thread.currentThread().sleep(500);
    System.out.println("线程组中活跃的线程组数量为===>> " +
 threadGroup.activeGroupCount());
    System.out.println("线程组中活跃的线程数量为===>> " + threadGroup.activeCount());
    System.out.println("主线程名称===>> " + Thread.currentThread().getName());
}
```

在上述代码中，首先创建了两个线程组，一个是父线程组，另一个是子线程组。然后创建两个子线程，一个子线程关联到父线程组，另一个子线程关联到子线程组。在主线程中分别启动两个线程后休眠 500 毫秒，最后输出父线程组中活跃的线程组和线程数量。

运行 main()方法，输出结果如下。

```
子线程名称===>> Thread-1
子线程名称===>> Thread-0
线程组中活跃的线程组数量为===>> 1
线程组中活跃的线程数量为===>> 2
主线程名称===>> main
```

在父线程组中存在 1 个活跃的线程组，2 个活跃的线程。

2.3.2　线程组自动归属

在创建一个线程组时，如果不指定线程组所属的线程组，则创建的线程组会自动归属到当前线程所属的线程组中，代码示例如下。

```java
public static void main(String[] args) throws InterruptedException {
    ThreadGroup threadGroup = new ThreadGroup("binghe-thread-group");
    ThreadGroup subThreadGroup = new ThreadGroup(threadGroup,
"binghe-sub-thread-group");
    Thread thread1 = new Thread(threadGroup, () -> {
        System.out.println("子线程名称===>> " + Thread.currentThread().getName());
        try {
            Thread.currentThread().sleep(1000);
        } catch (InterruptedException e) {
            e.printStackTrace();
        }
    });
    Thread thread2 = new Thread(subThreadGroup, () -> {
        System.out.println("子线程名称===>> " + Thread.currentThread().getName());
        //加入subThreadGroup线程组
        ThreadGroup thread2Group = new ThreadGroup("binghe-thread2-group");
        try {
            Thread.currentThread().sleep(1000);
        } catch (InterruptedException e) {
            e.printStackTrace();
        }
    });
    thread1.start();
    thread2.start();
    //保证子线程已经启动
    Thread.currentThread().sleep(500);
    System.out.println("threadGroup 线程组中活跃的线程组数量为===>> " +
threadGroup.activeGroupCount());
    System.out.println("threadGroup 线程组中活跃的线程数量为===>> " +
threadGroup.activeCount());
    System.out.println("subThreadGroup 线程组中活跃的线程组数量为===>> " +
                subThreadGroup.activeGroupCount());
    System.out.println("主线程名称===>> " + Thread.currentThread().getName());
}
```

上述代码在子线程 thread2 中创建了一个名为 binghe-thread2-group 的线程组对象 thread2Group，并且 thread2 线程在 subThreadGroup 线程组中。

运行 main()方法，输出结果如下。

```
子线程名称===>> Thread-0
子线程名称===>> Thread-1
threadGroup 线程组中活跃的线程组数量为===>> 2
threadGroup 线程组中活跃的线程数量为===>> 2
subThreadGroup 线程组中活跃的线程组数量为===>> 1
主线程名称===>> main
```

subThreadGroup 线程组中活跃的线程组数量为 1，说明在 thread2 线程中创建出来的 thread2Group 线程组已经自动归属到 thread2 线程所在的 subThreadGroup 线程组中。

2.3.3　顶级线程组

顶级线程组又叫根线程组，是 JVM 中顶级的线程组。获取顶级线程组的示例代码如下。

```
public static void main(String[] args) {
    System.out.println(Thread.currentThread().getThreadGroup().getName());
    System.out.println(Thread.currentThread().getThreadGroup().getParent().getName());
    System.out.println(Thread.currentThread().getThreadGroup().getParent().getParent().getName());
}
```

运行 main()方法，输出结果如下。

```
main
system
Exception in thread "main" java.lang.NullPointerException
at io.binghe.consurrent.chapter02.ThreadHandlerTest.main(ThreadHandlerTest.java:239)
```

main()方法所在的线程所属的线程组名称为 main，main 线程组的父级线程组名称为 system，system 线程组没有父级线程组，所以 JVM 中的顶级线程组，也就是根线程组为 system 线程组。

2.3.4　向线程组里添加线程组

线程组中支持添加新的线程组，示例程序如下。

```
public static void main(String[] args) {
    ThreadGroup threadGroup = new ThreadGroup("binghe-thread-group");
    System.out.println("threadGroup 线程组中活跃的线程组数量为===>> " +
                       threadGroup.activeGroupCount());
    ThreadGroup subThreadGroup = new ThreadGroup(threadGroup,
                       "binghe-sub-thread-group");
    System.out.println("threadGroup 线程组中活跃的线程组数量为===>> " +
                       threadGroup.activeGroupCount());
}
```

上述代码示例的逻辑比较简单，运行 main()方法，输出结果如下。

```
threadGroup 线程组中活跃的线程组数量为===>> 0
threadGroup 线程组中活跃的线程组数量为===>> 1
```

新创建的 subThreadGroup 线程组已经加入 threadGroup 线程组。

2.3.5　获取线程组内的对象

可以通过递归和非递归两种方式来获取线程组内的对象，示例程序如下。

```
public static void main(String[] args) {
    ThreadGroup mainGroup = Thread.currentThread().getThreadGroup();
    ThreadGroup threadGroup = new ThreadGroup(mainGroup, "threadGroup");
    ThreadGroup subThreadGroup1 = new ThreadGroup(threadGroup,"subThreadGroup1");
    ThreadGroup subThreadGroup2 = new ThreadGroup(threadGroup,"subThreadGroup2");
    ThreadGroup[] threadGroups1 = new ThreadGroup[mainGroup.activeGroupCount()];
```

```java
    //递归获取
    mainGroup.enumerate(threadGroups1, true);

    Stream.of(threadGroups1).forEach((tg) -> {
        if (tg != null){
            System.out.println("递归获取的线程组===>> " + tg.getName());
        }
    });

    ThreadGroup[] threadGroups2 = new ThreadGroup[mainGroup.activeGroupCount()];
    //非递归获取
    mainGroup.enumerate(threadGroups2, false);

    Stream.of(threadGroups2).forEach((tg) -> {
        if (tg != null){
            System.out.println("非递归获取的线程组===>> " + tg.getName());
        }
    });
}
```

上述程序首先获取 main 线程所在的线程组 mainGroup，然后创建 threadGroup 线程组，并将其加入 mainGroup 线程组，接下来创建了两个线程组，分别为 subThreadGroup1 和 subThreadGroup2，并将其添加到 threadGroup 线程组中。最后，分别以递归和非递归的形式遍历 mainGroup 线程组中活跃线程组的名称。

运行 main()方法，输出结果如下。

```
递归获取的线程组===>> threadGroup
递归获取的线程组===>> subThreadGroup1
递归获取的线程组===>> subThreadGroup2
非递归获取的线程组===>> threadGroup
```

在通过递归的形式遍历 mainGroup 线程组中的线程组时，会输出 threadGroup、subThreadGroup1 和 subThreadGroup2 三个线程组。在通过非递归的形式遍历 mainGroup 线程组中的线程组时，只输出了 threadGroup 线程组。

2.3.6 批量中断线程组内的线程

Java 支持通过 ThreadGroup 线程组批量中断线程组内的线程，示例程序如下。

```java
public static void main(String[] args) throws InterruptedException {
    ThreadGroup threadGroup = new ThreadGroup("threadGroup");
    System.out.println("创建并启动所有的线程===>>> " + new Date());
    IntStream.range(0, 5).forEach((i) -> {
        //将线程都添加到 threadGroup 线程组
        new Thread(threadGroup, () -> {
            while (!Thread.currentThread().isInterrupted()){

            }
```

```
            System.out.println("子线程" + Thread.currentThread().getName() + "被中断
===>>>
                          " + new Date());
        }).start();
    });
    //主线程休眠5s
    Thread.currentThread().sleep(5000);
    System.out.println("主线程中断子线程");
    //使用线程组批量中断线程
    threadGroup.interrupt();
}
```

上述程序首先创建一个名为 threadGroup 的线程组，输出创建并启动所有线程的时间；然后创建 5 个线程，并将线程统一添加到 threadGroup 线程组中，在每个线程的 run()方法中都使用 isInterrupted()方法检测当前线程是否被中断，如果被中断则输出子线程被中断的时间；接下来，主线程休眠 5s，输出"主线程中断子线程"的日志；最后，使用 threadGroup 线程组批量中断线程。

运行 main()方法，输出结果如下。

```
创建并启动所有的线程===>>> Sat Aug 20 22:44:16 CST 2022
主线程中断子线程
子线程Thread-1被中断===>>> Sat Aug 20 22:44:21 CST 2022
子线程Thread-0被中断===>>> Sat Aug 20 22:44:21 CST 2022
子线程Thread-3被中断===>>> Sat Aug 20 22:44:21 CST 2022
子线程Thread-2被中断===>>> Sat Aug 20 22:44:21 CST 2022
子线程Thread-4被中断===>>> Sat Aug 20 22:44:21 CST 2022
```

从输出结果可以看出，子线程启动后 5s，被 threadGroup 线程组批量中断。

核心工具篇

第 3 章

同步集合

JDK 提供了非常丰富的集合类、并发工具类、锁工具类、线程工具类，Java 8 中又新增了对并行流的操作和异步编程的工具类等，使用 JDK 提供的并发编程工具类能够极大地降低并发编程的复杂度。

3.1 Vector 同步集合类及案例

Vector 是从 JDK 1.0 开始提供的一个线程安全的同步集合类，在某些情况下需要通过对额外的客户端或者同步集合类加锁来保护在 Vector 集合中进行的复合操作。

当在 Vector 集合上执行循环遍历、根据当前元素获取下一个元素，或者根据某一条件执行后续逻辑等复合操作时，如果有其他线程并发修改 Vector 集合中的元素，就会出现线程安全问题。

Vector 类的类继承关系如图 3-1 所示。

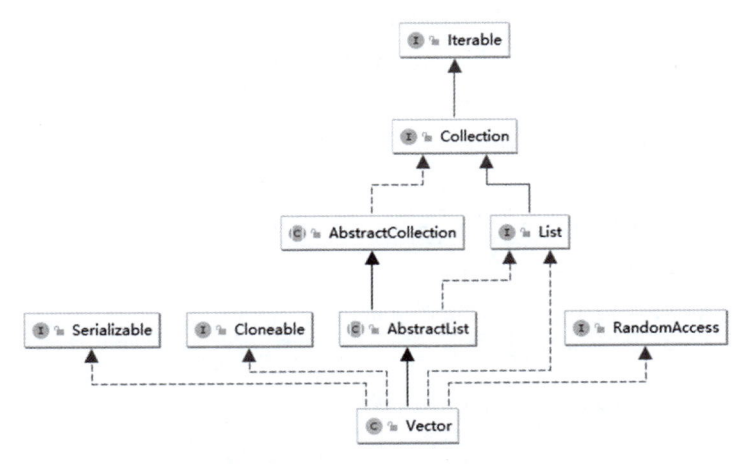

图 3-1 Vector 类的类继承关系

Vector 类实现了 List 接口，Vector 类的底层实际上是一个数组。从 JDK 中的 Vector 类的代码可以看出，Vector 类中的方法基本上都使用了 synchronized 关键字进行修饰，说明 Vector 中的方法都是线程安全的。换句话说，就是 Vector 类中的方法都是串行执行的，在单线程下的性能远远不及 ArrayList。Vector 类的大小支持手动设置。

Vector 类提供的部分方法说明如下。

- void setSize(int newSize)：设置当前 Vector 集合的大小。
- int capacity()：获取当前 Vector 集合的容量，返回的是 Vector 内部维护的数组的实际长度。
- int size()：获取当前 Vector 集合中元素的数量。
- boolean isEmpty()：判断当前 Vector 集合是否为空。
- Enumeration<E> elements()：获取 Vector 集合中所有值的集合。
- boolean contains(Object o)：查找 Vector 集合中是否包含对象 o。
- int indexOf(Object o)：查找对象 o 在 Vector 集合中首次出现的位置，若未找到，则返回 −1。
- int lastIndexOf(Object o)：查找对象 o 在 Vector 集合中最后一次出现的位置，若未找到，则返回 −1。
- E elementAt(int index)：查找 Vector 集合中 index 位置的值。
- E firstElement()：返回 Vector 集合中第 1 个位置的值。
- E lastElement()：返回 Vector 集合中最后一个位置的值。
- void setElementAt(E obj, int index)：将 Vector 集合中 index 位置的值设置为 obj。
- void removeElementAt(int index)：移除 Vector 集合中 index 位置的元素。
- void insertElementAt(E obj, int index)：将 obj 插入 Vector 集合中的 index 位置。
- void addElement(E obj)：将对象 obj 添加到 Vector 集合中。
- boolean removeElement(Object obj)：从 Vector 集合中移除对象 obj。
- void removeAllElements()：移除 Vector 集合中所有的元素。
- E get(int index)：获取 Vector 集合中 index 位置的值。
- E set(int index, E element)：将 Vector 集合中 index 位置的值设置为 elememt。
- boolean add(E e)：将元素 e 添加到 Vector 集合中。
- boolean remove(Object o)：从 Vector 集合中移除对象 o。
- void add(int index, E element)：将元素 element 添加到 Vector 集合的 index 位置。
- E remove(int index)：移除 Vector 集合中 index 位置的元素。
- void clear()：清空 Vector 集合中所有的元素。
- boolean containsAll(Collection<?> c)：判断 Vector 集合中是否包含 Collection 集合中的所有元素。

- boolean addAll(Collection<? extends E> c)：将 Collection 集合中的所有元素都添加到 Vector 集合中。
- boolean removeAll(Collection<?> c)：从 Vector 集合中移除 Collection 集合中包含的所有元素。
- boolean addAll(int index, Collection<? extends E> c)：将 Collection 集合中的所有元素插入 Vector 集合，具体的插入方式是将 Collection 集合中的第 1 个元素插入 Vector 集合中的 index 位置，将 Collection 集合中的第 2 个元素插入 Vector 集合中的 index+1 位置，直到将 Collection 集合中的所有元素都插入 Vector 集合。
- List<E> subList(int fromIndex, int toIndex)：截取 Vector 集合中 fromIndex 到 toIndex 之间的元素，并返回一个 List 集合。
- void removeRange(int fromIndex, int toIndex)：移除 Vector 集合中 fromIndex 到 toIndex 位置之间的元素。
- void forEach(Consumer<? super E> action)：遍历 Vector 集合中的元素。
- boolean removeIf(Predicate<? super E> filter)：删除 Vector 集合中符合某种条件的元素。
- void replaceAll(UnaryOperator<E> operator)：根据规则替换 Vector 集合中的所有元素。
- void sort(Comparator<? super E> c)：根据某种规则对 Vector 集合中的元素进行排序。

注意：同步集合的缺陷会在 3.5 节介绍，有关 synchronized 锁的底层核心原理，可以参考《深入理解高并发编程：核心原理与案例实战》一书。

在实际的工作场景中，Vector 类并不常用，本节简单给出一个关于 Vector 类的小案例，以加深读者对 Vector 类的理解。

使用 JMH 分别测试在 Vector 集合中添加元素、获取元素和删除元素所消耗的时间，具体的实现与测试步骤如下。

（1）在使用 JMH 对代码进行基准测试时，需要单独在项目的 pom.xml 文件中添加如下配置。

```
<dependency>
    <groupId>org.openjdk.jmh</groupId>
    <artifactId>jmh-core</artifactId>
    <version>1.19</version>
</dependency>
<dependency>
    <groupId>org.openjdk.jmh</groupId>
    <artifactId>jmh-generator-annprocess</artifactId>
    <version>1.19</version>
</dependency>
```

（2）在 mykit-concurrent-chapter03 子工程下的 io.binghe.concurrent.chapter03 包下创建 VectorTest 类，在 VectorTest 类上标注使用 JMH 进行基准测试的注解，代码如下。

```
@Fork(1)
@Measurement(iterations = 5)
@Warmup(iterations = 3)
@BenchmarkMode(Mode.AverageTime)
@OutputTimeUnit(TimeUnit.MICROSECONDS)
@Threads(value = 5)
@State(Scope.Benchmark)
public class VectorTest {
}
```

上述注解总体上表示在使用 JMH 进行基准测试时，同时开启 5 个线程，在线程共享实例的模式下预热 3 次、执行 5 次，并且输出每种方法的平均响应时间。

（3）在 VectorTest 类中，定义一个常量 ELEMENT_DATA 表示向 Vector 集合中插入的元素，并且在 VectorTest 类中创建一个标注有 JMH 的@Setup 注解的 setup()方法，用于创建 Vector 集合实例，同时向集合中添加一个 ELEMENT_DATA 常量元素，代码如下。

```
private static final String ELEMENT_DATA = "binghe";
private Vector<String> vector;
@Setup
public void setup(){
    this.vector = new Vector<>();
    vector.add(ELEMENT_DATA);
}
```

（4）在 VectorTest 类中分别创建添加元素的方法 addElement()、获取元素的方法 getElement()和移除元素的方法 removeElement()，并在每种方法上标注 JMH 的@Benchmark 注解，说明这 3 种方法是主要的基准测试方法，代码如下。

```
@Benchmark
public void addElement(){
    vector.add(ELEMENT_DATA);
}
@Benchmark
public void getElement(){
    vector.get(0);
}
@Benchmark
public void removeElement(){
    vector.remove(ELEMENT_DATA);
}
```

（5）创建 main()方法，执行 JMH 的基准测试，代码如下。

```
public static void main(String[] args) throws RunnerException {
    final Options opts = new
OptionsBuilder().include(VectorTest.class.getSimpleName()).build();
    new Runner(opts).run();
}
```

（6）在运行 main()方法后，会输出大量的基准测试信息，只需要关注最后的输出结果即可，如下所示。

```
Benchmark                    Mode    Cnt   Score       Error     Units
VectorTest.addElement        avgt    5     0.341  ±    0.125     us/op
VectorTest.getElement        avgt    5     0.293  ±    0.023     us/op
VectorTest.removeElement     avgt    5     0.367  ±    0.090     us/op
```

由输出结果可以看出，向 Vector 集合中添加一个元素平均耗时 0.341μs，误差范为±0.125μs。从 Vector 集合中获取一个元素平均耗时 0.293μs，误差为±0.023μs。从 Vector 集合中移除一个元素平均耗时 0.367μs，误差为±0.090μs。

注意：在不同的机器上运行上面的程序，得出的结果可能不同。

另外，JMH 是一款由 JVM 团队开发的、专门对代码进行基准测试的工具集，读者可以阅读"冰河技术"微信公众号的文章了解相关案例与原理剖析。

3.2　Stack 同步集合类及案例

Stack 与 Vector 一样，也是从 JDK 1.0 开始提供的一个线程安全的同步集合类，既然是同步集合类，就会存在同步集合类的缺陷。在对 Stack 集合进行复合操作时，需要注意线程安全的问题。

在 Stack 集合上进行循环遍历、根据当前元素获取下一个元素，或者根据某一条件执行后续逻辑等复合操作时，如果其他线程并发修改 Stack 集合中的元素，那么也会出现线程安全问题。

Stack 类的类继承关系如图 3-2 所示。

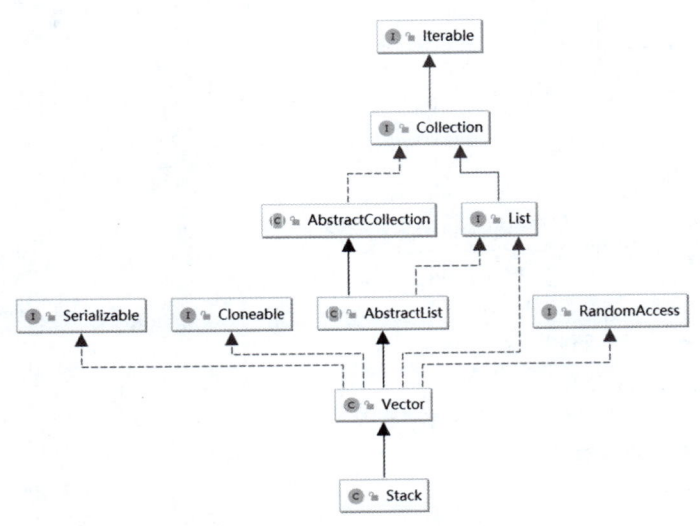

图 3-2　Stack 类的类继承关系

由图 3-2 可以看出，Stack 类继承自 Vector 类，所以 Stack 类具备 Vector 类对外提供的所有功能。Stack 类还实现了一个先进后出（First In Last Out，FILO）的栈结构，并对外提供了操作栈结构常用的方法。

- E push(E item)：向栈顶添加元素并返回添加的元素。
- E pop()：移除并返回栈顶的元素。
- E peek()：返回但不会移除栈顶的元素
- boolean empty()：判断当前栈是否为空。
- int search(Object o)：在栈中查找对象，并返回对象在栈中的位置。

注意：Stack 类继承自 Vector 类，Stack 类具备 Vector 类的所有功能，所以，从严格意义上讲，Stack 类不只是实现一个栈结构。

这里，继续使用 JMH 测试 Stack 类添加元素、获取元素和查找元素消耗的时间，对 Stack 类进行基准测试与对 Vector 类进行基准测试类似，只不过在对 Stack 进行基准测试时，是对 Stack 类的 push()方法、peek()方法和 search()方法进行基准测试。

在 mykit-concurrent-chapter03 工程下的 io.binghe.concurrent.chapter03 包下创建 StackTest 类，代码如下。

```
@Fork(1)
@Measurement(iterations = 5)
@Warmup(iterations = 3)
@BenchmarkMode(Mode.AverageTime)
@OutputTimeUnit(TimeUnit.MICROSECONDS)
@Threads(value = 5)
@State(Scope.Benchmark)
public class StackTest {
    private static final String ELEMENT_DATA = "binghe";
    private Stack<String> stack;
    @Setup
    public void setup(){
        this.stack = new Stack<>();
        stack.push(ELEMENT_DATA);
    }
    @Benchmark
    public void pushElement(){
        stack.push(ELEMENT_DATA);
    }
    @Benchmark
    public void peekElement(){
        stack.peek();
    }
    @Benchmark
    public void searchElement(){
```

```
        stack.search(ELEMENT_DATA);
    }
    public static void main(String[] args) throws RunnerException {
        final Options opts =
        new OptionsBuilder().include(StackTest.class.getSimpleName()).build();
        new Runner(opts).run();
    }
}
```

在使用 JMH 进行基准测试时，同时开启 5 个线程，在线程共享实例模式下预热 3 次，执行 5 次，并输出每种方法的平均响应时间，运行上面的程序，最终输出结果如下。

```
Benchmark                    Mode  Cnt  Score   Error   Units
StackTest.peekElement        avgt   5   0.409 ± 0.035   us/op
StackTest.pushElement        avgt   5   0.466 ± 0.177   us/op
StackTest.searchElement      avgt   5   0.408 ± 0.014   us/op
```

从 Stack 集合中获取一个元素但不删除元素的平均耗时为 0.409μs，误差为±0.035μs。向 Stack 集合中插入一个元素的平均耗时为 0.466μs，误差为±0.177μs。在 Stack 集合中搜索一个元素的平均耗时为 0.408μs，误差为±0.014μs。

注意：在不同的机器上运行上面的程序，得出的结果可能不同。

3.3　Hashtable 同步集合类及案例

Hashtable 也是从 JDK 1.0 开始提供的一个线程安全的同步集合类，但是后来逐渐被 ConcurrentHashMap 替代，在对 Hashtable 类进行复合操作时，同样需要注意线程安全的问题。

Hashtable 类对外提供的方法基本上都是使用 synchronized 关键字修饰的。也就是说，不管是在单线程环境下，还是在多线程环境下，在调用 Hashtable 类的方法读写数据时，都是串行执行的。

尽管 Hashtable 类存取数据的每种方法都使用了 synchronized 关键字修饰，但是在 Hashtable 集合上进行行循环遍历、根据当前元素获取下一个元素，或者根据某一条件执行后续逻辑等复合操作时，如果其他线程并发修改 Hashtable 集合中的元素，那么也会出现线程安全问题。

Hashtable 类的类继承关系如图 3-3 所示。

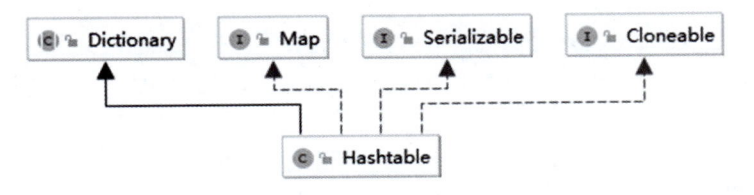

图 3-3　Hashtable 类的类继承关系

由图 3-3 可以看出，Hashtable 类继承自 Dictionary 类，并实现了 Map 接口，Dictionary 类是一个抽象类，没有实现具体的方法逻辑，只是简单定义了几个抽象方法。Map 接口中也只是简单定义了关于 Map 集合的方法。

对于 Hashtable 类提供的一些方法说明如下。

- int size()：获取 Hashtable 集合的大小，实际上是获取的 Hashtable 集合中 key 的数量。
- boolean isEmpty()：判断 Hashtable 集合是否为空。
- Enumeration<K> keys()：获取 Hashtable 集合中所有的 key。
- Enumeration<V> elements()：获取 Hashtable 集合中所有的 value。
- boolean contains(Object value)：判断 Hashtable 集合中是否包含 value 元素。
- boolean containsValue(Object value)：判断 Hashtable 集合中是否包含值为 value 的元素。
- boolean containsKey(Object key)：判断 Hashtable 集合中是否包含键为 key 的元素。
- V get(Object key)：获取 Hashtable 集合中键为 key 的 value 的值。
- V put(K key, V value)：向 Hashtable 集合中添加键为 key、值为 value 的元素。
- V remove(Object key)：移除 Hashtable 中键为 key 的元素。
- void putAll(Map<? extends K, ? extends V> t)：将 Map 集合 t 中的元素全部添加到 Hashtable 集合中。
- void clear()：清空 Hashtable 集合中的元素。
- Set<K> keySet()：获取 Hashtable 集合中所有的 key。
- Set<Map.Entry<K,V>> entrySet()：获取 Hashtable 集合中的所有键-值对集合。
- Collection<V> values()：获取 Hashtable 集合中所有的值。
- void forEach(BiConsumer<? super K, ? super V> action)：遍历 Hashtable 集合。
- V putIfAbsent(K key, V value)：向 Hashtable 集合中添加元素。
- boolean remove(Object key, Object value)：移除 Hashtable 集合中键与 key 相等、值与 value 相等的元素。
- boolean replace(K key, V oldValue, V newValue)：将 Hashtable 集合中键与 key 相等、值与 oldValue 相等的元素的值替换为 newValue。
- V replace(K key, V value)：将 Hashtable 集合中键为 key 的元素的值替换为 value。

这里，继续使用 JMH 测试在 Hashtable 集合中添加元素、获取元素和移除元素所消耗的时间。

在 mykit-concurrent-chapter03 工程下的 io.binghe.concurrent.chapter03 包下创建 HashTableTest 类，代码如下。

```
@Fork(1)
@Measurement(iterations = 5)
@Warmup(iterations = 3)
```

```
@BenchmarkMode(Mode.AverageTime)
@OutputTimeUnit(TimeUnit.MICROSECONDS)
@Threads(value = 5)
@State(Scope.Benchmark)
public class HashTableTest {
    private static final String ELEMENT_DATA = "binghe";
    private Hashtable<String, String> hashtable;
    @Setup
    public void setup(){
        this.hashtable = new Hashtable<>();
    }
    @Benchmark
    public void putElement(){
        hashtable.put(ELEMENT_DATA, ELEMENT_DATA);
    }
    @Benchmark
    public void getElement(){
        hashtable.get(ELEMENT_DATA);
    }
    @Benchmark
    public void removeElement(){
        hashtable.remove(ELEMENT_DATA);
    }
    public static void main(String[] args) throws RunnerException {
        final Options opts =
new OptionsBuilder().include(HashTableTest.class.getSimpleName()).build();
        new Runner(opts).run();
    }
}
```

使用 JMH 进行基准测试时，同时开启 5 个线程，在线程共享实例的模式下，预热 3 次，执行 5 次，并输出每种方法的平均响应时间，运行上面的程序，最终输出结果如下。

```
Benchmark                        Mode  Cnt   Score    Error   Units
HashTableTest.getElement         avgt    5   0.337 ±  0.008   us/op
HashTableTest.putElement         avgt    5   0.354 ±  0.007   us/op
HashTableTest.removeElement      avgt    5   0.314 ±  0.006   us/op
```

从 Hashtable 集合中获取一个元素的平均耗时为 0.337μs，误差为±0.008μs。向 Hashtable 中添加一个元素的平均耗时为 0.354μs，误差为±0.007μs。从 Hashtable 中移除一个元素的平均耗时为 0.314μs，误差为±0.006μs。

3.4 同步包装器及测试

Java 提供的集合类总体上分为 List、Set、Map、Queue 四大类，有些集合类是非线程安全的，例如 ArrayList 和 HashMap，如果需要将其转换成线程安全的类，则需要使用 Java 提供的同步包装器。

通过下面的方式可以在操作 HashSet 中的数据时变得线程安全。

```java
public class SafeHashSet<E> {
    private volatile Set<E> set = new HashSet<>();
    public synchronized boolean add(E e){
        return set.add(e);
    }
    public synchronized boolean isEmpty(){
        return set.isEmpty();
    }
    public synchronized boolean remove(Object o){
        return set.remove(o);
    }
    public synchronized void clear(){
        set.clear();
    }
}
```

上述代码创建了一个非线程安全集合的包装类,将一个非线程安全的集合作为这个包装类的成员变量。在这个类中,定义的操作非线程安全集合的方法都添加了 synchronized 关键字,通过这个包装类间接操作非线程安全集合中的元素,只要不涉及复合操作,就能够实现线程安全的效果。

但是,在实际的工作过程中,不可能为每个非线程安全的集合都手动创建一个包装类。JDK 的开发者已经考虑到这个问题,并在 java.util 包下的 Collections 类提供了 synchronizedXxx()方法将非线程安全的集合类转换成线程安全的同步集合类。例如,可以使用如下方法将非线程安全的 ArrayList、HashSet 和 HashMap 转换成线程安全的 List、Set 和 Map。

```java
List<Integer> list = Collections.synchronizedList(new ArrayList<Integer>());
Set<String> set = Collections.synchronizedSet(new HashSet<String>());
Map<String, String> map = Collections.synchronizedMap(new HashMap<String, String>());
```

Collections 类提供的 synchronizedXxx()方法创建的集合就是线程安全的同步集合。但是在这些集合上进行循环遍历、根据当前元素获取下一个元素,或者根据某一条件执行后续逻辑等复合操作时,如果有其他线程并发修改这些集合中的元素,则也会出现线程安全问题。

这里,以 Map 集合为例,开启 200 个线程同时执行任务,向 Map 集合中插入 5000 个键-值对。程序执行完毕,输出 Map 集合的大小,预期结果如下。

(1)当多个线程同时直接操作 HashMap 集合时,结果大概率会小于 5000。

(2)当多个线程同时操作使用 Collections.synchronizedMap()方法包装的 HashMap 集合时,结果等于 5000。

1. 测试多个线程同时直接操作 HashMap 集合

在 mykit-concurrent-chapter03 子工程下的 io.binghe.concurrent.chapter03 包下创建

HashMapTest 类进行测试，代码如下。

```java
public class HashMapTest {
    // 同时并发执行的线程数
    public static final int THREAD_COUNT = 200;
    //执行的总次数
    public static final int TOTAL_COUNT = 5000;
    // HashMap
    private static Map<Integer,Integer> map = new HashMap<>();
    public static void main(String[] args) throws InterruptedException {
        //创建一个线程池
        ExecutorService threadPool = Executors.newCachedThreadPool();
        final Semaphore semaphore = new Semaphore(THREAD_COUNT);
        final CountDownLatch countDownLatch = new CountDownLatch(TOTAL_COUNT);
        for (int i = 0; i < TOTAL_COUNT; i++) {
            final int count = i;
            threadPool.execute(() -> {
                try {
                    semaphore.acquire();//是否允许被执行
                    map.put(count, count);
                    semaphore.release();
                } catch (Exception e) {
                }
                countDownLatch.countDown();
            });
        }
        countDownLatch.await();
        threadPool.shutdown();
        System.out.println("map 的最终大小为====>>> " + map.size());
    }
}
```

注意：在上述程序中使用到了并发编程中经常使用的线程池、Semaphore 和 CountDownLatch 工具类，有关线程池的核心原理，读者可以参考《深入理解高并发编程：核心原理与案例实战》一书。

上述程序的运行结果如下。

```
map 的最终大小为====>>> 4812
```

从输出结果可以看出，多次运行上述程序，得出的结果大概率是小于 5000 的，说明 HashMap 集合不是线程安全的。

2. 测试多个线程同时操作使用 Collections.synchronizedMap()方法包装的 HashMap 集合

这里的测试程序与多个线程同时直接操作 HashMap 集合的测试程序基本相同，只是将 map 使用 Collections.synchronizedMap(new HashMap<>())包装成线程安全的同步集合类，代码如下。

```
//测试的map集合
private static Map<Integer,Integer> map = Collections.synchronizedMap(new
HashMap<>());
```

其他代码保持不变,运行结果如下。

```
map 的最终大小为====>>> 5000
```

可以看出,在多个线程同时操作使用 Collections.synchronizedMap()方法包装的 HashMap 集合时,输出了正确的结果。

3.5 同步集合的缺陷

本章多次讲述了同步集合存在的问题,从本质上讲,这些问题总体可以归为性能问题、竞态条件问题和使用迭代器遍历同步集合问题三大类。

1. 性能问题

同步集合对外提供的主要方法基本上都使用了 synchronized 关键字修饰,并且基本上会使用 synchronized 关键字修饰整个方法,synchronized 关键字在多线程并发环境下会涉及自旋锁、偏向锁、轻量级锁和重量级锁的竞争,以及锁的升级过程,使用 synchronized 关键字修饰的方法的执行性能会降低。

注意:有关 synchronized 锁的核心原理,读者可以参考《深入理解高并发编程:核心原理与案例实战》一书。

2. 竞态条件问题

同步集合可以保证每种方法单独操作的原子性,但是不能保证这些方法组合起来的复合操作的原子性,当程序中出现复合操作时,有可能出现竞态条件问题。

以 Vector 集合为例,下面的程序会出现竞态条件问题。

```
//以Vector集合为例测试竞态条件问题
public class RaceConditionVectorTest {
    //测试的元素
    private static final String ELEMENT_DATA = "binghe";
    // 同时并发执行的线程数
    public static final int THREAD_COUNT = 200;
    //执行的总次数
    public static final int TOTAL_COUNT = 5000;
    private static Vector<String> vector = new Vector<>();
    public static void addIfEmpty(){
        if (vector.isEmpty()){
            vector.add(ELEMENT_DATA);
        }
    }
    public static void main(String[] args) throws InterruptedException {
```

```java
        //创建一个线程池
        ExecutorService threadPool = Executors.newCachedThreadPool();
        final Semaphore semaphore = new Semaphore(THREAD_COUNT);
        final CountDownLatch countDownLatch = new CountDownLatch(TOTAL_COUNT);
        for (int i = 0; i < TOTAL_COUNT; i++) {
            threadPool.execute(() -> {
                try {
                    semaphore.acquire();//是否允许被执行
                    addIfEmpty();
                    semaphore.release();
                } catch (Exception e) {
                }
                countDownLatch.countDown();
            });
        }
        countDownLatch.await();
        threadPool.shutdown();
        System.out.println("vector 的最终大小为====>>> " + vector.size());
    }
}
```

上面的程序会并发调用 addIfEmpty() 方法向 Vector 集合中添加元素，addIfEmpty() 方法的逻辑比较简单：只有 Vector 集合为空，才会向 Vector 集合中添加元素。

由于 Vector 类中的 isEmpty() 方法和 add(E e) 方法都使用了 synchronized 关键字进行修饰，所以上面的程序看上去最终的运行结果为 Vector 集合中只有 1 个元素，但实际上由于竞态条件问题，运行上述程序得出的结果可能是 Vector 集合中存在多个相同的元素。

运行上述程序，输出结果如下。

```
vector 的最终大小为====>>> 2
```

从输出结果可以看出，最终 Vector 集合中存在 2 个元素，而不是 1 个元素，说明此时出现了竞态条件问题。

在多线程并发环境下，如果使用同步集合进行复合操作，就需要时刻注意是否会出现竞态条件问题。在使用同步集合进行复合操作时，对同步集合加锁，可以有效避免竞态条件问题。

修改 RaceConditionVectorTest 类的代码，在 addIfEmpty() 方法中，首先对 Vector 集合加锁，如下所示。

```java
public static void addIfEmpty(){
    synchronized (vector){
        if (vector.isEmpty()){
            vector.add(ELEMENT_DATA);
        }
    }
}
```

再次运行 RaceConditionVectorTest 类的代码，输出结果如下。

vector 的最终大小为====>>> 1

从输出结果可以看出，Vector 集合中只有 1 个元素，说明在使用同步集合进行复合操作时对同步集合加锁，能够有效避免竞态条件问题。

注意： 读者可以自行使用其他同步集合类进行验证，验证的效果与 Vector 集合相同。

3. 使用迭代器遍历同步集合问题

使用迭代器遍历同步集合也会出现线程安全问题，本质上使用迭代器遍历同步集合也是一种复合操作。

以 List 同步集合为例，程序代码如下。

```java
public class RaceConditionArrayListTest {
    private static final String ELEMENT_DATA = "binghe";
    private static List<String> list = Collections.synchronizedList(new ArrayList<>());
    static {
        IntStream.range(0, 100).forEach((i) -> {
            list.add(ELEMENT_DATA);
        });
    }
    //遍历集合
    private static void ergodicArrayList(){
        Iterator<String> iterator = list.iterator();
        while (iterator.hasNext()){
            printElement(iterator.next());
        }
    }
    private static void printElement(String str) {
        System.out.println("输出的元素信息====>>> " + str);
    }
    private static void removeElement(){
        list.remove(ELEMENT_DATA);
    }
    public static void main(String[] args) throws InterruptedException {
        new Thread(()->{ ergodicArrayList();}, "read-thread").start();
        new Thread(() -> {removeElement();}, "write-thread").start();
    }
}
```

上述程序首先定义了一个 list 同步集合，并在静态代码块中向 list 集合中添加了 10 个元素；然后创建了一个遍历集合的方法 ergodicArrayList()，该方法的逻辑是使用 Iterator 迭代器对 list 集合进行遍历，并调用 printElement()方法输出 list 集合中的每个元素；最后创建了一个 removeElement()方法移除 list 集合中的某个元素。

在 main()方法中，创建一个读线程和一个写线程，在读线程中调用 ergodicArrayList()方法遍历 list 集合并输出每个元素，在写线程中调用 removeElement()方法移除 list 集合中的某个元素。

运行结果如下。

```
输出的元素信息====>>> binghe
Exception in thread "read-thread" java.util.ConcurrentModificationException
    at java.util.ArrayList$Itr.checkForComodification(ArrayList.java:909)
    at java.util.ArrayList$Itr.next(ArrayList.java:859)
    at io.binghe.concurrent.chapter03.RaceConditionArrayListTest.
ergodicArrayList(RaceConditionArrayListTest.java:51)
    at io.binghe.concurrent.chapter03.RaceConditionArrayListTest.
lambda$main$1(RaceConditionArrayListTest.java:64)
    at java.lang.Thread.run(Thread.java:748)
```

当一个线程使用迭代器遍历同步集合时，如果有其他线程对同步集合进行写操作，则可能导致线程安全问题，抛出 ConcurrentModificationException 异常。

使用迭代器遍历集合问题的解决方案与竞态条件的解决方案类似，就是对同步集合加锁。

在 RaceConditionArrayListTest 类中，修改 ergodicArrayList() 方法，在 ergodicArrayList() 方法中对 list 同步集合加锁，修改后的 ergodicArrayList() 方法的代码如下。

```java
private static void ergodicArrayList(){
    synchronized (list){
        Iterator<String> iterator = list.iterator();
        while (iterator.hasNext()){
            printElement(iterator.next());
        }
    }
}
```

再次运行 RaceConditionArrayListTest 的代码，输出结果如下。

```
输出的元素信息====>>> binghe
输出的元素信息====>>> binghe
输出的元素信息====>>> binghe
输出的元素信息====>>> binghe
输出的元素信息====>>> binghe
输出的元素信息====>>> binghe
输出的元素信息====>>> binghe
输出的元素信息====>>> binghe
输出的元素信息====>>> binghe
输出的元素信息====>>> binghe
```

可以看出，此时已经正确地输出了遍历 list 同步集合的结果信息，说明对同步集合加锁能够有效地避免使用迭代器遍历集合时出现的线程安全问题。

注意：读者可以自行使用其他同步集合进行验证，效果与 List 同步集合相同。

第 4 章

并发 List 集合类

Java1.5 版本之前提供的线程安全集合基本上都是同步集合，主要通过在方法上添加 synchronized 关键字实现方法的同步操作，在使用这些同步集合进行并发编程时，除了性能较差，还要注意竞态条件问题和使用迭代器遍历集合的问题。

Java 1.5 及之后版本的 JDK 提供了性能更高的并发集合，主要包括 List、Set、Map 和 Queue 四大类。

4.1 CopyOnWriteArrayList 概述

Java 并发包中的并发 List 集合只有 CopyOnWriteArrayList，CopyOnWriteArrayList 使用了写时复制技术，添加、修改或删除 CopyOnWriteArrayList 中的数据都是基于底层数组的一个快照进行的，CopyOnWriteArrayList 类是线程安全的类。

每个 CopyOnWriteArrayList 类内部都有一个对象数组类型的成员变量 array 用来指向存储具体元素的数组，还会有一个 ReentrantLock 独占锁类型的成员变量 lock 用来保证在同一时刻只能有一个线程修改 array 数组中的数据。

CopyOnWriteArrayList 类的类继承关系如图 4-1 所示。

由图 4-1 可以看出，CopyOnWriteArrayList 类实现了 List 接口，CopyOnWriteArrayList 的底层本质上是一个数组。这个数组的读操作会直接返回原数组中的值；这个数组的写操作（包括添加元素、修改元素和删除元素）会首先获取 ReentrantLock 独占锁，然后复制一份底层数组 array 的副本，在 array 数组副本上进行写操作，在执行完毕后，再将 array 数组的副本赋值给 array 引用。

对 CopyOnWriteArrayList 进行写操作会复制一份内部数组，此时内存中同时存在两个内部数组。虽然原来的数组会在某个时刻被 JVM 进行 GC 操作，但是在对 CopyOnWriteArrayList

进行写操作的过程中，以及在写操作完成后、GC 操作前，会浪费将近一半的内存空间。

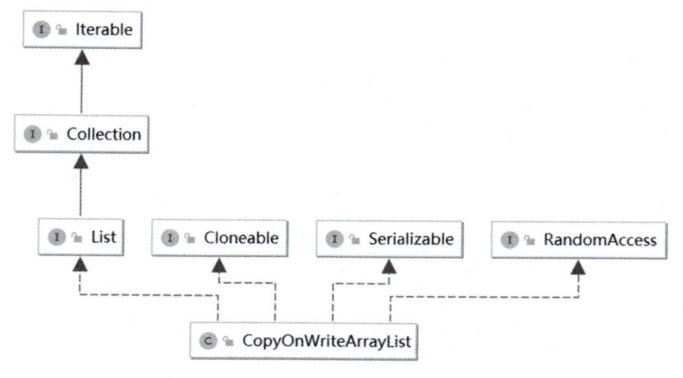

图 4-1　CopyOnWriteArrayList 类的类继承关系

另外，CopyOnWriteArrayList 并不能保证数据实时一致。例如，在线程 A 对 CopyOnWriteArrayList 进行写操作时，首先会将 CopyOnWriteArrayList 内部的数组复制一个副本，在副本上执行写操作，在执行完写操作且 array 引用未指向新的数组副本时，如果线程 B 读取 CopyOnWriteArrayList 中的数据，那么读取的还是原来数组中的数据。

所以，CopyOnWriteArrayList 适用于写操作比较少，并且能够容忍读写操作在短时间内不一致的场景。

4.2　写时复制技术

CopyOnWriteArrayList 内部使用了写时复制技术，所谓的写时复制技术，从字面意思理解就是写的时候复制一个副本，读的时候还是读原来的数据。这样做的优点就是在读取 CopyOnWriteArrayList 中的数据时，可以完全无锁化，能够极大地提升读取数据的性能。

在 CopyOnWriteArrayList 内部有一个指向内部数组的 array 引用，当发生读操作时，会直接读取 array 引用指向的原来数组中的数据。当发生写操作（添加元素、修改元素和删除元素）时，并不会在 array 引用指向的原来数组上操作，而是首先将 array 引用指向的数组复制一个副本，对副本进行写操作。写完之后，再将内部的 array 引用指向数组副本，这样可以确保对 CopyOnWriteArrayList 中元素的写操作不会影响读操作，数据的读取操作完全可以做到无锁化。

CopyOnWriteArrayList 实现写时复制技术最核心的逻辑是在 CopyOnWriteArrayList 内部维护一个数组，成员变量 array 指向这个数组，其核心代码如下。

```
private transient volatile Object[] array;
final Object[] getArray() {
    return array;
}
```

```
final void setArray(Object[] a) {
    array = a;
}
```

读操作都是基于成员变量 array 指向的这个内部数组进行的，例如在使用 E get(int index)方法获取 CopyOnWriteArrayList 中的数据时，会按照图 4-2 所示的方式读取 array 引用指向的数组。

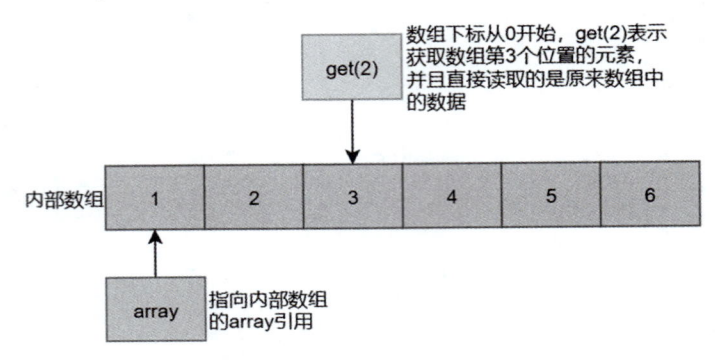

图 4-2　获取 CopyOnWriteArrayList 中的数据

由图 4-2 可以看出，CopyOnWriteArrayList 中的成员变量 array 指向内部数组的首地址，在使用 CopyOnWriteArrayList 类中的 get()方法读取数据时，会从 array 引用指向的原来的内部数组中读取数据。

在使用 Iterator 迭代器遍历 CopyOnWriteArrayList 时，会按照图 4-3 所示的方式遍历数组。

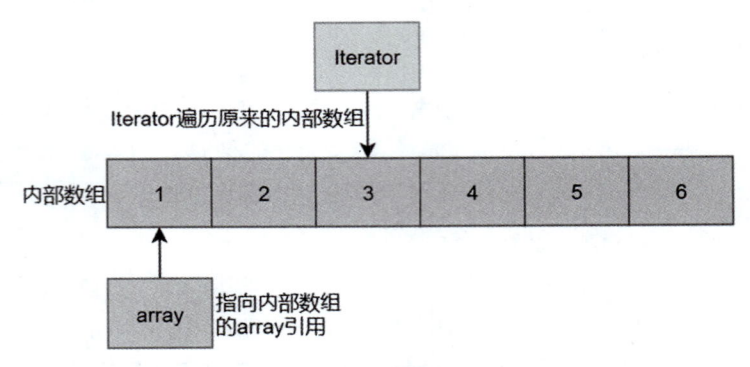

图 4-3　使用 Iterator 迭代器遍历 CopyOnWriteArrayList

由图 4-3 可以看出，在使用 Iterator 迭代器遍历 CopyOnWriteArrayList 时，实际上遍历的还是 CopyOnWriteArrayList 中 array 引用指向的原来的内部数组。

在使用 E get(int index)方法获取 CopyOnWriteArrayList 中的数据时，另一个线程向 CopyOnWriteArrayList 中添加元素，此时的处理逻辑如图 4-4 所示。

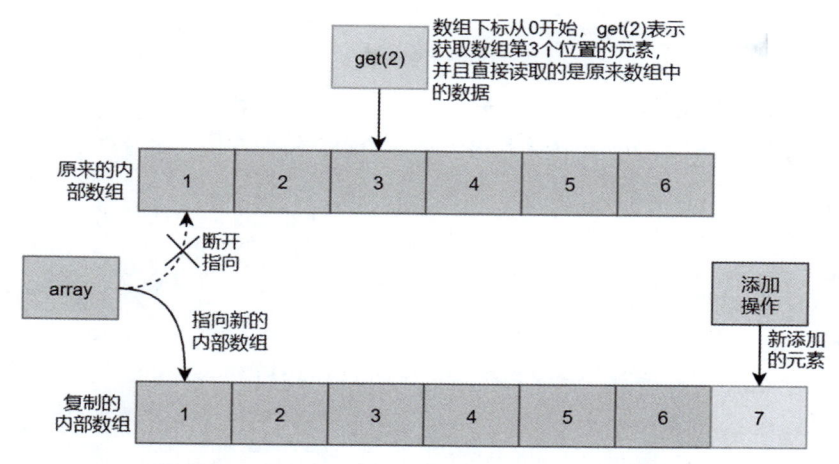

图 4-4　向 CopyOnWriteArrayList 中添加数据

由图 4-4 可以看出，如果此时有其他线程对 CopyOnWriteArrayList 进行写操作，则 get() 方法还是从原来的数组中读取数据，而在其他线程向 CopyOnWriteArrayList 中添加数据时，会先将原来的数组复制一份副本，在副本中添加数据，再将 array 引用指向数组副本。

在使用 Iterator 迭代器遍历 CopyOnWriteArrayList 时，另一个线程删除 CopyOnWriteArrayList 中的数据，处理逻辑如图 4-5 所示。

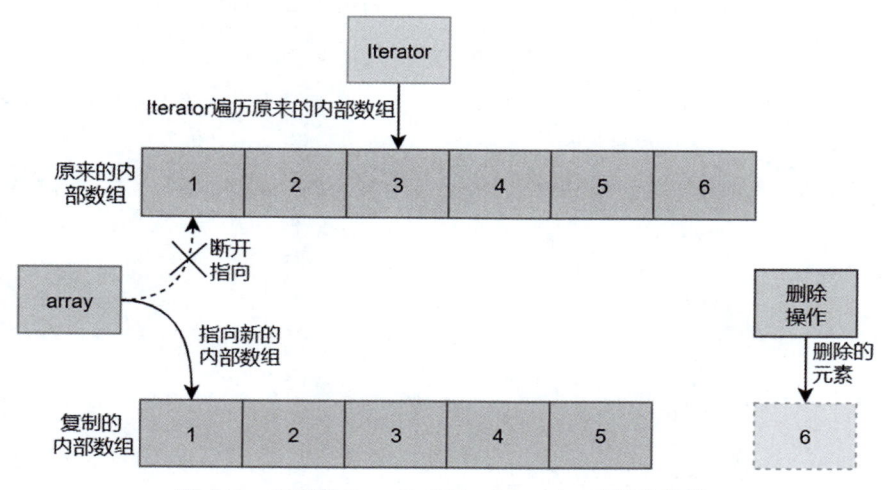

图 4-5　在遍历 CopyOnWriteArrayList 时删除数据

由图 4-5 可以看出，当使用 Iterator 遍历 CopyOnWriteArrayList 时，会直接在 array 引用指向的原来的数组上进行遍历操作，如果此时有其他线程删除 CopyOnWriteArrayList 中的元素，则会先将原来的数组复制一份副本，在副本中删除对应的元素，再将 array 引用指向数组副本。

4.3 初始化

CopyOnWriteArrayList 对外提供了 3 种构造方法来执行初始化操作，包含一个无参构造方法和两个有参构造方法，其中一个传入数组类型的参数，另一个传入集合类型的参数。

首先，查看 CopyOnWriteArrayList 的无参构造方法，代码如下。

```
public CopyOnWriteArrayList() {
    setArray(new Object[0]);
}
```

在 CopyOnWriteArrayList 中的无参构造方法中，调用 setArray()方法创建了一个大小为 0 的 Object 数组作为 array 的初始值，其中 setArray()方法的代码如下。

```
final void setArray(Object[] a) {
    array = a;
}
```

然后查看包含一个数组类型参数的构造方法，代码如下。

```
public CopyOnWriteArrayList(E[] toCopyIn) {
    setArray(Arrays.copyOf(toCopyIn, toCopyIn.length, Object[].class));
}
```

在 CopyOnWriteArrayList 中，有一个数组类型的参数的构造方法也会调用 setArray()方法初始化 array 引用，但是在初始化 array 引用之前，会将传入的泛型类型的数组中的数据复制到一个 Object 类型的数组中，随后将 array 引用指向 Object 类型的数组。

最后，再来查看包含一个集合类型参数的构造方法，代码如下。

```
public CopyOnWriteArrayList(Collection<? extends E> c) {
    Object[] elements;
    if (c.getClass() == CopyOnWriteArrayList.class)
        elements = ((CopyOnWriteArrayList<?>)c).getArray();
    else {
        elements = c.toArray();
        // c.toArray might (incorrectly) not return Object[] (see 6260652)
        if (elements.getClass() != Object[].class)
            elements = Arrays.copyOf(elements, elements.length, Object[].class);
    }
    setArray(elements);
}
```

在 CopyOnWriteArrayList 包含一个集合类型参数的构造方法中，首先会定义一个 Object 类型数组的局部变量 elements。然后，如果传入的集合类型是 CopyOnWriteArrayList，则将传入的参数 c 强转为 CopyOnWriteArrayList 类型，并调用 getArray()方法将获取的数组赋值给 elements 局部变量。否则，先将传入的参数 c 转换成数组，并赋值给局部变量 elements，再进行判断：如果 elements 数组类型不是 Object，则将 elements 数组中的元素复制到 Object 类型的数组中，并将 Object 类型的数组再次赋值给 elements 局部变量；如果 elements 数组类型是 Object，则不做任何处理。

最后调用 setArray() 方法将数组类型的局部变量 elements 赋值给 array 引用。

4.4 添加数据

CopyOnWriteArrayList 提供了多个添加数据的方法，常见的方法如下。

```
boolean add(E e)
void add(int index, E element)
boolean addIfAbsent(E e)
int addAllAbsent(Collection<? extends E> c)
boolean addAll(Collection<? extends E> c)
boolean addAll(int index, Collection<? extends E> c)
```

这里，以 boolean add(E e) 方法为例进行说明，boolean add(E e) 方法的代码如下。

```
public boolean add(E e) {
    final ReentrantLock lock = this.lock;
    lock.lock();
    try {
        Object[] elements = getArray();
        int len = elements.length;
        Object[] newElements = Arrays.copyOf(elements, len + 1);
        newElements[len] = e;
        setArray(newElements);
        return true;
    } finally {
        lock.unlock();
    }
}
```

在 boolean add(E e) 方法中，首先获取 ReentrantLock 锁；然后调用 getArray() 方法获取 array 引用，并将其赋值给 Object 数组类型的局部变量 elements，获取 elements 数组的长度；随后将 elements 数组的元素复制到一个比 elements 数组长度多 1 的数组中，并将结果赋值给 Object 类型的数组 newElements；接下来，将传入的参数 e 添加到 newElements 数组的末尾；随后调用 setArray() 方法将 newElements 数组赋值给 array 引用；最后返回 true 并释放 ReentrantLock 锁。

通过上述分析可以得知，当多个线程同时执行 boolean add(E e) 方法时，只有获取到 ReentrantLock 锁的线程才能执行 boolean add(E e) 方法的逻辑。在添加元素时，首先会将 array 指向的原来的数组复制一份副本，在副本中执行添加元素的操作，添加成功后将数组副本赋值给 array 引用。整个添加元素的过程不会影响从 CopyOnWriteArrayList 中读取数据。

4.5 读取数据

CopyOnWriteArrayList 提供了 E get(Object[] a, int index) 和 E get(int index) 两种读取数据的方

法。在 E get(int index)方法内部，直接调用 E get(Object[] a, int index)方法实现数据的读取操作，如果传入的 index 下标超出内部数组的范围，则会抛出 IndexOutOfBoundsException 异常。代码如下。

```
private E get(Object[] a, int index) {
    return (E) a[index];
}
public E get(int index) {
    return get(getArray(), index);
}
```

无论是 E get(Object[] a, int index)方法还是 E get(int index)方法，都没有加锁和释放锁的操作。说明读取 CopyOnWriteArrayList 中的数据时不需要获取锁，并且每次都从 array 引用指向的原来的数组中读取数据。

4.6 修改数据

CopyOnWriteArrayList 提供了 E set(int index, E element)和 void replaceAll(UnaryOperator<E> operator)两种修改数据的方法。

1. E set(int index, E element)方法

E set(int index, E element)方法主要用于修改指定位置的数据，如果传入的 index 下标超出数组的下标范围，则会抛出 IndexOutOfBoundsException 异常，代码如下。

```
public E set(int index, E element) {
    final ReentrantLock lock = this.lock;
    lock.lock();
    try {
        Object[] elements = getArray();
        E oldValue = get(elements, index);

        if (oldValue != element) {
            int len = elements.length;
            Object[] newElements = Arrays.copyOf(elements, len);
            newElements[index] = element;
            setArray(newElements);
        } else {
            // Not quite a no-op; ensures volatile write semantics
            setArray(elements);
        }
        return oldValue;
    } finally {
        lock.unlock();
    }
}
```

在 E set(int index, E element)方法中，首先获取 ReentrantLock 锁，然后调用 getArray()方法获取内部数组的 array 引用并将其赋值给 Object 数组类型的 elements 局部变量。随后调用 get()方法并且传入 elements 和 index 下标获取数组中 index 位置的元素，将其赋值给 oldValue。

接下来，判断 oldValue 的值是否等于传入的 element 的值，此时有两种情况。

如果 oldValue 的值不等于 element 的值，则将 elements 数组的中的元素复制到 newElements 数组中，并将 newElements 数组中下标为 index 的值设置为传入的 element，然后调用 setArray()方法将 newElement 数组赋值给 array 引用。

如果 oldValue 的值等于 element 的值，则不做任何处理，直接调用 setArray()方法将 elements 再次赋值给 array 引用。

最后返回数组中 index 处修改之前的值并释放 ReentrantLock 锁。

2. void replaceAll(UnaryOperator<E> operator)

void replaceAll(UnaryOperator<E> operator)方法会根据一定的规则修改数组中的所有数据，代码如下。

```
public void replaceAll(UnaryOperator<E> operator) {
    if (operator == null) throw new NullPointerException();
    final ReentrantLock lock = this.lock;
    lock.lock();
    try {
        Object[] elements = getArray();
        int len = elements.length;
        Object[] newElements = Arrays.copyOf(elements, len);
        for (int i = 0; i < len; ++i) {
            @SuppressWarnings("unchecked")
            E e = (E) elements[i];
            newElements[i] = operator.apply(e);
        }
        setArray(newElements);
    } finally {
        lock.unlock();
    }
}
```

从上述代码可以看出，在 replaceAll()方法中传入的是一个 UnaryOperator 类型的接口实例，UnaryOperator 接口是一个函数式接口，UnaryOperator 接口的代码如下。

```
@FunctionalInterface
public interface UnaryOperator<T> extends Function<T, T> {
    static <T> UnaryOperator<T> identity() {
        return t -> t;
    }
}
```

UnaryOperator 接口继承了 Function 接口，Function 接口的代码如下。

```
@FunctionalInterface
public interface Function<T, R> {
    R apply(T t);
    default <V> Function<V, R> compose(Function<? super V, ? extends T> before) {
        Objects.requireNonNull(before);
        return (V v) -> apply(before.apply(v));
    }
    default <V> Function<T, V> andThen(Function<? super R, ? extends V> after) {
        Objects.requireNonNull(after);
        return (T t) -> after.apply(apply(t));
    }
    static <T> Function<T, T> identity() {
        return t -> t;
    }
}
```

在 Function 接口中，定义了一个 apply() 方法，接收一个泛型 T 类型的参数，返回一个泛型 R 类型的结果。

UnaryOperator 接口继承自 Function 接口，在调用 UnaryOperator 接口中的 apply() 方法时，会传入一个泛型 T 类型的参数，返回一个泛型 T 类型的结果。

注意：UnaryOperator 接口和 Function 接口都是 Java 8 提供的函数式接口，有关 Java 8 函数式接口和其他 Java 8 新特性的知识，读者可以关注"冰河技术"微信公众号阅读相关文章。

回到 void replaceAll(UnaryOperator<E> operator) 方法的代码，在 void replaceAll (UnaryOperator<E> operator) 方法中，首先会判断传入的 operator 是否为空，如果 operator 为空，则直接抛出 NullPointerException 异常。

然后获取 ReentrantLock 锁，调用 getArray() 方法获取数组的引用，将其赋值给 Object 数组类型的局部变量 elements，并获取数组的长度。

接下来，将 elements 数组的元素赋值到一个新的 Object 数组类型的 newElements 数组中并遍历 elements 数组，取出 elements 数组中的每一个元素都传入 UnaryOperator 接口的 apply() 方法中进行处理，将处理后的接口数据添加到 newElements 数组中。

最后调用 setArray() 方法将 newElements 数组赋值给 array 引用，并释放 ReentrantLock 锁。

无论是 E set(int index, E element) 方法，还是 void replaceAll(UnaryOperator<E> operator) 方法，在执行方法体的逻辑时，都要先获取 ReentrantLock 锁，在执行完方法体的逻辑后，都会释放 ReentrantLock 锁，也就是说同一时刻只能有一个线程修改 CopyOnWriteArrayList 中的数据。

在执行方法体的逻辑时，都要将 array 引用指向的数组复制一份副本，在副本中进行写操作，在写操作执行完毕后，再将数组副本赋值给 array 引用。整个过程不影响其他线程对数组中数据的读取操作。

4.7 删除数据

CopyOnWriteArrayList 提供了如下删除数据的方法。

```
E remove(int index)
boolean remove(Object o)
boolean removeAll(Collection<?> c)
boolean removeIf(Predicate<? super E> filter)
void clear()
```

这里以 E remove(int index) 方法为例进行说明。E remove(int index) 方法会删除数组中 index 位置的元素，如果 index 下标超出了数组下标的范围，则会抛出 IndexOutOfBoundsException 异常，E remove(int index) 方法的代码如下。

```
public E remove(int index) {
    final ReentrantLock lock = this.lock;
    lock.lock();
    try {
        Object[] elements = getArray();
        int len = elements.length;
        E oldValue = get(elements, index);
        int numMoved = len - index - 1;
        if (numMoved == 0)
            setArray(Arrays.copyOf(elements, len - 1));
        else {
            Object[] newElements = new Object[len - 1];
            System.arraycopy(elements, 0, newElements, 0, index);
            System.arraycopy(elements, index + 1, newElements, index,
                             numMoved);
            setArray(newElements);
        }
        return oldValue;
    } finally {
        lock.unlock();
    }
}
```

在 E remove(int index) 方法中，首先获取 ReentrantLock 锁，然后调用 getArray() 方法获取数组引用并赋值给 Object 数组类型的 elements 局部变量，随后获取 elements 数组的长度 len 和 elements 数组中 index 位置的元素，并将其赋值给 oldValue。接下来计算删除数组中的元素后需要移动的元素数量 numMoved。

如果 numMoved 的值为 0，也就是删除的是数组中的最后一个元素（位置为 len−1 的元素），则直接将 elements 数组中的元素复制到一个比 elements 数组长度 len 少 1 的数组中，并将其赋值给 array 引用。

如果 numMoved 的值大于 0，则创建一个长度为 len-1 的数组 newElements，将 elements 数组中从位置 0 到位置 index-1 的元素复制到 newElements 数组的位置 0 到位置 index-1，再将 elements 数组中自位置 index+1 之后的 numMoved 个元素依次复制到 newElements 数组的 index 位置到 index+numMoved-1 位置。

在 E remove(int index)方法中，删除数组中的元素是通过复制数组中的元素实现的。同一时刻只能有一个线程获取 ReentrantLock 锁，获取锁成功后才能执行方法体内的逻辑，执行完毕后释放 ReentrantLock 锁，通过复制数组和复制数据元素的方式删除数组中的元素，不影响其他线程读取数组中的数据。

4.8 遍历数据

在 CopyOnWriteArrayList 中实现了自己的迭代器，在迭代遍历内部的 array 数组的过程中不会检查修改的状态。这是因为当某个线程使用迭代器遍历 array 数组时，array 数组在当前线程是只读的，其他线程修改了数组中的数据，对当前线程不可见，当前线程还是在遍历原来的数据。

CopyOnWriteArrayList 提供了如下遍历数据的方法。

```
Iterator<E> iterator()
ListIterator<E> listIterator()
ListIterator<E> listIterator(int index)
void forEach(Consumer<? super E> action)
```

这里，以 Iterator<E> iterator()方法为例，代码如下。

```
public Iterator<E> iterator() {
    return new COWIterator<E>(getArray(), 0);
}
```

在 Iterator<E> iterator()方法中，直接调用了 COWIterator 类的构造方法，并传入了内部的 array 数组引用和 0。

COWIterator 类的代码如下。

```
static final class COWIterator<E> implements ListIterator<E> {
    //array 的快照副本
    private final Object[] snapshot;
    //数组下标
    private int cursor;
    //构造方法
    private COWIterator(Object[] elements, int initialCursor) {
        cursor = initialCursor;
        snapshot = elements;
    }
    //是否存在下一个元素
    public boolean hasNext() {
```

```
        return cursor < snapshot.length;
    }
    //是否存在上一个元素
    public boolean hasPrevious() {
        return cursor > 0;
    }
    //获取快照副本中的下一个元素
    public E next() {
        if (! hasNext())
            throw new NoSuchElementException();
        return (E) snapshot[cursor++];
    }
    //获取快照副本中的上一个元素
    public E previous() {
        if (! hasPrevious())
            throw new NoSuchElementException();
        return (E) snapshot[--cursor];
    }
    //获取下一个元素的下标
    public int nextIndex() {
        return cursor;
    }
    //获取上一个元素的下标
    public int previousIndex() {
        return cursor-1;
    }
    //删除数据直接抛出异常
    public void remove() {
        throw new UnsupportedOperationException();
    }

    //修改数据直接抛出异常
    public void set(E e) {
        throw new UnsupportedOperationException();
    }
    //添加数据直接抛出异常
    public void add(E e) {
        throw new UnsupportedOperationException();
    }
    //遍历数据
    @Override
    public void forEachRemaining(Consumer<? super E> action) {
        Objects.requireNonNull(action);
        Object[] elements = snapshot;
        final int size = elements.length;
        for (int i = cursor; i < size; i++) {
            @SuppressWarnings("unchecked") E e = (E) elements[i];
            action.accept(e);
        }
        cursor = size;
```

    }
}
```

笔者已经为 COWIterator 类的方法添加了注释，当调用 CopyOnWriteArrayList 类的 iterator() 方法时，实际上返回的是 COWIterator 类的对象。在 COWIterator 类中保存了当前 array 数组的副本。

在 COWIterator 类的构造方法中，直接将 array 复制给了 snapshot，如果在遍历数组中数据的过程中，没有其他线程修改数组中的数据，则 snapshot 与 array 是引用关系，否则 snapshot 就是 array 的副本。当其他线程修改数据完毕时，snapshot 引用的是修改前的数组，而 array 引用的是修改后的数组。

## 4.9 使用案例

CopyOnWriteArrayList 是线程安全的，在每次添加、修改和删除数据时，同一时刻只有获取到 ReentrantLock 锁的线程才能执行添加、修改和删除数据的操作，并且在添加、修改和删除数据时，会复制一份副本，在副本上执行添加、修改和删除数据的操作，执行完毕会释放 ReentrantLock 锁。整个执行过程不影响其他线程对数据的读操作。

接下来，为读者列举一个在使用迭代器遍历 CopyOnWriteArrayList 数据时，其他线程修改 CopyOnWriteArrayList 数据的案例。

案例期望：在获取 CopyOnWriteArrayList 迭代器后、使用迭代器遍历 CopyOnWriteArrayList 数据前，开启一个新的线程向 CopyOnWriteArrayList 中添加数据，当添加数据的线程执行完毕后，使用迭代器遍历 CopyOnWriteArrayList 中的数据，发现遍历出的数据还是修改前的数据。实现代码如下。

```java
public class CopyOnWriteArrayListTest {
    private static final CopyOnWriteArrayList<Integer> LIST = new CopyOnWriteArrayList<>();
    static {
        //向 CopyOnWriteArrayList 中添加数据 1~5
        IntStream.rangeClosed(1, 5).forEach((i) -> LIST.add(i));
    }
    public static void main(String[] args) throws InterruptedException {
        //在新线程中向 CopyOnWriteArrayList 中添加数据 6~10
        Thread thread = new Thread(()->{
            IntStream.rangeClosed(6, 10).forEach((i) -> LIST.add(i));
        });
        //在启动线程前获取 CopyOnWriteArrayList 的迭代器
        Iterator<Integer> iterator = LIST.iterator();
        //启动线程
        thread.start();
        //等待线程执行完毕

```
        thread.join();
        //迭代 CopyOnWriteArrayList 中的数据
        while (iterator.hasNext()){
            System.out.println("遍历出的数据为===>>> " + iterator.next());
        }
    }
}
```

在 CopyOnWriteArrayListTest 类中定义了一个 CopyOnWriteArrayList 类型并且泛型是 Integer 的常量 LIST，在静态代码块中向 LIST 中添加了数据 1~5。

在 main()方法中创建了一个线程 Thread 对象，并在 Thread 的 run()方法中向 LIST 添加数据 6~10，在启动 thread 线程前获取 LIST 的迭代器 iterator。接下来，启动 thread 线程，并在主线程中等待 thread 线程执行完毕后，使用迭代器 iterator 遍历 LIST 中的元素。如果输出结果中只包含数据 1~5，则符合预期。

运行 CopyOnWriteArrayListTest 类的代码，输出结果如下。

```
遍历出的数据为===>>> 1
遍历出的数据为===>>> 2
遍历出的数据为===>>> 3
遍历出的数据为===>>> 4
遍历出的数据为===>>> 5
```

从输出结果可以看出，输出的数据中只包含 1~5，并不包含向 LIST 中新添加的数据 6~10，符合预期。

第 5 章

并发 Set 集合类

从 Java 1.5 版本开始，JDK 提供了高并发的 CopyOnWriteArraySet 集合。为了进一步提升 Set 集合的高并发性能，从 Java 1.6 版本开始，JDK 提供了并发性能更高的 ConcurrentSkipListSet 集合。

5.1 CopyOnWriteArraySet 集合类

CopyOnWriteArraySet 是线程安全的无序集合，可以将其理解为线程安全的 HashSet。

5.1.1 概述

从代码角度来看，CopyOnWriteArraySet 与 HashSet 都继承自 AbstractSet，但是 CopyOnWriteArraySet 是线程安全的 Set 集合实现，可以理解为线程安全的 HashSet。在底层实现上，CopyOnWriteArraySet 与 HashSet 是有区别的，HashSet 底层是基于散列表实现的，而 CopyOnWriteArraySet 底层是基于动态数组实现的。

CopyOnWriteArraySet 与 CopyOnWriteArrayList 的区别是 CopyOnWriteArraySet 是无序的，CopyOnWriteArrayList 是有序的。

CopyOnWriteArraySet 类的类继承关系如图 5-1 所示。

由图 5-1 可以看出，CopyOnWriteArraySet 继承了 AbstractSet 类，实现了 Set 接口。

在 CopyOnWriteArraySet 底层，数据的添加、遍历和删除操作都是通过 CopyOnWriteArrayList 实现的，所以多个线程在对 CopyOnWriteArraySet 进行数据添加和删除操作时，也会获取 ReentrantLock 锁，只有成功获取锁的线程才能执行数据的添加和删除操作，并且在添加和删除数据时，也会复制一份 array 引用的数组的副本，在副本上执行数据的添加和删除操作，最后将数组的副本赋值给 array 引用，并释放 ReentrantLock 锁。整个添加数据和删除数据的过程不影响其他线程对 CopyOnWriteArraySet 中数据的读取。

第 5 章　并发 Set 集合类

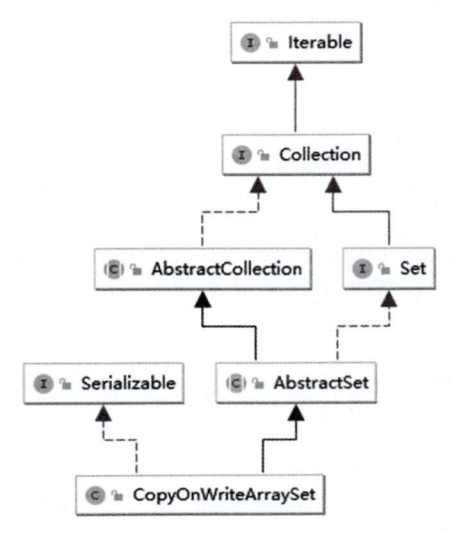

图 5-1　CopyOnWriteArraySet 类的类继承关系

与 CopyOnWriteArrayList 一样，在线程对 CopyOnWriteArraySet 进行写操作时，也会复制一个数组的副本，所以，会出现内存浪费的现象。CopyOnWriteArraySet 也不能保证数据实时一致。

综上所述，CopyOnWriteArraySet 与 CopyOnWriteArrayList 一样，适用于写操作比较少，读操作比较多，并且能够容忍读写操作在短时间内不一致的场景。

注意：CopyOnWriteArraySet 与 CopyOnWriteArrayList 都使用了写时复制技术，关于写时复制技术，读者可以参考 4.2 节的内容。

5.1.2　初始化

CopyOnWriteArraySet 底层维护了一个 CopyOnWriteArrayList 类型的成员变量 al。数据的添加、删除和遍历操作都是通过 al 进行的。

CopyOnWriteArraySet 类提供了两种构造方法进行初始化，一种是无参构造方法，一种是有一个 Collection 集合类型参数的构造方法。构造方法的主要作用是实例化 CopyOnWriteArraySet 类型的对象，同时实例化并初始化 CopyOnWriteArrayList 类型的成员变量 al，CopyOnWriteArraySet 中定义的 CopyOnWriteArrayList 类型的成员变量 al 的代码如下。

```
private final CopyOnWriteArrayList<E> al;
```

首先来看 CopyOnWriteArraySet 类中的无参构造方法，代码如下。

```
public CopyOnWriteArraySet() {
    al = new CopyOnWriteArrayList<E>();
}
```

在 CopyOnWriteArraySet 类的无参构造方法中，只是简单地创建一个 CopyOnWriteArrayList 类型的对象并赋值给成员变量 al。

CopyOnWriteArraySet 类中包含一个 Collection 集合类型参数的构造方法，代码如下。

```java
public CopyOnWriteArraySet(Collection<? extends E> c) {
    if (c.getClass() == CopyOnWriteArraySet.class) {
        @SuppressWarnings("unchecked")
        CopyOnWriteArraySet<E> cc = (CopyOnWriteArraySet<E>)c;
        al = new CopyOnWriteArrayList<E>(cc.al);
    }
    else {
        al = new CopyOnWriteArrayList<E>();
        al.addAllAbsent(c);
    }
}
```

CopyOnWriteArraySet 类中包含一个 Collection 集合类型参数的构造方法中，首先判断传入的 Collection 集合类型的参数 c 是否是 CopyOnWriteArraySet 类型的。

如果传入的 Collection 集合类型的参数 c 是 CopyOnWriteArraySet 类型的，则将参数 c 强转为 CopyOnWriteArraySet 类型，并赋值给 CopyOnWriteArraySet 类型的局部变量 cc。接下来调用 CopyOnWriteArrayList 的有参构造方法，传入 cc 的成员变量 al 来创建 CopyOnWriteArrayList 类型的对象，并将其赋值给成员变量 al。

如果传入的 Collection 集合类型的参数 c 不是 CopyOnWriteArraySet 类型的，则先创建 CopyOnWriteArrayList 类型的对象，将其赋值给成员变量 al，再调用 al 的 int addAllAbsent(Collection<? extends E> c)方法将 Collection 集合类型 c 中的元素添加到数组中，并在 al 的 int addAllAbsent(Collection<? extends E> c)方法中对添加的元素进行去重处理。

CopyOnWriteArrayList 类中的 int addAllAbsent(Collection<? extends E> c)方法的代码如下。

```java
public int addAllAbsent(Collection<? extends E> c) {
    Object[] cs = c.toArray();
    if (cs.length == 0)
        return 0;
    final ReentrantLock lock = this.lock;
    lock.lock();
    try {
        Object[] elements = getArray();
        int len = elements.length;
        int added = 0;
        // uniquify and compact elements in cs
        for (int i = 0; i < cs.length; ++i) {
            Object e = cs[i];
            if (indexOf(e, elements, 0, len) < 0 &&
                indexOf(e, cs, 0, added) < 0)
                cs[added++] = e;
```

```
        }
        if (added > 0) {
            Object[] newElements = Arrays.copyOf(elements, len + added);
            System.arraycopy(cs, 0, newElements, len, added);
            setArray(newElements);
        }
        return added;
    } finally {
        lock.unlock();
    }
}
```

在上述代码中,首先调用传入的 Collection 集合类型参数 c 的 toArray()方法,将其转换成数组并赋值给 Object 数组类型的局部变量 cs,如果 cs 的长度为 0,则直接返回 0。

获取 ReentrantLock 锁,然后调用 getArray()方法获取 array 数组引用并将其赋值给 Object 数组类型的局部变量 elements,同时获取 elements 数组的长度 len。定义成功添加到数组中元素数量的局部变量 added。

接下来,遍历 cs 数组,对 cs 数组中的数据进行去重,遍历 cs 数组中的每个元素。当当前遍历的元素 e 不在 elements 数组中且不在 cs 数组的前 added 个元素中时,会将 cs 数组的第 added 个位置的元素设置为当前遍历的元素 e。当遍历完 cs 数组时,cs 数组下标从 0 到 added 位置的元素都是经过去重的。

随后判断 added 的值是否大于 0,如果大于 0,则说明 cs 数组进行了去重处理。接下来将 elements 数组中的元素复制到一个长度为 len+added 的数组中,并将数组赋值给 Object 数组类型的 newElements 局部变量,再将 cs 数组中 0 到 added 位置的元素复制到 newElements 数组中 len 到 len+added−1 位置。

最后调用 setArray()方法将 newElements 数组赋值给 array 引用,并返回 added,释放 ReentrantLock 锁。

5.1.3 添加数据

CopyOnWriteArraySet 提供了 boolean add(E e)和 boolean addAll(Collection<? extends E> c)两种方法添加数据。其中,boolean addAll(Collection<? extends E> c)方法底层调用的 CopyOnWriteArrayList 中的 int addAllAbsent(Collection<? extends E> c)方法与 CopyOnWriteArraySet 构造方法中调用的 CopyOnWriteArrayList 中的 int addAllAbsent(Collection<? extends E> c)方法相同。

这里,以 boolean add(E e)方法为例进行说明,boolean add(E e)方法的代码如下。

```
public boolean add(E e) {
    return al.addIfAbsent(e);
}
```

在 boolean add(E e)方法中，直接调用了 al 成员变量的 addIfAbsent()方法，并将传入的 e 传递给了 addIfAbsent()方法。其中，addIfAbsent()方法位于 CopyOnWriteArrayList 类中。

查看 CopyOnWriteArrayList 类中的 boolean addIfAbsent(E e)方法，代码如下。

```java
public boolean addIfAbsent(E e) {
    Object[] snapshot = getArray();
    return indexOf(e, snapshot, 0, snapshot.length) >= 0 ? false :
        addIfAbsent(e, snapshot);
}
```

在 boolean addIfAbsent(E e)方法中，首先调用 getArray()方法获取 array 数组的引用并复制给 Object 数组类型的 snapshot 局部变量，然后判断当前要添加的元素是否在 snapshot 数组中，如果在 snapshot 数组中，则直接返回 false，否则调用 boolean addIfAbsent(E e, Object[] snapshot)方法。

boolean addIfAbsent(E e, Object[] snapshot)方法的代码如下。

```java
private boolean addIfAbsent(E e, Object[] snapshot) {
    final ReentrantLock lock = this.lock;
    lock.lock();
    try {
        Object[] current = getArray();
        int len = current.length;
        if (snapshot != current) {
            // Optimize for lost race to another addXXX operation
            int common = Math.min(snapshot.length, len);
            for (int i = 0; i < common; i++)
                if (current[i] != snapshot[i] && eq(e, current[i]))
                    return false;
            if (indexOf(e, current, common, len) >= 0)
                return false;
        }
        Object[] newElements = Arrays.copyOf(current, len + 1);
        newElements[len] = e;
        setArray(newElements);
        return true;
    } finally {
        lock.unlock();
    }
}
```

boolean addIfAbsent(E e, Object[] snapshot)方法的逻辑比较简单，首先获取 ReentrantLock 锁，然后调用 getArray()方法获取 array 引用并赋值给 Object 数组类型的 current 局部变量，获取 current 数组的长度 len，当传入的 snapshot 与 current 不是同一个数组时，判断要添加的元素 e 是否在 current 数组中，如果要添加的元素 e 在 current 数组中，则直接返回 false。

接下来，将 current 数组中的元素复制到一个长度为 len+1 的数组中，并将其赋值给

newElements 数组，将 newElements 数组的最后一个元素，也就是下标为 len 的元素设置为 e。调用 setArray()方法将 newElements 数组赋值给 array 引用，返回 true。最后释放 ReentrantLock 锁。向 CopyOnWriteArraySet 中添加数据的过程不影响其他线程遍历 CopyOnWriteArraySet 中的数据。

CopyOnWriteArraySet 一方面在添加数据的方法中保证了数据的唯一性，另一方面，在一定程度上保证了数据在 hashCode()方法和 equals()方法中的唯一性。

hashCode()方法在 CopyOnWriteArraySet 类的父类 AbstractSet 类中，代码如下。

```java
public int hashCode() {
    int h = 0;
    Iterator<E> i = iterator();
    while (i.hasNext()) {
        E obj = i.next();
        if (obj != null)
            h += obj.hashCode();
    }
    return h;
}
```

equals()方法在 CopyOnWriteArraySet 类中，代码如下。

```java
public boolean equals(Object o) {
    if (o == this)
        return true;
    if (!(o instanceof Set))
        return false;
    Set<?> set = (Set<?>)(o);
    Iterator<?> it = set.iterator();
    Object[] elements = al.getArray();
    int len = elements.length;
    boolean[] matched = new boolean[len];
    int k = 0;
    outer: while (it.hasNext()) {
        if (++k > len)
            return false;
        Object x = it.next();
        for (int i = 0; i < len; ++i) {
            if (!matched[i] && eq(x, elements[i])) {
                matched[i] = true;
                continue outer;
            }
        }
        return false;
    }
    return k == len;
}
```

hashCode()方法与 equals()方法的代码比较简单。

5.1.4 删除数据

CopyOnWriteArraySet 提供如下删除数据的方法。

```
void clear()
boolean remove(Object o)
boolean removeAll(Collection<?> c)
```

这里，以 boolean remove(Object o)方法为例进行说明。boolean remove(Object o)方法的代码如下。

```
public boolean remove(Object o) {
    return al.remove(o);
}
```

在 CopyOnWriteArraySet 的 boolean remove(Object o)方法中，直接调用的是 al 的 remove()方法。接下来，查看 CopyOnWriteArrayList 中的 boolean remove(Object o)方法，代码如下。

```
public boolean remove(Object o) {
    Object[] snapshot = getArray();
    int index = indexOf(o, snapshot, 0, snapshot.length);
    return (index < 0) ? false : remove(o, snapshot, index);
}
```

在 CopyOnWriteArrayList 的 boolean remove(Object o)方法中，首先调用 getArray()方法获取 array 引用，将其赋值给 Object 数组类型的 snapshot 局部变量，并获取传入的对象 o 在 snapshot 数组中的位置。如果 snapshot 数组中不包含传入的对象 o，则直接返回 false。

如果 snapshot 数组中包含传入的对象 o，则调用 boolean remove(Object o, Object[] snapshot, int index)方法删除元素。boolean remove(Object o, Object[] snapshot, int index)方法的代码如下。

```
private boolean remove(Object o, Object[] snapshot, int index) {
    final ReentrantLock lock = this.lock;
    lock.lock();
    try {
        Object[] current = getArray();
        int len = current.length;
        if (snapshot != current) findIndex: {
            int prefix = Math.min(index, len);
            for (int i = 0; i < prefix; i++) {
                if (current[i] != snapshot[i] && eq(o, current[i])) {
                    index = i;
                    break findIndex;
                }
            }
            if (index >= len)
                return false;
            if (current[index] == o)
                break findIndex;
            index = indexOf(o, current, index, len);
            if (index < 0)
```

```
                return false;
            }
            Object[] newElements = new Object[len - 1];
            System.arraycopy(current, 0, newElements, 0, index);
            System.arraycopy(current, index + 1,
                        newElements, index,
                        len - index - 1);
            setArray(newElements);
            return true;
        } finally {
            lock.unlock();
        }
    }
}
```

由 boolean remove(Object o, Object[] snapshot, int index)方法的代码可以看出，首先获取 ReentrantLock 锁，然后调用 getArray()方法获取 array 引用赋值给 Object 数组类型的 current 局部变量，并将获取的 current 数组的长度赋值给 len。

如果 snapshot 不等于 current，则获取传入的 index 与 len 中的较小值赋给 prefix，随后的逻辑如下。

（1）遍历 prefix 的值，将每次遍历的值赋给 i，比较 current 数组下标为 i 处的值与 snapshot 数组下标为 i 处的值，如果二者不相等且传入的对象 o 等于 current 数组下标为 i 处的值，则将 i 的值赋给 index 并退出最外层的 if 判断。

（2）当不满足 for 循环中的条件时，比较 index 与 len 的值。如果 index 的值大于或等于 len 的值，则说明对象 o 不在 current 数组中，直接返回 false。

（3）如果 current 数组中下标为 index 的值等于传入的对象 o，则直接退出最外层 if 判断。

如果（1）（2）（3）都不满足，则获取对象 o 在 current 数组中下标从 index 到 len−1 区间的位置，将其再次赋值给 index。如果 index 小于 0，则说明在 current 数组中不存在对象 o，直接返回 false。

接下来创建一个长度为 len−1 的数组 newElements，将 current 数组中下标从 0 到 index−1 的元素依次复制到 newElements 数组下标从 0 到 index−1 处，将 current 数组中下标自 index+1 位置之后的 len−index−1 个元素依次复制到 newElements 数组下标自 index 之后的位置。

这样就删除了对象 o。接下来，调用 setArray()方法将 newElements 数组赋值给 array 引用，返回 true 并释放 ReentrantLock 锁。

从 CopyOnWriteArraySet 中删除数据的过程不影响其他线程遍历 CopyOnWriteArraySet 中的数据。

5.1.5 遍历数据

CopyOnWriteArraySet 提供了如下遍历数据的方法。

```
Iterator<E> iterator()
void forEach(Consumer<? super E> action)
```

在 CopyOnWriteArraySet 的 Iterator<E> iterator()方法中，调用的是 CopyOnWriteArrayList 的 Iterator<E> iterator()方法。4.8 节已经介绍过 CopyOnWriteArrayList 的 Iterator<E> iterator() 方法。

这里重点分析一下 CopyOnWriteArraySet 的 void forEach(Consumer<? super E> action)方法，代码如下。

```
public void forEach(Consumer<? super E> action) {
    al.forEach(action);
}
```

CopyOnWriteArraySet 的 void forEach(Consumer<? super E> action)方法也是直接调用了 CopyOnWriteArrayList 的 void forEach(Consumer<? super E> action)方法。CopyOnWriteArrayList 的 void forEach(Consumer<? super E> action)方法代码如下。

```
public void forEach(Consumer<? super E> action) {
    if (action == null) throw new NullPointerException();
    Object[] elements = getArray();
    int len = elements.length;
    for (int i = 0; i < len; ++i) {
        @SuppressWarnings("unchecked")
        E e = (E) elements[i];
        action.accept(e);
    }
}
```

CopyOnWriteArrayList 的 void forEach(Consumer<? super E> action)方法的参数 Consumer 是 Java 8 提供的消费数据的函数式接口，这个接口中包含一个 void accept(T t)方法，可以对传入的数据进行处理。

在 CopyOnWriteArrayList 的 void forEach(Consumer<? super E> action)方法中，首先判断传入的函数式接口对象 action 是否为空，如果为空，则直接抛出空指针异常。

随后通过 getArray()方法获取 array 引用，将其赋值给 Object 数组类型的 elements 局部变量，并获取 elements 数组的长度 len。最后遍历 elements 数组，获取数组中的每一个元素 e，调用函数式接口对象 action 的 accept()方法，将元素 e 传入进行处理。

整个遍历过程不需要加锁处理，能够极大地提升遍历数据的性能。

5.1.6　使用案例

CopyOnWriteArraySet 底层是通过 CopyOnWriteArrayList 实现的，对数据的增加、删除和修改操作不会影响正在进行的数据遍历操作。

案例期望：获取 CopyOnWriteArraySet 迭代器后，在使用迭代器遍历 CopyOnWriteArraySet 中的数据前，开启一个新的线程从 CopyOnWriteArraySet 中删除数据，当删除数据的线程执行完毕后，使用迭代器遍历 CopyOnWriteArraySet 中的数据，发现遍历出的数据还是删除之前的数据。

案例代码如下。

```java
public class CopyOnWriteArraySetTest {
    private static final CopyOnWriteArraySet<Integer> SET = new CopyOnWriteArraySet<>();
    static {
        //向 CopyOnWriteArraySet 中添加数据 1~10
        IntStream.rangeClosed(1, 10).forEach((i) -> SET.add(i));
    }
    public static void main(String[] args) throws InterruptedException {
        //在新线程中删除 CopyOnWriteArraySet 中的数据 6~10
        Thread thread = new Thread(()->{
            IntStream.rangeClosed(6, 10).forEach((i) -> SET.remove(i));
        });
        //在启动线程之前获取 CopyOnWriteArraySet 的迭代器
        Iterator<Integer> iterator = SET.iterator();
        //启动线程
        thread.start();
        //等待线程执行完毕
        thread.join();
        //遍历 CopyOnWriteArraySet 中的数据
        while (iterator.hasNext()){
            System.out.println("遍历出的数据为===>>> " + iterator.next());
        }
    }
}
```

上述代码的主要逻辑与 4.9 节的案例的主要逻辑相同，只不过 4.9 节的案例是在遍历数据前向 CopyOnWriteArrayList 中添加数据，此处的案例是在遍历数据前删除 CopyOnWriteArraySet 中的部分数据，这里不再赘述上述案例程序的执行逻辑。

运行程序代码，输出结果如下。

```
遍历出的数据为===>>> 1
遍历出的数据为===>>> 2
遍历出的数据为===>>> 3
遍历出的数据为===>>> 4
遍历出的数据为===>>> 5
遍历出的数据为===>>> 6
```

```
遍历出的数据为===>>> 7
遍历出的数据为===>>> 8
遍历出的数据为===>>> 9
遍历出的数据为===>>> 10
```

可以看出，在获取 CopyOnWriteArraySet 迭代器之后、遍历数据之前，虽然开启线程删除了 CopyOnWriteArraySet 中的数据，但是在使用 CopyOnWriteArraySet 迭代器遍历数据时，输出结果仍然为删除之前的数据，符合预期。

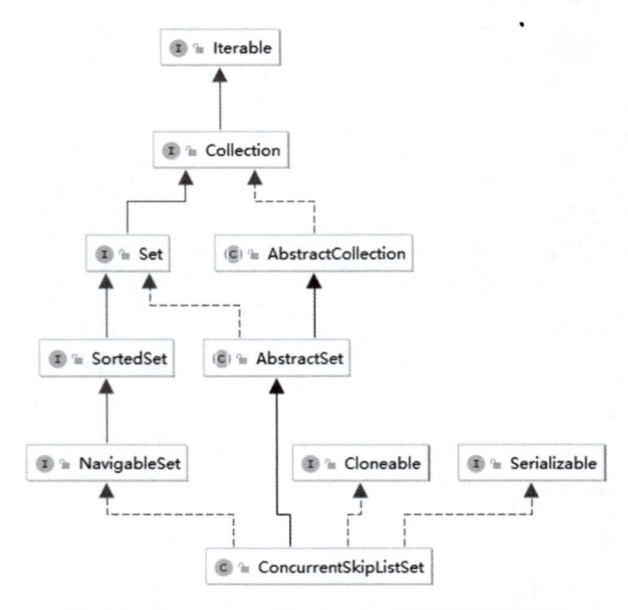

5.2 ConcurrentSkipListSet 集合类

ConcurrentSkipListSet 是 JDK 从 Java 1.6 版本开始提供的线程安全的有序 Set 集合类，适用于高并发的场景。

5.2.1 概述

从代码角度看，ConcurrentSkipListSet 与 CopyOnWriteArraySet 和 HashSet 一样，都继承了 AbstractSet 类，ConcurrentSkipListSet 与 CopyOnWriteArraySet 一样，都是线程安全的 Set 集合。与 CopyOnWriteArraySet 不同的是，ConcurrentSkipListSet 是有序的。ConcurrentSkipListSet 类的类继承关系如图 5-2 所示。

图 5-2 ConcurrentSkipListSet 类的类继承关系

可以看出，ConcurrentSkipListSet 类继承了 AbstractSet，说明 ConcurrentSkipListSet 类本质上是一个 Set 集合。ConcurrentSkipListSet 类实现了 NavigableSet 接口，而 NavigableSet 接口继

承了 SortedSet 接口，说明 ConcurrentSkipListSet 类是一个有序 Set 集合。

在 ConcurrentSkipListSet 内部，维护了一个 ConcurrentNavigableMap 类型的成员变量 m，在初始化时，实际上创建的是 ConcurrentSkipListMap 对象，在 ConcurrentSkipListMap 中维护的是一个键-值对集合，而 ConcurrentSkipListSet 中只使用了 ConcurrentSkipListMap 中的键（key）。

注意：ConcurrentSkipListSet 底层所使用的 ConcurrentSkipListMap 集合在第 6 章并发 Map 集合类中进行详细的介绍。

5.2.2 跳表

在 ConcurrentSkipListSet 的实现中使用了跳表这种数据结构，很多读者对它比较陌生，我们先来看一下链表。

在链表中查询某个特定的数据时，需要从头节点 header 开始，逐个对比链表中的数据与当前要查询的数据是否相等，如图 5-3 所示。

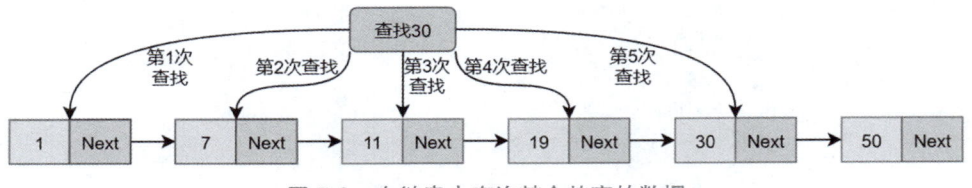

图 5-3　在链表中查询某个特定的数据

由图 5-3 可以看出，在链表中，每一个节点都存储了指向下一个节点的指针。如果要查找数值为 30 的节点，那么一共需要查找 5 次。

也就是说，在链表中查找数据时，如果数据正好在链表的头部，则查找数据的时间复杂度为 $O(1)$，如果数据正好在链表的尾部，则查找数据的时间复杂度为 $O(n)$。

为了提高数据的查找效率，可以在链表上添加索引，通过增加数据的层级来提高查找数据的效率，此时就形成了跳表，如图 5-4 所示。

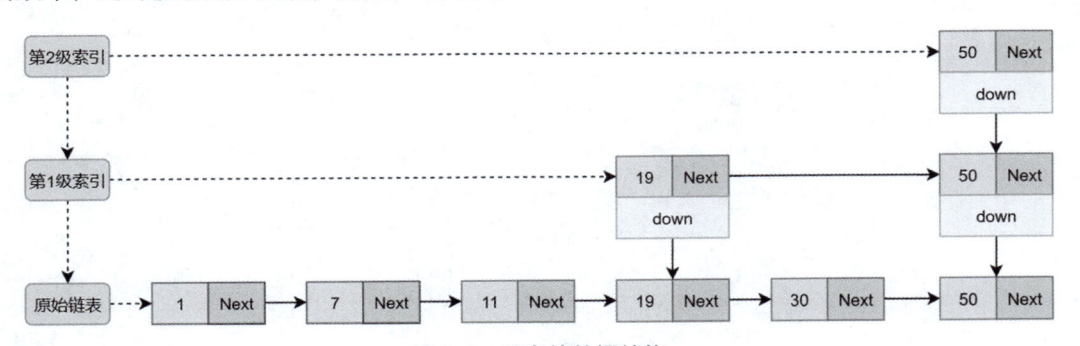

图 5-4　跳表的数据结构

跳表在链表的基础上多了几级索引，每一级索引也是一个链表，并且高层级索引中的节点会存在一个指向低层级索引中节点的指针。

在图 5-4 所示的跳表数据结构中查找数值为 30 的节点，整体流程如下。

（1）在第 2 级索引中查找，发现第 2 级索引中节点的数值（50）大于 30，则继续查找第 1 级索引中的节点。

（2）第 1 级索引中的第 1 个节点的数值（19）小于 30，第 2 个节点的数值（50）大于 30，通过数值为 19 的节点定位原始链表中数值为 19 的节点。

（3）在原始链表中，通过数值为 19 的节点找到数值为 30 的节点，整个查找过程完毕。

使用跳表查找数据的时间复杂度是 $O(\log n)$。另外，因为在向跳表中插入数据时，已经根据某种规则对数据进行了排序，所以在查找数据时不需要再次对数据进行排序。

注意： Redis 底层的主要数据结构之一就是跳表。有关跳表的更多知识，读者可以关注"冰河技术"微信公众号阅读相关文章。

5.2.3 初始化

ConcurrentSkipListSet 类内部定义了一个 ConcurrentNavigableMap 类型的成员变量 m，代码如下。

```
private final ConcurrentNavigableMap<E,Object> m;
```

在初始化 ConcurrentSkipListSet 类时，为成员变量 m 赋了一个 ConcurrentSkipListMap 类型的对象，后续的数据读写操作都是使用成员变量 m 来完成的，并且主要使用 ConcurrentSkipListMap 中的键（key）。

ConcurrentSkipListSet 类主要提供了如下构造方法进行初始化。

```
public ConcurrentSkipListSet() {
    m = new ConcurrentSkipListMap<E,Object>();
}
public ConcurrentSkipListSet(Comparator<? super E> comparator) {
    m = new ConcurrentSkipListMap<E,Object>(comparator);
}
public ConcurrentSkipListSet(Collection<? extends E> c) {
    m = new ConcurrentSkipListMap<E,Object>();
    addAll(c);
}
public ConcurrentSkipListSet(SortedSet<E> s) {
    m = new ConcurrentSkipListMap<E,Object>(s.comparator());
    addAll(s);
}
ConcurrentSkipListSet(ConcurrentNavigableMap<E,Object> m) {
    this.m = m;
}
```

ConcurrentSkipListSet 类提供的构造方法主要是为成员变量赋值。如果需要在 ConcurrentSkipListSet 集合中添加元素，则需要调用 ConcurrentSkipListMap 类的构造方法，或者调用 ConcurrentSkipListSet 类的 boolean add(E e)方法，代码如下。

```
public boolean add(E e) {
    return m.putIfAbsent(e, Boolean.TRUE) == null;
}
```

ConcurrentSkipListSe t 类的 boolean add(E e)方法最终通过调用 ConcurrentSkipListMap 类的 V putIfAbsent(K key, V value)方法将元素添加到 ConcurrentSkipListMap 集合中。而 ConcurrentSkipListSet 主要使用了 ConcurrentSkipListMap 集合中的键（key）。

5.2.4 添加数据

ConcurrentSkipListSet 类提供了 boolean add(E e)方法来添加数据，代码如下。

```
public boolean add(E e) {
    return m.putIfAbsent(e, Boolean.TRUE) == null;
}
```

此处添加元素的方法与在 ConcurrentSkipListSet 构造方法中调用的添加数据的方法相同，最终都会调用 ConcurrentSkipListMap 类的 V putIfAbsent(K key, V value)方法。

5.2.5 删除数据

ConcurrentSkipListSet 类主要提供了 boolean remove(Object o)方法、boolean removeAll(Collection<?> c)方法和 void clear()方法删除数据，boolean remove(Object o)方法用于删除 ConcurrentSkipListSet 集合中的对象 o，boolean removeAll(Collection<?> c)方法用于删除 ConcurrentSkipListSet 集合中包含在 Collection 集合 c 中的所有数据，void clear()方法用于清空 ConcurrentSkipListSet 集合中的所有数据。

ConcurrentSkipListSet 类中删除数据的代码如下。

```
//删除单个数据
public boolean remove(Object o) {
    return m.remove(o, Boolean.TRUE);
}
//删除所有包含在集合 c 中的所有数据
public boolean removeAll(Collection<?> c) {
    boolean modified = false;
    for (Object e : c)
        if (remove(e))
            modified = true;
    return modified;
}
//清空所有数据
public void clear() {
```

```
    m.clear();
}
```

boolean removeAll(Collection<?> c)方法内部在遍历集合 c 的过程中，调用 boolean remove(Object o)方法删除 ConcurrentSkipListSet 集合中包含在 Collection 集合 c 中的数据。

boolean remove(Object o)方法内部调用 ConcurrentSkipListMap 类的 boolean remove(Object key, Object value)方法删除数据。

void clear()方法内部调用 ConcurrentSkipListMap 类的 void clear()方法清除数据。

5.2.6 遍历数据

ConcurrentSkipListSet 类主要提供了 Iterator<E> iterator()和 Iterator<E> descendingIterator()两种方法来获取 ConcurrentSkipListSet 的迭代器 Iterator。其中，Iterator<E> iterator()方法表示获取升序遍历 ConcurrentSkipListSet 集合中数据的迭代器，Iterator<E> descendingIterator()方法表示获取降序遍历 ConcurrentSkipListSet 集合中数据的迭代器。

获取了 ConcurrentSkipListSet 的 Iterator 迭代器，就可以使用 Iterator 迭代器来遍历 ConcurrentSkipListSet 集合中的数据。

ConcurrentSkipListSet 类中获取遍历数据迭代器的代码如下。

```
//获取升序遍历迭代器
public Iterator<E> iterator() {
    return m.navigableKeySet().iterator();
}
//获取降序遍历迭代器
public Iterator<E> descendingIterator() {
    return m.descendingKeySet().iterator();
}
```

无论是获取升序遍历迭代器的 Iterator<E> iterator()方法，还是获取降序遍历迭代器的 Iterator<E> descendingIterator()方法，内部都是调用 ConcurrentSkipListMap 类的相关方法实现的。

5.2.7 关系运算

ConcurrentSkipListSet 类提供了一系列的关系运算方法，其内部也是主要通过 ConcurrentSkipListMap 类的相关方法实现的，代码如下。

```
//获取ConcurrentSkipListSet集合中小于e的最大数据
public E lower(E e) {
    return m.lowerKey(e);
}
//获取ConcurrentSkipListSet集合中小于或等于e的最大数据
public E floor(E e) {
    return m.floorKey(e);
}
```

```java
//获取ConcurrentSkipListSet集合中大于或等于e的最小数据
public E ceiling(E e) {
    return m.ceilingKey(e);
}
//获取ConcurrentSkipListSet集合中大于e的最小数据
public E higher(E e) {
    return m.higherKey(e);
}
//移除ConcurrentSkipListSet集合中的最小数据并返回
public E pollFirst() {
    Map.Entry<E,Object> e = m.pollFirstEntry();
    return (e == null) ? null : e.getKey();
}
//移除ConcurrentSkipListSet集合中的最大数据并返回
public E pollLast() {
    Map.Entry<E,Object> e = m.pollLastEntry();
    return (e == null) ? null : e.getKey();
}
```

上述方法都是通过ConcurrentSkipListMap类的相关方法实现的，实现逻辑比较简单，代码中也给出了相应的注释。

5.2.8　有序集合操作

ConcurrentSkipListSet类提供了一系列针对有序集合的操作方法，其内部也是主要通过ConcurrentSkipListMap类的相关方法实现的，代码如下。

```java
//获取比较器
public Comparator<? super E> comparator() {
    return m.comparator();
}
//获取ConcurrentSkipListSet集合中最小的数据
public E first() {
    return m.firstKey();
}
//获取ConcurrentSkipListSet集合中最大的数据
public E last() {
    return m.lastKey();
}
//获取fromElement与toElement之间的数据并返回一个NavigableSet集合
public NavigableSet<E> subSet(E fromElement,
                    boolean fromInclusive,
                    E toElement,
                    boolean toInclusive) {
    return new ConcurrentSkipListSet<E>
        (m.subMap(fromElement, fromInclusive,
              toElement,   toInclusive));
}
//获取ConcurrentSkipListSet集合中从第1个数据到toElement的集合
```

```java
//返回NavigableSet集合
public NavigableSet<E> headSet(E toElement, boolean inclusive) {
    return new ConcurrentSkipListSet<E>(m.headMap(toElement, inclusive));
}
//获取ConcurrentSkipListSet集合中从fromElement到最后一个数据的集合
//返回NavigableSet集合
public NavigableSet<E> tailSet(E fromElement, boolean inclusive) {
    return new ConcurrentSkipListSet<E>(m.tailMap(fromElement, inclusive));
}
//获取fromElement与toElement之间的数据
//包含fromElement，不包含toElement，返回NavigableSet集合
public NavigableSet<E> subSet(E fromElement, E toElement) {
    return subSet(fromElement, true, toElement, false);
}
//获取ConcurrentSkipListSet集合中从第1个数据到toElement的集合，不包含toElement
//返回NavigableSet集合
public NavigableSet<E> headSet(E toElement) {
    return headSet(toElement, false);
}
//获取ConcurrentSkipListSet集合中从fromElement到最后一个数据的集合，包含fromElement
//返回NavigableSet集合
public NavigableSet<E> tailSet(E fromElement) {
    return tailSet(fromElement, true);
}
//获取降序NavigableSet集合
public NavigableSet<E> descendingSet() {
    return new ConcurrentSkipListSet<E>(m.descendingMap());
}
//获取可分割的迭代器
@SuppressWarnings("unchecked")
public Spliterator<E> spliterator() {
    if (m instanceof ConcurrentSkipListMap)
        return ((ConcurrentSkipListMap<E,?>)m).keySpliterator();
    else
        return (Spliterator<E>)((ConcurrentSkipListMap.SubMap<E,?>)m).keyIterator();
}
//使用原子方式更新map，主要提供给clone()方法使用
private void setMap(ConcurrentNavigableMap<E,Object> map) {
    UNSAFE.putObjectVolatile(this, mapOffset, map);
}
```

上述代码中方法的主体逻辑比较简单，基本上都是通过调用ConcurrentSkipListMap类的方法实现的。

5.2.9 使用案例

案例期望：证明ConcurrentSkipListSet集合是有序的，并且集合中不存在相同的元素。在向ConcurrentSkipListSet集合中添加数据后，使用升序迭代器遍历ConcurrentSkipListSet集合中

的数据时，按照从小到大的顺序输出结果，在使用降序迭代器遍历 ConcurrentSkipListSet 集合中的数据时，按照从大到小的顺序输出结果。

完整的案例程序代码如下。

```java
public class ConcurrentSkipListSetTest {
    private static final ConcurrentSkipListSet<Integer> SET = new ConcurrentSkipListSet<>();
    static {
        //向 ConcurrentSkipListSet 中随机添加数据
        IntStream.rangeClosed(1, 10).forEach((i) -> {
            int randomNum = new Random().nextInt(10);
            //向 ConcurrentSkipListSet 中重复添加随机获取的数据
            //检查 ConcurrentSkipListSet 中是否会存储重复的数据
            SET.add(randomNum);
            SET.add(randomNum);
        });
    }
    public static void main(String[] args){
        //升序遍历 ConcurrentSkipListSet
        Iterator<Integer> iterator = SET.iterator();
        while (iterator.hasNext()){
            System.out.println("升序遍历输出的结果数据===>>> " + iterator.next());
        }
        System.out.println("===========================");
        //降序遍历 ConcurrentSkipListSet
        iterator = SET.descendingIterator();
        while (iterator.hasNext()){
            System.out.println("降序遍历输出的结果数据===>>> " + iterator.next());
        }
    }
}
```

在上述案例程序代码中，首先定义了一个 ConcurrentSkipListSet 类型的集合常量 SET，在静态代码块中向 SET 集合中随机添加 10 次数字 1~10，并且每次会添加两次获取的随机数。

在 main()方法中，先升序遍历 SET 集合中的数据，再降序遍历 SET 集合中的数据，运行程序输出结果如下。

```
升序遍历输出的结果数据===>>> 0
升序遍历输出的结果数据===>>> 2
升序遍历输出的结果数据===>>> 4
升序遍历输出的结果数据===>>> 5
升序遍历输出的结果数据===>>> 6
升序遍历输出的结果数据===>>> 9
===========================
降序遍历输出的结果数据===>>> 9
降序遍历输出的结果数据===>>> 6
降序遍历输出的结果数据===>>> 5
降序遍历输出的结果数据===>>> 4
```

```
降序遍历输出的结果数据===>>> 2
降序遍历输出的结果数据===>>> 0
```

从输出结果可以看出，ConcurrentSkipListSet 集合是有序的，并且不会存储重复的数据。使用升序迭代器遍历 ConcurrentSkipListSet 中的数据，会按照从小到大的顺序输出结果，使用降序迭代器遍历 ConcurrentSkipListSet 中的数据，会按照从大到小的顺序输出结果，符合预期。

第 6 章

并发 Map 集合类

Map 集合类一直是 Java 编程中使用频率较高的数据结构，在多线程并发环境下，HashMap 是非线程安全的。HashTable 虽然是线程安全的，但是它解决线程安全问题的方式是在每种方法上都添加 synchronized 锁，这在高并发环境下会影响程序的性能，除此以外，存在竞态条件和使用迭代器遍历集合的问题。

为解决以上问题，JDK 提供了并发 Map 集合类，并发 Map 集合类是线程安全的，并且在高并发环境下具备更高的数据读写性能。

6.1 ConcurrentHashMap 集合类

随着 JDK 版本的不断迭代，ConcurrentHashMap 也会不断升级和优化。ConcurrentHashMap 读取数据的 get() 方法基本上是无锁的，而添加数据的 put() 方法会将锁粒度控制在很小的范围内，所以，ConcurrentHashMap 非常适用于多线程高并发的场景。

在多线程并发环境下使用 HashMap 的 put() 方法添加数据可能导致 HashMap 的 Entry 链表形成环形数据结构。一旦形成环形数据结构，Entry 节点的 Next 指针指向的下一个节点永远不为空，在获取 Entry 时，就会引起死循环，导致 CPU 利用率飙升，甚至出现 CPU 使用率接近 100%的问题。

HashTable 中的方法使用 synchronized 锁来保证线程的安全性，在多线程高并发场景下，程序的执行性能低下，可能引起激烈的锁竞争，进而引发锁升级。在 HashTable 中，读数据和写数据是互斥的，一个线程在读数据时，其他线程无论是要读数据还是要写数据，都会被阻塞，直到读数据的线程释放 synchronized 锁。除此以外，HashTable 还存在竞态条件和使用迭代器遍历集合的问题。

注意：有关 synchronized 锁的核心原理和升级过程，读者可以参考《深入理解高并发编程：核心原理与案例实战》一书。

6.1.1 概述

ConcurrentHashMap 是 JDK 从 Java 1.5 版本开始提供的适用于多线程高并发环境的线程安全的 Map 集合类。随着 JDK 的不断迭代，ConcurrentHashMap 类也在不断升级和优化。

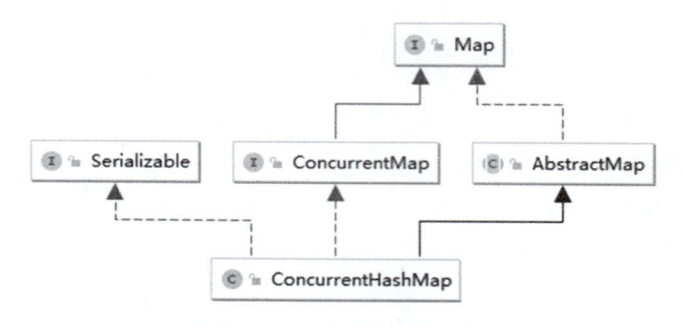

Java 7 以及之前版本的 ConcurrentHashMap 使用分段锁技术将数据分段存储，然后为每一段数据单独分配锁，此时一个线程占有锁访问其中一个段的数据不影响其他线程访问其他段的数据，多个线程能够并发访问 ConcurrentHashMap 中不同段的数据。

Java 8 版本对 ConcurrentHashMap 内部结构进行了优化和升级，使用 CAS 和 synchronized 锁的方式保证数据的一致性，在高并发场景下锁住数组的节点，在性能上得到进一步提升。

ConcurrentHashMap 类的类继承关系如图 6-1 所示。

图 6-1　ConcurrentHashMap 类的类继承关系

由图 6-1 可以看出，ConcurrentHashMap 类实现了 Map 接口和 ConcurrentMap 接口，说明它是一个线程安全的 Map 集合类。

随着不断迭代和升级，ConcurrentHashMap 在高并发场景下的性能也越来越高。在高并发场景下，可以将 HashMap 和 HashTable 直接替换为 ConcurrentHashMap。

6.1.2 结构

Java 7 及之前版本的 ConcurrentHashMap 的结构与 Java 8 及之后版本的 ConcurrentHashMap 的结构差异还是比较大的，本节将分别进行介绍。

1. Java 7 及之前版本的 ConcurrentHashMap 的结构

Java 7 及之前版本的 ConcurrentHashMap 是使用 Segment 组、HashEntry 数组和链表实现的，整体结构如图 6-2 所示。

此时的 ConcurrentHashMap 使用了分段锁技术，也就是 Segment 数组保证线程安全，将数据分段存储。Segment 数组中的每一个元素都是一个段，都会存储一个 HashEntry 数组。从结构上看，每个单独的 HashEntry 数组都相当于一个 HashMap。

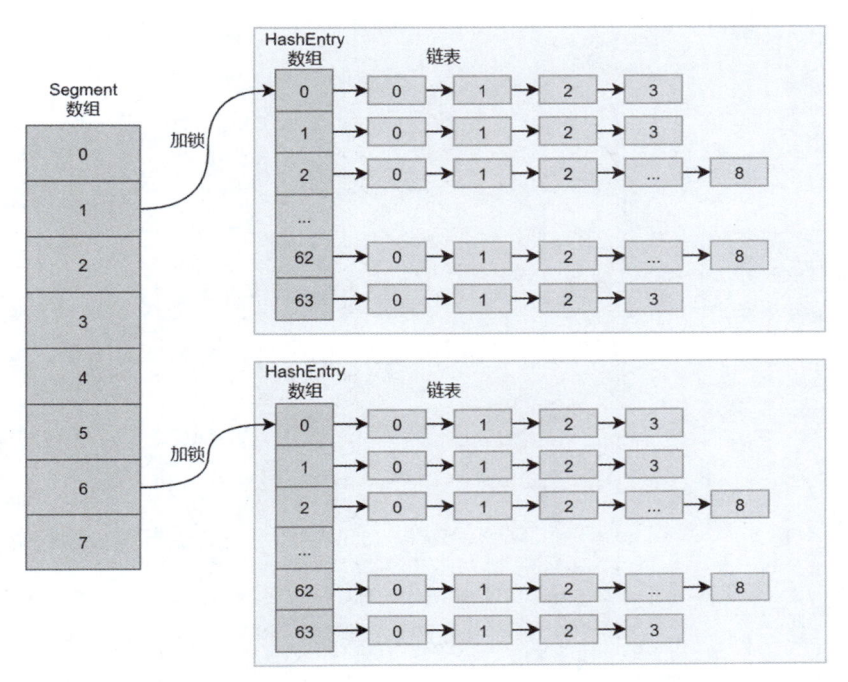

图 6-2　Java 7 及之前版本的 ConcurrentHashMap 结构

在高并发场景下修改 ConcurrentHashMap 中的数据时，只会针对 Segment 数组中的每个数据加锁，也就是只会锁住 Segment 数组中的所有数据，它们分别负责自身对应的锁。采用这种分段锁技术向 ConcurrentHashMap 的不同段插入和更新数据，不同段之间不会有任何影响。所以，ConcurrentHashMap 真正做到了并发插入和更新数据。

2. Java 8 及之后版本的 ConcurrentHashMap 的结构

Java 8 及之后版本的 ConcurrentHashMap 去除了 Segment 数组和分段锁方案，使用和 HashMap 相同的结构，也就是数组、链表和红黑树的结构，并使用 CAS+synshronized 锁的方式保证线程的安全性。

在 ConcurrentHashMap 类中，存在一个 Node 类型的数组 table。table 是一个哈希桶数组，数组中的每个节点都可以看作一个哈希桶。如果不发生哈希冲突，则每个元素都保存在哈希桶数组中。如果发生哈希冲突，则使用标准的链地址（拉链法）解决哈希冲突。为了防止拉链，也就是链表的长度过长，ConcurrentHashMap 会根据一定的规则将链表转换成红黑树。table 数组中的每个元素实际上存储的都是单链表的头节点或者红黑树的根节点，当向 ConcurrentHashMap 中插入键-值对时，首先要定位插入的桶，也就是定位 table 数组的某个索引下标。

当 ConcurrentHashMap 中只存在 table 数组和链表时，整体结构如图 6-3 所示。

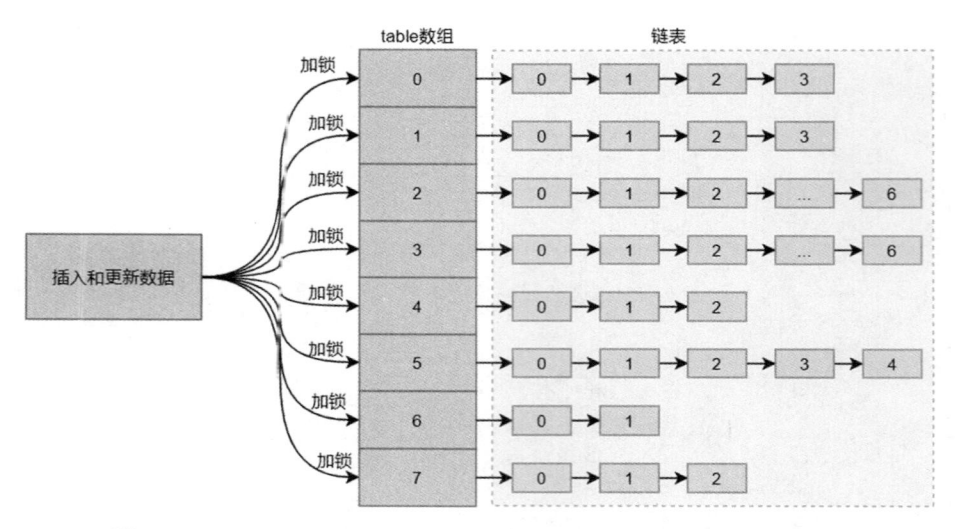

图 6-3　ConcurrentHashMap 中只存在 table 数组和链表时的整体结构

由图 6-3 可以看出，在向 ConcurrentHashMap 中插入和更新数据时，会对 table 数组中的每一个节点都单独加锁，并且在数组中的每个节点下存储一个链表。

在使用链表存储数据时，会从链表的头部向后遍历数据，如果要查找的数据恰好在链表的尾部，则每次获取数据都要遍历整个链表。如果链表长度过长，就会极大地影响获取数据的效率。为解决这个问题，在 Java 8 及之后版本的 JDK 中，ConcurrentHashMap 会在一定条件下将内部的链表自动转换为红黑树，如图 6-4 所示。

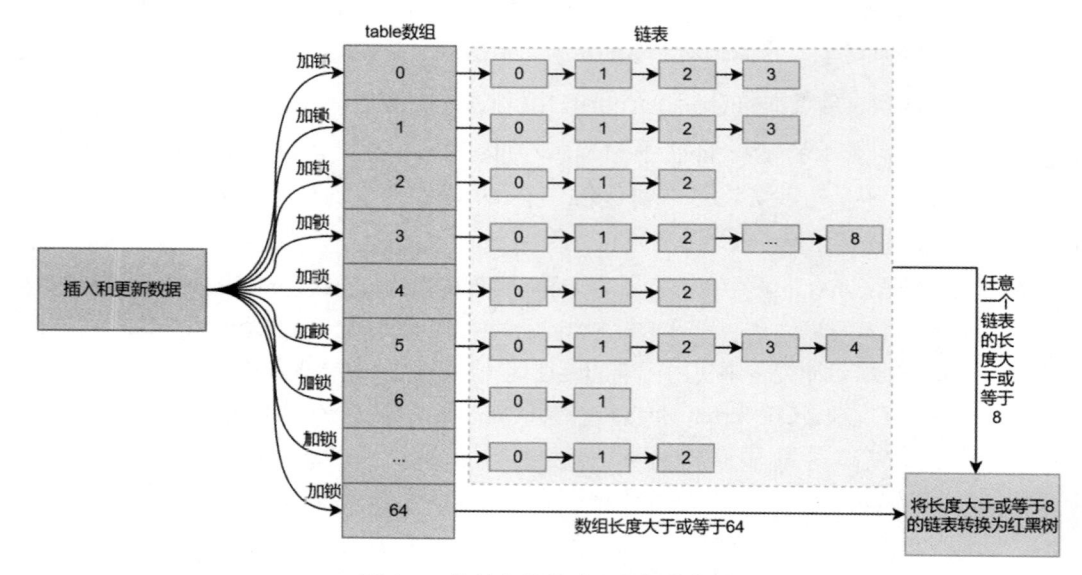

图 6-4　将链表转换成红黑树的条件

由图 6-4 可以看出，当 ConcurrentHashMap 中的数组长度大于或等于 64、table 数组中任意一个链表的长度大于或等于 8 时，会将长度大于或等于 8 的链表转换为红黑树，数组中其他位置的链表保持不变。图 6-4 所示的结构可转换成图 6-5 所示的结构。

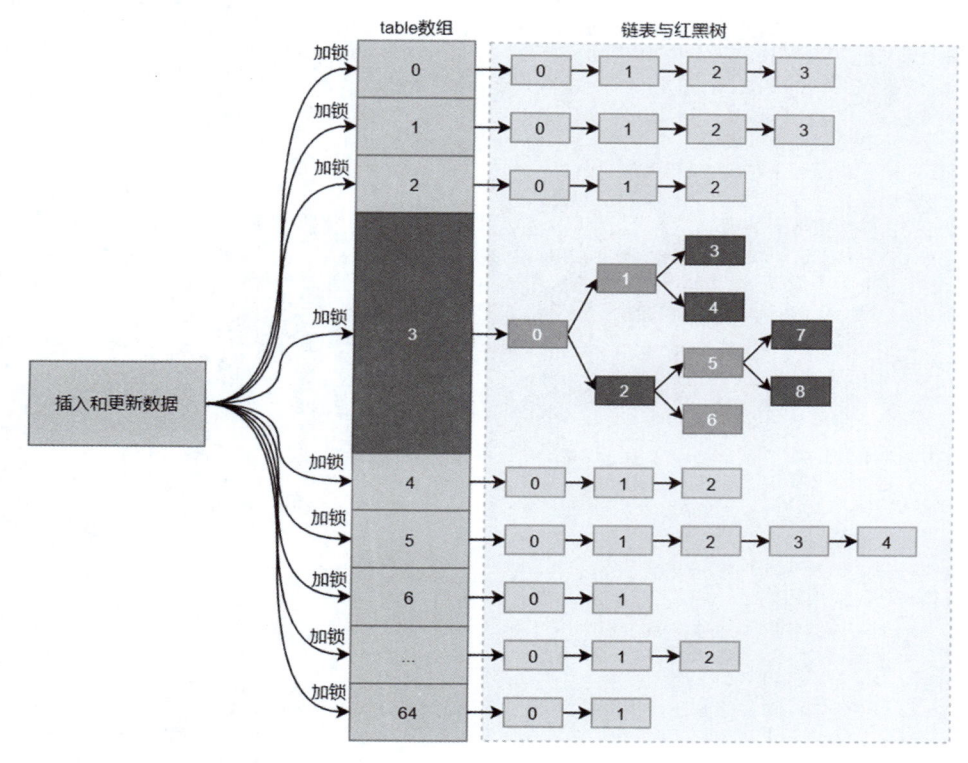

图 6-5　将链表转换成红黑树

注意：本书后续章节中的 ConcurrentHashMap 都基于 Java 8 版本进行介绍。如果读者对 Java 7 版本的 ConcurrentHashMap 的更多细节感兴趣，可以关注"冰河技术"微信公众号阅读相关技术文章。

6.1.3　成员变量

ConcurrentHashMap 定义了一些非常重要的成员变量，有些成员变量决定了 ConcurrentHashMap 的扩容和缩容，有些成员变量表示链表与红黑树之间的转换阈值，有些成员变量则用来控制 table 数组的初始化和扩容过程。ConcurrentHashMap 的主要成员变量的代码如下。

```
public class ConcurrentHashMap<K,V> extends AbstractMap<K,V>
    implements ConcurrentMap<K,V>, Serializable {
    //数组的最大容量
```

```
private static final int MAXIMUM_CAPACITY = 1 << 30;
//初始化时默认的数组容量
private static final int DEFAULT_CAPACITY = 16;
//数组的最大容量
static final int MAX_ARRAY_SIZE = Integer.MAX_VALUE - 8;
//默认的并发级别，兼容之前版本的ConcurrentMap
private static final int DEFAULT_CONCURRENCY_LEVEL = 16;
//默认的负载因子0.75(目前在容量 ≥ 负载因子*总容量时进行resize扩容)
private static final float LOAD_FACTOR = 0.75f;
//将链表转换成红黑树的阈值，当数组容量大于或等于64时
//将长度大于或等于8的链表转换成红黑树
static final int TREEIFY_THRESHOLD = 8;
//将红黑树转换成链表的阈值，当数组中链表的长度小于6时，红黑树会转换成链表
static final int UNTREEIFY_THRESHOLD = 6;
//当数组长度大于或等于64时，会将长度大于或等于8的链表转换成红黑树
//当数组长度小于64时，链表长度大于或等于8时进行数组扩容
static final int MIN_TREEIFY_CAPACITY = 64;
//正在转移
static final int MOVED     = -1;
//正在转换成树结构
static final int TREEBIN   = -2;
static final int RESERVED  = -3;
static final int HASH_BITS = 0x7fffffff;
//保存元素的数组，大小是2的幂次方
transient volatile Node<K,V>[] table;
//数据转移时使用的数组
private transient volatile Node<K,V>[] nextTable;
//主要用来控制table数组的初始化和扩容
private transient volatile int sizeCtl;
//##############省略其他成员变量####################
}
```

这里简单介绍sizeCtl这个成员变量，它主要用来控制table数组的初始化和扩容操作。

当sizeCtl的值为正时，包括以下两种情况。

（1）如果table数组还未初始化，则sizeCtl表示table数组需要初始化的大小。

（2）如果table数组已经初始化，则sizeCtl表示table数组的容量，默认是table数组大小的0.75倍。

当sizeCtl的值为负时，包括以下两种情况。

（1）当sizeCtl的值为–1时，表示table数组正在进行初始化。

（2）当sizeCtl的值为–N，也就是小于–1时，表示有N–1个线程正在扩容。

注意： 上述代码中给出了详细的注释，关于其他成员变量的含义，这里不再赘述。

6.1.4 内部类

ConcurrentHashMap 针对 table 数组的节点定义了 5 个内部类，分别是 Node、TreeNode、TreeBin、ForwardingNode 和 ReservationNode，其中 ReservationNode 类在 compute()方法和 computeIfAbsent()方法中作为占位节点使用。

接下来，重点介绍 Node、TreeNode、TreeBin 和 ForwardingNode 这 4 个类。

（1）Node 类

Node 类实现了 Map.Entry 接口，主要存放键-值对。Node 类 value 和 next 属性使用 volatile 关键字进行修饰，保证了数据的可见性，比 HashMap 的 Node 类少了 V setValue(V newValue)方法修改 value 的值，Node 类的部分代码如下。

```
static class Node<K,V> implements Map.Entry<K,V> {
    final int hash;
    final K key;
    volatile V val;
    volatile Node<K,V> next;
    Node(int hash, K key, V val, Node<K,V> next) {
        this.hash = hash;
        this.key = key;
        this.val = val;
        this.next = next;
    }
    //##############其他代码省略################
}
```

注意：有关 volatile 核心原理与内存屏障的知识，读者可以参考《深入理解高并发编程：核心原理与案例实战》一书。

（2）TreeNode 类

TreeNode 类继承自 Node 类，主要在 TreeBin 类中使用，部分代码如下。

```
static final class TreeNode<K,V> extends Node<K,V> {
    TreeNode<K,V> parent;
    TreeNode<K,V> left;
    TreeNode<K,V> right;
    TreeNode<K,V> prev;
    boolean red;
    TreeNode(int hash, K key, V val, Node<K,V> next,
             TreeNode<K,V> parent) {
        super(hash, key, val, next);
        this.parent = parent;
    }
    //###########省略其他代码##############
}
```

（3）TreeBin 类

TreeBin 类继承自 Node 类，在 TreeBin 类中，并没有直接封装键-值对信息，而是封装了 TreeNode 节点，在 ConcurrentHashMap 的哈希桶数组 table 中，实际上存储的是 TreeBin 类的对象，不是 TreeNode 类的对象。TreeBin 类的部分代码如下。

```
static final class TreeBin<K,V> extends Node<K,V> {
    TreeNode<K,V> root;
    volatile TreeNode<K,V> first;
    volatile Thread waiter;
    volatile int lockState;
    static final int WRITER = 1;
    static final int WAITER = 2;
    static final int READER = 4;
    #####################其他代码省略##########################
}
```

（4）ForwardingNode 类

ForwardingNode 类继承自 Node 类，ForwardingNode 类节点主要在 ConcurrentHashMap 扩容时使用，其内部会存储一个 Node 数组类型的 nextTable 引用指向新的 table 数组。ForwardingNode 类的部分代码如下。

```
static final class ForwardingNode<K,V> extends Node<K,V> {
    final Node<K,V>[] nextTable;
    ForwardingNode(Node<K,V>[] tab) {
        super(MOVED, null, null, null);
        this.nextTable = tab;
    }
    //############省略其他代码#################
}
```

6.1.5 构造方法

ConcurrentHashMap 主要提供了 5 种构造方法用于实例化 ConcurrentHashMap 对象。构造方法的代码如下。

```
//构造方法①
//使用默认大小(16)创建一个新的空映射
public ConcurrentHashMap() {
}
//构造方法②
public ConcurrentHashMap(int initialCapacity) {
    if (initialCapacity < 0)
        throw new IllegalArgumentException();
    int cap = ((initialCapacity >= (MAXIMUM_CAPACITY >>> 1)) ?
            MAXIMUM_CAPACITY :
            tableSizeFor(initialCapacity + (initialCapacity >>> 1) + 1));
    this.sizeCtl = cap;
```

```
}
//构造方法③
public ConcurrentHashMap(Map<? extends K, ? extends V> m) {
    this.sizeCtl = DEFAULT_CAPACITY;
    putAll(m);
}
//构造方法④
public ConcurrentHashMap(int initialCapacity, float loadFactor) {
    this(initialCapacity, loadFactor, 1);
}
//构造方法⑤
public ConcurrentHashMap(int initialCapacity,
                        float loadFactor, int concurrencyLevel) {
    if (!(loadFactor > 0.0f) || initialCapacity < 0 || concurrencyLevel <= 0)
        throw new IllegalArgumentException();
    if (initialCapacity < concurrencyLevel)
        initialCapacity = concurrencyLevel;
    long size = (long)(1.0 + (long)initialCapacity / loadFactor);
    int cap = (size >= (long)MAXIMUM_CAPACITY) ?
        MAXIMUM_CAPACITY : tableSizeFor((int)size);
    this.sizeCtl = cap;
}
```

接下来，分别对以上代码中的 5 种构造方法进行简要的说明。

（1）构造方法①为默认的无参构造方法。在调用无参构造方法后，控制 table 数组初始化和扩容的 sizeCtl 的值为 0，初始化操作会在 put 操作中完成。

（2）构造方法②会传入初始容量。如果传入的初始容量 initialCapacity 小于 0，则抛出 IllegalArgumentException 异常。如果传入的初始容量 initialCapacity 大于或等于允许的最大容量 MAXIMUM_CAPACITY，则 initialCapacity 取值为 MAXIMUM_CAPACITY，如果小于 MAXIMUM_CAPACITY，则对 initialCapacity 使用 tableSizeFor(initialCapacity + (initialCapacity >>> 1) + 1))进行处理，也就是向上取最近的 2 的幂次方，然后将结果赋值给 sizeCtl。

（3）构造方法③传入一个 Map 集合，并创建一个与传入的 Map 集合具有相同键-值对集合的 ConcurrentHashMap。

（4）构造方法④传入初始容量和负载因子，并调用构造方法⑤。传入的并发度为 1。

（5）构造方法⑤传入初始容量、负载因子及并发度。如果负载因子小于或等于 0，或者初始容量小于 0，或者并发度小于或等于 0，则直接抛出 IllegalArgumentException 异常。如果初始容量小于并发度，则将并发度的值赋给初始容量。接下来计算 sizeCtl 的值，主要逻辑与构造方法②相同。

6.1.6 初始化

ConcurrentHashMap 提供了一个私有方法 Node<K,V>[] initTable()，用于 ConcurrentHashMap

的初始化。Node<K,V>[] initTable()方法会在第一次进入 V putVal(K key, V value, boolean onlyIfAbsent)方法、循环发现 table 数组为空时被调用,也就是在使用默认的无参构造方法创建 ConcurrentHashMap 时被调用。Node<K,V>[] initTable()方法最主要的逻辑就是初始化一个大小合适的数组并设置 sizeCtl 的值。

Node<K,V>[] initTable()方法的代码如下。

```
//初始化table 数组,大小为 sizeCtl
private final Node<K,V>[] initTable() {
    Node<K,V>[] tab; int sc;
    //如果 table 数组为空或者 table 数组的长度为 0,则一直初始化,直到完成
    while ((tab = table) == null || tab.length == 0) {
        //将 sizeCtl 的值赋给 sc,
        //如果 sc 的值小于 0,则说明有其他线程正在进行初始化或者扩容操作
        //当前线程让出 CPU 使用权
        if ((sc = sizeCtl) < 0)
            Thread.yield();
        //如果 sc 的值大于或等于 0,则使用 CAS 操作将 sc 的值修改为-1
        //表示 table 数组正在初始化
        else if (U.compareAndSwapInt(this, SIZECTL, sc, -1)) {
            try {
                //再次确认 table 数组为空或者 table 数组的长度为 0
                //这个判断绝对不能少,非常关键
                if ((tab = table) == null || tab.length == 0) {
                    //如果 sc 大于 0,则 n 取值为 sc
                    //如果 sc 小于或等于 0,则 n 取值为 DEFAULT_CAPACITY
                    int n = (sc > 0) ? sc : DEFAULT_CAPACITY;
                    //创建一个长度为 n 的 Node 数组
                    @SuppressWarnings("unchecked")
                    Node<K,V>[] nt = (Node<K,V>[])new Node<?,?>[n];
                    table = tab = nt;
                    //计算 sc 阈值,大于这个阈值就会扩容
                    sc = n - (n >>> 2);
                }
            } finally {
                //将 sc 的值赋给 sizeCtl
                sizeCtl = sc;
            }
            break;
        }
    }
    return tab;
}
```

由 Node<K,V>[] initTable()方法的代码可以看出,table 数组的初始化操作并没有加锁,在初始化时会通过 CAS 操作将 sizeCtl 的值设置为 -1,表示 table 数组正在进行初始化。sizeCtl 使用 volatile 关键字进行修饰,修饰后的 sizeCtl 值立刻对其他线程可见。后续如果有其他线程进入 Node<K,V>[] initTable()方法,检测到 sizeCtl 的值小于 0,则说明有线程正在对 table 数组进

行初始化或扩容，当前线程会调用 Thread.yield()方法让出 CPU 的使用权。

整个 table 数组的初始化过程做到了无锁并发。

6.1.7 扩容

ConcurrentHashMap 在扩容过程中主要使用 sizeCtl 和 transferIndex 这两个使用 volatile 关键字修饰的成员变量协调多个线程之间的并发操作，能够在扩容过程中保证访问的大部分数据不会发生阻塞。

总体来说，在调用 V put(K key, V value)方法或者 void putAll(Map<? extends K, ? extends V> m)方法向 ConcurrentHashMap 中添加数据时，最终都会调用 V putVal(K key, V value, boolean onlyIfAbsent)方法。扩容的几种情况如下。

（1）在向 ConcurrentHashMap 中添加数据、对数据进行合并和计算时，如果所在链表的元素数量大于或等于 8，则会调用 void treeifyBin(Node<K,V>[] tab, int index)方法将链表转换成红黑树，以添加数据为例，此逻辑在 ConcurrentHashMap 的 V putVal(K key, V value, boolean onlyIfAbsent)方法中，代码如下。

```
//代码位于 java.util.concurrent.ConcurrentHashMap#putVal 方法中
if (binCount >= TREEIFY_THRESHOLD)
    treeifyBin(tab, i);
```

在 void treeifyBin(Node<K,V>[] tab, int index)方法中，会检测 table 数组的长度，如果 table 数组的长度大于或等于 MIN_TREEIFY_CAPACITY，也就是大于或等于 64，则会将链表转换成红黑树。如果 table 数组的长度小于 MIN_TREEIFY_CAPACITY，则调用 void tryPresize(int size)方法将 table 数组的长度扩容至原来的 2 倍，并触发 void transfer(Node<K,V>[] tab, Node<K,V>[] nextTab)方法，调整 table 数组各个节点的位置。代码如下。

```
//代码位于 java.util.concurrent.ConcurrentHashMap#treeifyBin 方法中
if ((n = tab.length) < MIN_TREEIFY_CAPACITY)
    tryPresize(n << 1);
```

（2）在 V putVal(K key, V value, boolean onlyIfAbsent)方法的最后会调用 void addCount(long x, int check)方法记录 table 数组中元素的数量，代码如下。

```
//代码位于 java.util.concurrent.ConcurrentHashMap#putVal 方法中
addCount(1L, binCount);
```

在 addCount(1L, binCount)方法中，记录 table 数组中元素的数量并检查 table 数组是否需要扩容。如果 table 数组需要扩容，则触发 void transfer(Node<K,V>[] tab, Node<K,V>[] nextTab)方法调整 table 数组各节点的位置。代码如下。

```
//代码位于 java.util.concurrent.ConcurrentHashMap#addCount 方法中
if (check >= 0) {
    Node<K,V>[] tab, nt; int n, sc;
    //判断当前 table 数组的元素数量是否大于或等于 sizeCtl，并且数组不为空
```

```
        //同时数组的长度是否小于MAXIMUM_CAPACITY
        while (s >= (long)(sc = sizeCtl) && (tab = table) != null &&
            (n = tab.length) < MAXIMUM_CAPACITY) {
            int rs = resizeStamp(n);
            //table数组正处于扩容状态
            if (sc < 0) {
                //判断table数组扩容是否完成，或者并发线程数是否达到最大值
                //如果table数组扩容完成或者并发线程数达到最大值，就会退出while循环
                if ((sc >>> RESIZE_STAMP_SHIFT) != rs || sc == rs + 1 ||
                    sc == rs + MAX_RESIZERS || (nt = nextTable) == null ||
                    transferIndex <= 0)
                    break;
                //table数组扩容还未完成，当前线程会参与table数组的扩容，并且将sc的计数加1
                if (U.compareAndSwapInt(this, SIZECTL, sc, sc + 1))
                    transfer(tab, nt);
            }
            //table数组还未处于扩容状态，此时第1个线程会执行此方法初始化table数组
            else if (U.compareAndSwapInt(this, SIZECTL, sc,
                                    (rs << RESIZE_STAMP_SHIFT) + 2))
                transfer(tab, null);
            s = sumCount();
        }
    }
}
```

（3）在当前ConcurrentHashMap处于扩容状态时，如果其他线程在ConcurrentHashMap中进行添加、修改、删除数据等操作，或者对ConcurrentHashMap中的数据进行合并或计算等操作，则会调用Node<K,V>[] helpTransfer(Node<K,V>[] tab, Node<K,V> f)方法判断当前节点是否是ForwardingNode节点。如果是，则触发扩容操作，调用void transfer(Node<K,V>[] tab, Node<K,V>[] nextTab)方法，调整table数组中各节点的位置。Node<K,V>[] helpTransfer(Node<K,V>[] tab, Node<K,V> f)方法的代码如下。

```
//代码位于java.util.concurrent.ConcurrentHashMap#helpTransfer方法中
final Node<K,V>[] helpTransfer(Node<K,V>[] tab, Node<K,V> f) {
    Node<K,V>[] nextTab; int sc;
    if (tab != null && (f instanceof ForwardingNode) &&
        (nextTab = ((ForwardingNode<K,V>)f).nextTable) != null) {
        int rs = resizeStamp(tab.length);
        while (nextTab == nextTable && table == tab &&
            (sc = sizeCtl) < 0) {
            if ((sc >>> RESIZE_STAMP_SHIFT) != rs || sc == rs + 1 ||
                sc == rs + MAX_RESIZERS || transferIndex <= 0)
                break;
            if (U.compareAndSwapInt(this, SIZECTL, sc, sc + 1)) {
                transfer(tab, nextTab);
                break;
            }
        }
        return nextTab;
```

```
    }
    return table;
}
```

（4）在调用 void putAll(Map<? extends K, ? extends V> m)方法向 ConcurrentHashMap 添加数据时，最终会在 void treeifyBin(Node<K,V>[] tab, int index)方法中检测 table 数组的长度。如果 table 数组的长度小于 MIN_TREEIFY_CAPACITY，那么还会调用 void tryPresize(int size)方法将 table 数组的长度扩容至原来的 2 倍，在 void putAll(Map<? extends K, ? extends V> m)方法的开始位置直接调用 void tryPresize(int size)方法，在 void tryPresize(int size)方法中检测是否会真正扩容，代码如下。

```
//代码位于 java.util.concurrent.ConcurrentHashMap#putAll 方法中
public void putAll(Map<? extends K, ? extends V> m) {
    tryPresize(m.size());
    for (Map.Entry<? extends K, ? extends V> e : m.entrySet())
        putVal(e.getKey(), e.getValue(), false);
}
```

注意：当 table 数组中某个索引下标处的链表长度大于或等于 8 时，会有两种情况。

（1）如果 table 数组的长度小于 64，则对 table 数组进行扩容。

（2）如果 table 数组的长度大于或等于 64，则将长度大于或等于 8 的链表转换为红黑树，其他索引下标处的链表保持不变。

综上，在 ConcurrentHashMap 中，涉及扩容的方法主要如下。

```
void treeifyBin(Node<K,V>[] tab, int index)
void tryPresize(int size)
void addCount(long x, int check)
Node<K,V>[] helpTransfer(Node<K,V>[] tab, Node<K,V> f)
void transfer(Node<K,V>[] tab, Node<K,V>[] nextTab)
```

为了加深读者的理解，给出上述方法代码的详细注释。

（1）void treeifyBin(Node<K,V>[] tab, int index)方法。void treeifyBin(Node<K,V>[] tab, int index)方法的主要逻辑是判断 table 数组的长度与 MIN_TREEIFY_CAPACITY（64）的关系，如果 table 数组的长度小于 MIN_TREEIFY_CAPACITY，则调用 void tryPresize(int size)方法将 table 数组的长度扩容至原来的 2 倍，代码如下。

```
private final void treeifyBin(Node<K,V>[] tab, int index) {
    Node<K,V> b; int n, sc;
    if (tab != null) {
        //如果 table 数组的长度小于 64，则对数组进行扩容
        if ((n = tab.length) < MIN_TREEIFY_CAPACITY)
            //扩容至原来的 2 倍
            tryPresize(n << 1);
        //将 table 数组中长度大于或等于 8 的链表转换成红黑树
        else if ((b = tabAt(tab, index)) != null && b.hash >= 0) {
```

```
            synchronized (b) {
                if (tabAt(tab, index) == b) {
                    TreeNode<K,V> hd = null, tl = null;
                    for (Node<K,V> e = b; e != null; e = e.next) {
                        TreeNode<K,V> p =
                            new TreeNode<K,V>(e.hash, e.key, e.val,
                                              null, null);
                        if ((p.prev = tl) == null)
                            hd = p;
                        else
                            tl.next = p;
                        tl = p;
                    }
                    setTabAt(tab, index, new TreeBin<K,V>(hd));
                }
            }
        }
    }
}
```

（2）void tryPresize(int size)方法。void tryPresize(int size)方法主要的逻辑是处理 table 数组的初始化和扩容，代码如下。

```
//在向ConcurrentHash 插入数据时，table 数组中链表的长度大于 8
private final void tryPresize(int size) {
    int c = (size >= (MAXIMUM_CAPACITY >>> 1)) ? MAXIMUM_CAPACITY :
        tableSizeFor(size + (size >>> 1) + 1);
    int sc;
    //如果 sizeCtl 大于或等于 0，说明没有其他线程正在扩容或初始化
    //则执行扩容或初始化的逻辑
    //如果 sizeCtl 小于 0，说明有其他线程正在扩容或初始化，则当前线程不需要扩容或初始化
    while ((sc = sizeCtl) >= 0) {
        Node<K,V>[] tab = table; int n;
        //如果数组为空，则进行 table 数组的初始化操作
        if (tab == null || (n = tab.length) == 0) {
            n = (sc > c) ? sc : c;
            //对 table 数组的初始化会将 sizeCtl 的值修改为-1，
            //保证只有一个线程进行初始化
            if (U.compareAndSwapInt(this, SIZECTL, sc, -1)) {
                try {
                    if (table == tab) {
                        @SuppressWarnings("unchecked")
                        Node<K,V>[] nt = (Node<K,V>[])new Node<?,?>[n];
                        table = nt;
                        sc = n - (n >>> 2);
                    }
                } finally {
                    //初始化完成后，sizeCtl 主要记录的是当前数组扩容的阈值
                    sizeCtl = sc;
                }
```

```
        }
    }
    //如果扩容后的容量小于或等于 sc, 或者 n 大于或等于最大容量, 则直接退出循环
    else if (c <= sc || n >= MAXIMUM_CAPACITY)
        break;
    //添加数据后, table 数组中的链表长度大于或等于 8, table 数组长度小于 64
    //会触发下面的逻辑
    else if (tab == table) {
        int rs = resizeStamp(n);
        if (sc < 0) {
            Node<K,V>[] nt;
            //如果扩容已经结束或者并发扩容线程数已经达到最大, 则退出 while 循环
            if ((sc >>> RESIZE_STAMP_SHIFT) != rs || sc == rs + 1 ||
                sc == rs + MAX_RESIZERS || (nt = nextTable) == null ||
                transferIndex <= 0)
                break;
            //如果 table 数组还在扩容, 则其他线程进入方法后会将 sc 计数加 1
            if (U.compareAndSwapInt(this, SIZECTL, sc, sc + 1))
                transfer(tab, nt);
        }
        //如果 table 数组还未扩容, 则第 1 个线程会将 sizeCtl 的值设置为
        //rs << RESIZE_STAMP_SHIFT) + 2, 后续线程执行方法时
        //可以根据这个值判断是不是最后一个线程
        else if (U.compareAndSwapInt(this, SIZECTL, sc,
                                     (rs << RESIZE_STAMP_SHIFT) + 2))
            transfer(tab, null);
    }
}
```

（3）void addCount(long x, int check)方法。在 V putVal(K key, V value, boolean onlyIfAbsent)方法的最后会调用 void addCount(long x, int check)方法记录 table 数组中元素的数量, 还会处理 table 扩容的逻辑, 代码如下。

```
private final void addCount(long x, int check) {
    //省略其他代码, 以下代码是与 table 数组扩容相关的
    if (check >= 0) {
        Node<K,V>[] tab, nt; int n, sc;
        //判断当前 table 数组的元素数量是否大于或等于 sizeCtl, 并且数组不为空
        //同时数组的长度要小于 MAXIMUM_CAPACITY
        while (s >= (long)(sc = sizeCtl) && (tab = table) != null &&
               (n = tab.length) < MAXIMUM_CAPACITY) {
            int rs = resizeStamp(n);
            //table 数组正处于扩容状态中
            if (sc < 0) {
                //判断 table 数组扩容是否完成或者并发线程是否达到最大值
                //如果 table 数组扩容完成, 或者并发线程达到了最大值
                //就会退出 while 循环
                if ((sc >>> RESIZE_STAMP_SHIFT) != rs || sc == rs + 1 ||
                    sc == rs + MAX_RESIZERS || (nt = nextTable) == null ||
```

```
                transferIndex <= 0)
                break;
            //table 数组扩容还未完成，当前线程会参与 table 数组的扩容
            //并且将 sc 的计数加 1
            if (U.compareAndSwapInt(this, SIZECTL, sc, sc + 1))
                transfer(tab, nt);
        }
        //table 数组还未处于扩容状态, 此时第 1 个线程会执行此方法初始化 table 数组
        else if (U.compareAndSwapInt(this, SIZECTL, sc,
                    (rs << RESIZE_STAMP_SHIFT) + 2))
            transfer(tab, null);
        s = sumCount();
    }
}
```

（4）Node<K,V>[] helpTransfer(Node<K,V>[] tab, Node<K,V> f)方法。Node<K,V>[] helpTransfer(Node<K,V>[] tab, Node<K,V> f)方法在情况（3）中有完整的代码展示。

（5）void transfer(Node<K,V>[] tab, Node<K,V>[] nextTab)方法。void transfer(Node<K,V>[] tab, Node<K,V>[] nextTab)方法会在 void addCount(long x, int check)方法、Node<K,V>[] helpTransfer(Node<K,V>[] tab, Node<K,V> f)方法和 void tryPresize(int size)方法中调用，主要对 table 数组进行扩容和转移数据操作。

由于 void transfer(Node<K,V>[] tab, Node<K,V>[] nextTab)方法的代码整体逻辑过长，这里将其拆分成几部分进行说明，如下所示。

① 初始化核心数据，代码如下。

```
//代码位于java.util.concurrent.ConcurrentHashMap#transfer 方法中
int n = tab.length, stride;
//计算每个线程处理的哈希桶数量, 也就是处理的 table 数组中元素的数量
//每个线程处理的桶数量一样, 如果 CPU 为单核, 则使用一个线程处理 table 数组中的所有桶
//每个线程至少处理 16 个桶, 如果计算结果小于 16, 则使用一个线程处理 16 个桶
if ((stride = (NCPU > 1) ? (n >>> 3) / NCPU : n) < MIN_TRANSFER_STRIDE)
    stride = MIN_TRANSFER_STRIDE;
//传入的 nextTab 数组为空
if (nextTab == null) {
    try {
        //初始化一个长度是原来 nextTab 数组 2 倍的新数组
        @SuppressWarnings("unchecked")
        Node<K,V>[] nt = (Node<K,V>[])new Node<?,?>[n << 1];
        nextTab = nt;
    } catch (Throwable ex) {
        //初始化 nextTab 数组抛出异常, 将 sizeCtl 设置为 Integer 的最大值
        sizeCtl = Integer.MAX_VALUE;
        //退出方法
        return;
    }
```

```
        nextTable = nextTab;
        //将 transferIndex 指向 table 数组索引下标最大的位置
        //也就是 table 数组中最后一个元素的位置
        transferIndex = n;
}
int nextn = nextTab.length;
//新建一个占位 ForwardingNode 对象
ForwardingNode<K,V> fwd = new ForwardingNode<K,V>(nextTab);
//该标识用于控制是否继续处理 table 数组中的下一个元素
//值为 true 表示已经处理完当前元素，可以继续迁移下一个元素中的数据
boolean advance = true;
//该标识用于控制扩容是否结束
//从代码逻辑上看，这个标识的另外一个作用是
//最后一个扩容线程会负责重新检查一遍 table 数组，查看是否有未处理的元素
boolean finishing = false;
```

② 执行具体扩容操作前的各项判断逻辑，代码如下。

```
//代码位于 java.util.concurrent.ConcurrentHashMap#transfer 方法中
//在循环中，分配任务和控制当前线程的任务进度
for (int i = 0, bound = 0;;) {
    Node<K,V> f; int fh;
    while (advance) {
        int nextIndex, nextBound;
        //每处理完 table 数组中的一个元素就将 bound 值减 1
        if (--i >= bound || finishing)
            advance = false;
        //transferIndex ≤ 0 说明 table 数组中的元素已被线程分配完毕，没有待分配的元素
        else if ((nextIndex = transferIndex) <= 0) {
            //将 i 设置为-1，后续代码根据这个数值退出当前线程的扩容操作
            i = -1;
            advance = false;
        }
        //只有首次进入 for 循环才会进入这个条件判断
        else if (U.compareAndSwapInt(this, TRANSFERINDEX, nextIndex, nextBound =
        (nextIndex > stride ? nextIndex - stride : 0))) {
            //设置 bound 和 i 的值，也就是迁移任务的数组区间
            bound = nextBound;
            i = nextIndex - 1;
            advance = false;
        }
    }
    if (i < 0 || i >= n || i + n >= nextn) {
        int sc;
        //扩容结束
        if (finishing) {
            //将 nextTable 设置为 null，表示扩容已结束
            nextTable = null;
            //将 nextTab 数组赋值给 table 数组
            table = nextTab;
            //将 sizeCtl 设置为扩容阈值
```

```
                sizeCtl = (n << 1) - (n >>> 1);
                return;
            }
            //每当一个线程扩容结束就更新一次 sizeCtl 的值，进行减 1 操作
            if (U.compareAndSwapInt(this, SIZECTL, sc = sizeCtl, sc - 1)) {
                //判断当前线程是否为扩容 table 数组的最后一个线程
                if ((sc - 2) != resizeStamp(n) << RESIZE_STAMP_SHIFT)
                    //直接回到上层 while 循环
                    return;
                //(sc - 2) == resizeStamp(n) << RESIZE_STAMP_SHIFT
                //说明这个线程是最后一条扩容线程
                //将扩容结束的标识修改为 true
                finishing = advance = true;
                //设置 i = n；以便最后一个线程重新检查一遍数组
                //防止有迁移失败的数组元素
                i = n;
            }
        }
        //如果数组当前位置为空
        else if ((f = tabAt(tab, i)) == null)
            //则在数组的空位置填充一个 ForwardingNode 类型的占位对象
            advance = casTabAt(tab, i, null, fwd);
        else if ((fh = f.hash) == MOVED)
            //遇到哈希值为 MOVED，也就是数组下标为-1 的位置
            //说明该位置已经被其他线程迁移了
            //将 advance 设置为 true，以便继续迁移下一个元素中的数据
            advance = true;
}
```

③ 根据扩容数组和数据迁移的细节不同，分为链表迁移和红黑树迁移两种情况，代码如下。

```
//代码位于 java.util.concurrent.ConcurrentHashMap#transfer 方法中
//完成 table 数组的扩容和数据迁移操作
private final void transfer(Node<K,V>[] tab, Node<K,V>[] nextTab) {
    else {
        synchronized (f) {
            if (tabAt(tab, i) == f) {
                Node<K,V> ln, hn;
                //table 数组为链表结构
                if (fh >= 0) {
                    int runBit = fh & n;
                    Node<K,V> lastRun = f;
                    //遍历整个链表
                    for (Node<K,V> p = f.next; p != null; p = p.next) {
                        int b = p.hash & n;
                        //查找 b 不等于 runBit 的节点
                        if (b != runBit) {
                            runBit = b;
                            lastRun = p;
                        }
                    }
```

```java
            //将 lastRun 设置为 ln 链表的末尾节点
            if (runBit == 0) {
                ln = lastRun;
                hn = null;
            }
            //将 lastRun 设置为 hn 链表的末尾节点
            else {
                hn = lastRun;
                ln = null;
            }
            //使用 hn 和 ln 两个链表进行迁移，使用头插法拼接链表
            for (Node<K,V> p = f; p != lastRun; p = p.next) {
                int ph = p.hash; K pk = p.key; V pv = p.val;
                if ((ph & n) == 0)
                    ln = new Node<K,V>(ph, pk, pv, ln);
                else
                    hn = new Node<K,V>(ph, pk, pv, hn);
            }
            setTabAt(nextTab, i, ln);
            setTabAt(nextTab, i + n, hn);
            //将当前数组标记为迁移完成
            setTabAt(tab, i, fwd);
            //将 advance 设置为 true
            //表示当前数组元素已处理完，可以继续处理下一个数组元素
            advance = true;
        }
        //当前节点为红黑树结构
        else if (f instanceof TreeBin) {
            TreeBin<K,V> t = (TreeBin<K,V>)f;
            //lo 为 ln 链表头节点，loTail 为 ln 链表尾节点
            TreeNode<K,V> lo = null, loTail = null;
            //hi 为 hn 链表头节点，hiTail 为 hn 链表尾节点
            TreeNode<K,V> hi = null, hiTail = null;
            int lc = 0, hc = 0;
            //使用 for 循环以链表方式遍历整棵红黑树
            for (Node<K,V> e = t.first; e != null; e = e.next) {
                int h = e.hash;
                TreeNode<K,V> p = new TreeNode<K,V>
                    (h, e.key, e.val, null, null);
                if ((h & n) == 0) {
                    if ((p.prev = loTail) == null)
                        lo = p;
                    else
                        loTail.next = p;
                    loTail = p;
                    ++lc;
                }
                else {
                    if ((p.prev = hiTail) == null)
                        hi = p;
```

```
                    else
                        hiTail.next = p;
                    hiTail = p;
                    ++hc;
                }
            }
            //lc 小于或等于 6，将红黑树转换成链表
            ln = (lc <= UNTREEIFY_THRESHOLD) ? untreeify(lo) :
                (hc != 0) ? new TreeBin<K,V>(lo) : t;
            //hc 小于或等于 6，将红黑树转换成链表
            hn = (hc <= UNTREEIFY_THRESHOLD) ? untreeify(hi) :
                (lc != 0) ? new TreeBin<K,V>(hi) : t;
            setTabAt(nextTab, i, ln);
            setTabAt(nextTab, i + n, hn);
            //迁移完成后使用 ForwardingNode 类型的对象占位
            setTabAt(tab, i, fwd);
            //将 advance 设置为 true
            //表示当前数组元素已处理完，可以继续处理下一个数组元素
            advance = true;
        }
    }
}
```

6.1.8 再谈 sizeCtl 成员变量

从 ConcurrentHashMap 的构造方法、初始化和扩容可以看出，sizeCtl 成员变量是非常重要的，控制着 table 数组的初始化和扩容操作。下面总结一下 sizeCtl 的变化过程。

（1）在未初始化阶段，在 ConcurrentHashMap 的构造方法 ConcurrentHashMap(int initialCapacity)中为 sizeCtl 赋值，此时创建了 ConcurrentHashMap 对象，还未对 table 数组进行初始化。此时的 sizeCtl 记录了 table 数组的初始容量，为 sizeCtl 赋值的代码如下。

```
int cap = ((initialCapacity >= (MAXIMUM_CAPACITY >>> 1)) ?
        MAXIMUM_CAPACITY :
        tableSizeFor(initialCapacity + (initialCapacity >>> 1) + 1));
this.sizeCtl = cap;
```

（2）在初始化过程中，在 Node<K,V>[] initTable()方法中使用 CAS 操作将 sizeCtl 的值修改为 -1，当其他线程进入 Node<K,V>[] initTable()方法后检测到 sizeCtl 的值为 -1 时，会调用 Thread.yield()方法让出 CPU 的使用权，此时修改 sizeCtl 值的代码如下。

```
U.compareAndSwapInt(this, SIZECTL, sc, -1)
```

（3）在初始化完成后，Node<K,V>[] initTable()方法中的 sizeCtl 会记录当前 table 数组扩容的阈值，此时修改 sizeCtl 的代码如下。

```
Node<K,V>[] nt = (Node<K,V>[])new Node<?,?>[n];
table = tab = nt;
sc = n - (n >>> 2);
sizeCtl = sc;
```

（4）在 table 数组整个扩容过程中，sizeCtl 用于记录当前扩容的线程并发数量，ConcurrentHashMap 的 void tryPresize(int size)方法和 void transfer(Node<K,V>[] tab, Node<K,V>[] nextTab)方法都会修改 sizeCtl 的值。

在 void tryPresize(int size)方法中，会增加 sizeCtl 的值，代码如下。

```
//第 1 个进入方法扩容 table 数组的线程设置 sc 的值，也就是设置 sizeCtl 的值
U.compareAndSwapInt(this, SIZECTL, sc,(rs << RESIZE_STAMP_SHIFT) + 2)
//后续每个进入方法扩容 table 数组的线程都会对 sc 的值进行加 1 操作
U.compareAndSwapInt(this, SIZECTL, sc, sc + 1)
```

在 void transfer(Node<K,V>[] tab, Node<K,V>[] nextTab)方法中，会减少 sizeCtl 的值，代码如下。

```
//每次完成 table 数组扩容的线程都会对 sc 执行减 1 操作，也就是对 sizeCtl 进行减 1 操作
U.compareAndSwapInt(this, SIZECTL, sc = sizeCtl, sc - 1)
```

6.1.9　添加数据

ConcurrentHashMap 主要提供了如下方法添加数据。

```
//添加键-值对
V put(K key, V value)
//将 Map 集合添加到 ConcurrentHashMap 中
void putAll(Map<? extends K, ? extends V> m)
//如果 ConcurrentHashMap 中不存在 key，则添加键-值对
V putIfAbsent(K key, V value)
```

这里以 V put(K key, V value)方法为例进行说明，代码如下。

```
public V put(K key, V value) {
    return putVal(key, value, false);
}
```

在 V put(K key, V value)方法中直接调用了 V putVal(K key, V value, boolean onlyIfAbsent)方法。V putVal(K key, V value, boolean onlyIfAbsent)方法的代码如下。

```
final V putVal(K key, V value, boolean onlyIfAbsent) {
    //如果 key 或 value 为空，则直接抛出 NullPointerException 异常
    if (key == null || value == null) throw new NullPointerException();
    int hash = spread(key.hashCode());
    int binCount = 0;
    //遍历数组
    for (Node<K,V>[] tab = table;;) {
        //f 表示当前计算出的 hashcode 值对应的数组下标元素
        //f 可能是链表，也可能是红黑树
        //n 表示数组的长度
```

```java
//i 表示数组的下标
//fh 表示数组中元素的 hashcode
Node<K,V> f; int n, i, fh;
//数组为空，进行数组的初始化操作
if (tab == null || (n = tab.length) == 0)
    tab = initTable();
//如果数组当前下标位置的元素为空
else if ((f = tabAt(tab, i = (n - 1) & hash)) == null) {
    //则创建一个 Node 节点，添加到数组当前为空的下标处
    if (casTabAt(tab, i, null,
            new Node<K,V>(hash, key, value, null)))
        break;
}
//判断 table 数组当前元素的第 1 个节点的 hashcode 是否为-1
else if ((fh = f.hash) == MOVED)
    //调用 helpTransfer()方法处理
    tab = helpTransfer(tab, f);
else {
    V oldVal = null;
    synchronized (f) {
        //在加锁的过程中，可能有其他线程操作过数组中当前下标的元素
        //所以，要增加这个判断
        if (tabAt(tab, i) == f) {
            //如果 table 数组中第 1 个元素的 hashCode 大于或等于 0
            //则说明是链表结构
            if (fh >= 0) {
                binCount = 1;
                //遍历整个链表
                for (Node<K,V> e = f;; ++binCount) {
                    K ek;
                    //找到与传递的 key 相同的节点数据
                    if (e.hash == hash &&
                        ((ek = e.key) == key ||
                         (ek != null && key.equals(ek)))) {
                        //修改 key 对应的 value 的值
                        oldVal = e.val;
                        if (!onlyIfAbsent)
                            e.val = value;
                        break;
                    }
                    //未找到与传递的 key 相同的节点数据
                    //将当前传递的 key value 构造成 Node 添加到链表尾部
                    Node<K,V> pred = e;
                    if ((e = e.next) == null) {
                        pred.next = new Node<K,V>(hash, key,
                                                  value, null);
                        break;
                    }
                }
            }
```

```
                //如果是红黑树，则直接将红黑树放入对应的位置
                else if (f instanceof TreeBin) {
                    Node<K,V> p;
                    binCount = 2;
                    if ((p = ((TreeBin<K,V>)f).putTreeVal(hash, key,
                                                   value)) != null) {
                        oldVal = p.val;
                        if (!onlyIfAbsent)
                            p.val = value;
                    }
                }
            }
            if (binCount != 0) {
                //链表的长度大于或等于8
                if (binCount >= TREEIFY_THRESHOLD)
                    //调用treeifyBin()方法处理链表转换成红黑树的逻辑
                    treeifyBin(tab, i);
                if (oldVal != null)
                    return oldVal;
                break;
            }
        }
    }
    //添加元素的数量，检测是否扩容
    addCount(1L, binCount);
    return null;
}
```

上述代码的注释比较详细，这里不再赘述实现细节。值得注意的是，在 V putVal(K key, V value, boolean onlyIfAbsent) 方法中，当链表长度大于或等于 8 时，调用 void treeifyBin(Node<K,V>[] tab, int index) 方法实现 table 数组扩容或者将链表转换成红黑树，调用 void addCount(long x, int check) 方法检测 table 数组是否扩容。

6.1.10 读取数据

ConcurrentHashMap 主要提供了如下方法读取数据。

```
//获取key对应的value的值
V get(Object key)
//获取key对应的value的值
//如果获取的value的值为空，则传递默认值defaultValue
V getOrDefault(Object key, V defaultValue)
```

在 V getOrDefault(Object key, V defaultValue) 方法的内部，首先会调用 V get(Object key) 方法获取数据，如果获取的数据为空，就会返回传入的默认值 defaultValue。重点看一下 V get(Object key) 方法，该方法是无锁的，当多个线程读取数据时不会阻塞，代码如下。

```
public V get(Object key) {
```

```
    Node<K,V>[] tab; Node<K,V> e, p; int n, eh; K ek;
    //计算传入的 key 的哈希值
    int h = spread(key.hashCode());
    //table 数组不为空，数组的长度大于 0，数组元素的头节点不为空
    if ((tab = table) != null && (n = tab.length) > 0 &&
        (e = tabAt(tab, (n - 1) & h)) != null) {
        //数组元素的头节点哈希值等于传入的 key 的哈希值
        if ((eh = e.hash) == h) {
            //数组元素的头节点的 key 与传入的 key 相等
            if ((ek = e.key) == key || (ek != null && key.equals(ek)))
                //返回 key 对应的 value
                return e.val;
        }
        //如果数组元素头节点的哈希值小于 0
        //则表示当前数组正在扩容或者当前数组元素是红黑树
        else if (eh < 0)
            return (p = e.find(h, key)) != null ? p.val : null;
        //查找的数据在链表上，遍历链表获取数据
        while ((e = e.next) != null) {
            if (e.hash == h &&
                ((ek = e.key) == key || (ek != null && key.equals(ek))))
                return e.val;
        }
    }
    return null;
}
```

通过上述代码可以看出，V get(Object key)方法是无锁的，Node 节点的 value 的值和指向下一个节点的 next 引用都使用 volatile 关键字保证了可见性。具体参见 6.1.4 节中 Node 类的代码。

6.1.11 修改数据

ConcurrentHashMap 主要提供了如下方法修改数据。

```
//将 ConcurrentHashMap 中键等于 key、值等于 oldValue 的元素的值修改为 newValue
boolean replace(K key, V oldValue, V newValue)
//将 ConcurrentHashMap 中键等于 key 的元素的值修改为 value
V replace(K key, V value)
//根据一定的规则修改 ConcurrentHashMap 中的键-值对
void replaceAll(BiFunction<? super K, ? super V, ? extends V> function)
```

这里，以 V replace(K key, V value)方法为例进行说明，V replace(K key, V value)方法的代码如下：

```
public V replace(K key, V value) {
    if (key == null || value == null)
        throw new NullPointerException();
```

```
        return replaceNode(key, value, null);
}
```

从 V replace(K key, V value)方法的代码可以看出，当传入的 key 或 value 为空时，会直接抛出 NullPointerException 异常，否则返回直接调用 V replaceNode(Object key, V value, Object cv)方法的结果。V replaceNode(Object key, V value, Object cv)方法的代码如下。

```
final V replaceNode(Object key, V value, Object cv) {
    //计算 key 的哈希值
    int hash = spread(key.hashCode());
    //遍历数组
    for (Node<K,V>[] tab = table;;) {
        Node<K,V> f; int n, i, fh;
        //table 数组为空、table 数组长度为 0，或者 key 对应的 table 数组为空
        if (tab == null || (n = tab.length) == 0 ||
            (f = tabAt(tab, i = (n - 1) & hash)) == null)
            //退出循环
            break;
        //判断 table 数组当前元素的第 1 个节点的 hashcode 是否为-1
        else if ((fh = f.hash) == MOVED)
            tab = helpTransfer(tab, f);
        else {
            V oldVal = null;
            boolean validated = false;
            synchronized (f) {
                //table 数组中当前元素的第 1 个节点无变化
                if (tabAt(tab, i) == f) {
                    //哈希值大于或等于 0
                    if (fh >= 0) {
                        validated = true;
                        //循环遍历链表
                        for (Node<K,V> e = f, pred = null;;) {
                            K ek;
                            //节点 key 的哈希值与传入的 key 的哈希值相等
                            //或者节点 key 与传入的 key 相等
                            if (e.hash == hash &&
                                ((ek = e.key) == key ||
                                 (ek != null && key.equals(ek)))) {
                                //设置 key 对应的 value 的值
                                V ev = e.val;
                                if (cv == null || cv == ev ||
                                    (ev != null && cv.equals(ev))) {
                                    oldVal = ev;
                                    if (value != null)
                                        e.val = value;
                                    else if (pred != null)
                                        pred.next = e.next;
                                    else
                                        setTabAt(tab, i, e.next);
                                }
```

```java
                    break;
                }
                pred = e;
                if ((e = e.next) == null)
                    break;
            }
        }
        //结构为红黑树
        else if (f instanceof TreeBin) {
            validated = true;
            TreeBin<K,V> t = (TreeBin<K,V>)f;
            TreeNode<K,V> r, p;
            //根节点不为空,并且根据 hash 和 key 能获取不为空的节点
            if ((r = t.root) != null &&
                (p = r.findTreeNode(hash, key, null)) != null) {
                //修改 value 对应的值
                V pv = p.val;
                if (cv == null || cv == pv ||
                    (pv != null && cv.equals(pv))) {
                    oldVal = pv;
                    if (value != null)
                        p.val = value;
                    else if (t.removeTreeNode(p))
                        setTabAt(tab, i, untreeify(t.first));
                }
            }
        }
    }
    if (validated) {
        if (oldVal != null) {
            if (value == null)
                //计数减 1
                addCount(-1L, -1);
            return oldVal;
        }
        break;
    }
  }
 }
 return null;
}
```

V replaceNode(Object key, V value, Object cv)方法代码的逻辑是根据一定的规则寻找键与 key 相同的节点,值与 cv 相同的节点,将其对应的值修改成 value。

6.1.12 删除数据

ConcurrentHashMap 主要提供了如下方法删除数据。

```
//根据 key 删除数据
V remove(Object key)
//清空所有数据
void clear()
//删除键与 key 相等、值与 value 相等的数据
boolean remove(Object key, Object value)
```

这里，以 V remove(Object key)方法为例进行说明，代码如下。

```
public V remove(Object key) {
    return replaceNode(key, null, null);
}
```

V remove(Object key)方法会调用 V replaceNode(Object key, V value, Object cv)方法将键为 key 的数据删除。

6.1.13 遍历数据

ConcurrentHashMap 主要提供了如下方法遍历数据。

```
//获取 key 的集合
KeySetView<K,V> keySet()
//获取所有的 value 集合
Collection<V> values()
//获取键-值对集合
Set<Map.Entry<K,V>> entrySet()
//根据一定的规则遍历 ConcurrentHashMap
void forEach(BiConsumer<? super K, ? super V> action)
//获取所有的 key
Enumeration<K> keys()
//获取所有的 value
Enumeration<V> elements()
```

这里，以 Set<Map.Entry<K,V>> entrySet()方法为例进行说明，代码如下。

```
public Set<Map.Entry<K,V>> entrySet() {
    EntrySetView<K,V> es;
    return (es = entrySet) != null ? es : (entrySet = new EntrySetView<K,V>(this));
}
```

从 Set<Map.Entry<K,V>> entrySet()方法的代码定义了一个 EntrySetView 类型的局部变量 es，将 entrySet 赋值给 es，如果 es 不为空则返回 es，否则在构造方法中传入 this 创建 EntrySetView 对象并返回。

EntrySetView 是 ConcurrentHashMap 中定义的一个可以遍历 ConcurrentHashMap 中键-值对的 Set 集合类，EntrySetView 的代码比较简单。

6.1.14 使用案例

案例期望：使用 ConcurrentHashMap 实现并发计数功能，具体实现逻辑为循环 100000 次，

每次向线程池提交一个任务，每 100 个任务为一个批次，同时对 ConcurrentHashMap 中指定 key 的 value 进行累加。如果 ConcurrentHashMap 是线程安全的，则无论运行多少次，最终的结果都是 100000；如果 ConcurrentHashMap 不是线程安全的，则运行多次后，最终的结果可能小于 100000。

案例完整代码如下。

```java
public class ConcurrentHashMapTest {
    //map 中的 key
    private static final String KEY = "binghe";
    // 同时并发执行的线程数
    public static final int THREAD_COUNT = 100;
    //执行的总次数
    public static final int TOTAL_COUNT = 100000;
    public static void main(String[] args) throws InterruptedException {
        Map<String, AtomicInteger> map = new ConcurrentHashMap<>();
        ExecutorService threadPool = Executors.newCachedThreadPool();
        final Semaphore semaphore = new Semaphore(THREAD_COUNT);
        final CountDownLatch countDownLatch = new CountDownLatch(TOTAL_COUNT);
        for (int i = 0; i < TOTAL_COUNT; i++) {
            threadPool.execute(() -> {
                try {
                    semaphore.acquire();//是否允许被执行
                    AtomicInteger value = map.get(KEY);;
                    if (value == null){
                        AtomicInteger zeroValue = new AtomicInteger(0);
                        value = map.putIfAbsent(KEY, zeroValue);
                        if (value == null){
                            value = zeroValue;
                        }
                    }
                    value.incrementAndGet();
                    semaphore.release();
                } catch (Exception e) {
                }
                countDownLatch.countDown();
            });
        }
        countDownLatch.await();
        threadPool.shutdown();
        System.out.println("执行完毕输出结果为====>>> " + map.get(KEY).get());
    }
}
```

上述案例代码的实现逻辑比较简单，先定义了一个 ConcurrentHashMap 的 key，同时定义了并发执行的线程数和执行的总次数。在 main() 方法中，创建了一个 ConcurrentHashMap 对象 map，使用线程池创建线程，并使用 Semaphore 和 CountDownLatch 控制并发。循环 100000 次，每次向线程池提交一个任务，每 100 个任务作为一个批次，同时对 ConcurrentHashMap 中指定 key 的 value 进行累加。

多次运行 ConcurrentHashMapTest 类的代码，输出结果如下。

执行完毕输出结果为====>>> 100000

最终的结果为 100000，说明 ConcurrentHashMap 是线程安全的。

注意：案例中使用的 Semaphore 和 CountDownLatch 会在第 9 章进行详细的介绍。

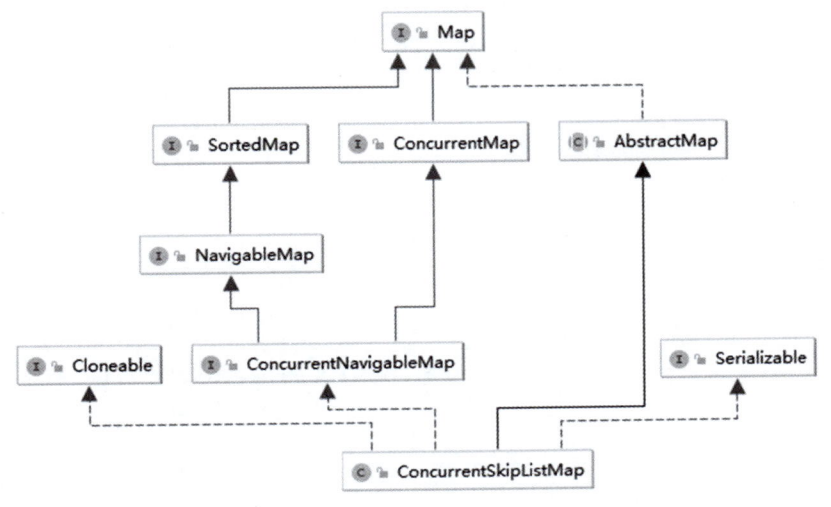

6.2 ConcurrentSkipListMap 集合类

ConcurrentSkipListMap 是线程安全的并发 Map 集合类，ConcurrentSkipListMap 中的元素是有序的。

6.2.1 概述

ConcurrentSkipListMap 是 JDK 1.6 中新增的线程安全的、支持并发访问的、有序 Map 集合类，内部是基于跳表数据结构实现的。也就是说，在 ConcurrentSkipListMap 内部对于数据的添加、修改、查找和删除等操作都是基于跳表实现的。ConcurrentSkipListMap 读写数据的时间是 $\log n$，与线程数无关。在调用 size() 方法获取元素数量时，支持多个线程同时对集合进行操作，集合需要遍历整个链表才能获取元素数量，所以调用 size() 方法的时间复杂度为 $O(\log n)$。

ConcurrentSkipListMap 类的类继承关系如图 6-6 所示。

图 6-6　ConcurrentSkipListMap 类的类继承关系

由图 6-6 可以看出，ConcurrentSkipListMap 类实现了 ConcurrentNavigableMap 接口，而 ConcurrentNavigableMap 接口继承了 ConcurrentMap 接口和 NavigableMap 接口，说明 ConcurrentSkipListMap 类既保证了线程的安全性，又保证了 key 的有序性。

在高并发环境下，如果需要对 Map 中的 key 进行排序，可以使用 ConcurrentSkipListMap 提高并发度。ConcurrentSkipListMap 适用于高并发且需要对键-值对排序的场景。

注意：ConcurrentSkipListMap 内部使用了跳表数据结构，关于跳表数据结构，读者可以参见 5.2.2 节的内容。

6.2.2　内部类

ConcurrentSkipListMap 中定义了两个非常重要的成员变量，一个是表示底层链表的头指针 BASE_HEADER，一个是表示最上层链表的头指针 header，代码如下。

```
//底层链表的头指针
private static final Object BASE_HEADER = new Object();
//最上层链表的头指针
private transient volatile HeadIndex<K,V> head;
```

在 ConcurrentSkipListMap 中，定义了三个非常重要的内部类，分别是 Node 普通节点、Index 索引节点和 HeadIndex 头索引节点，如下所示。

（1）Node 普通节点内部类

Node 普通节点是 ConcurrentSkipListMap 的底层节点，其中保存着实际的键-值对数据，并且保存的数据是按照键-值对中的 key 排序的，核心代码如下。

```
//普通节点Node
static final class Node<K,V> {
    final K key;
    volatile Object value;
    volatile Node<K,V> next;
    Node(K key, Object value, Node<K,V> next) {
        this.key = key;
        this.value = value;
        this.next = next;
    }
    Node(Node<K,V> next) {
        this.key = null;
        this.value = this;
        this.next = next;
    }
    //#############省略其他代码###############
}
```

在内部类 Node 中，key 使用 final 关键字修饰，说明 key 一旦被赋值后是不可变的，value 和 next 使用 volatile 关键字修饰，说明修改 value 和 next 的值对其他线程立刻可见。从整体结果上看，Node 是一个单向链表。

（2）Index 索引节点内部类

Index 索引节点是除底层链表外各层链表的非头节点，核心代码如下。

```
static class Index<K,V> {
    final Node<K,V> node;
    final Index<K,V> down;
    volatile Index<K,V> right;
    Index(Node<K,V> node, Index<K,V> down, Index<K,V> right) {
        this.node = node;
        this.down = down;
        this.right = right;
    }
    //################省略其他代码##################
}
```

Index 索引节点内部类包含 node、down 和 right 3 个指针，node 指针和 down 指针使用 final 关键字修饰，说明 node 指针和 down 指针一旦被赋值是不可变的。right 指针使用 volatile 关键字修饰，说明修改 right 指针的值后，立刻对其他线程可见。

其中，node 指针指向底部的 node 节点，down 指针指向下层链表的节点，right 指针指向当前链表的后继节点。

（3）HeadIndex 头索引节点内部类

HeadIndex 头索引节点是每个层级中链表的头节点，是 Index 索引节点内部类的子类，核心代码如下。

```
static final class HeadIndex<K,V> extends Index<K,V> {
    final int level;
    HeadIndex(Node<K,V> node, Index<K,V> down, Index<K,V> right, int level) {
        super(node, down, right);
        this.level = level;
    }
}
```

可以看出，HeadIndex 头索引节点内部类继承了 Index 索引节点内部类，并在 Index 索引节点内部类的基础上增加了 level 字段，level 字段使用 final 关键字修饰，说明 level 字段一旦被赋值，就不再变化。level 字段表示当前链表的级别，上层链表的级别更高，level 值更大。

6.2.3 初始化

ConcurrentSkipListMap 提供了 4 种构造方法创建 ConcurrentSkipListMap 对象，代码如下。

```
//默认无参构造方法，创建一个空的 ConcurrentSkipListMap 集合
public ConcurrentSkipListMap() {
    this.comparator = null;
```

```
        initialize();
}
//传入一个比较器,创建一个空的ConcurrentSkipListMap集合
public ConcurrentSkipListMap(Comparator<? super K> comparator) {
        this.comparator = comparator;
        initialize();
}
//根据指定的Map创建一个新的ConcurrentSkipListMap集合
public ConcurrentSkipListMap(Map<? extends K, ? extends V> m) {
        this.comparator = null;
        initialize();
        putAll(m);
}
//根据指定的SortedMap创建一个新的ConcurrentSkipListMap集合
//key的顺序与传入的SortedMap中key的顺序保持一致
public ConcurrentSkipListMap(SortedMap<K, ? extends V> m) {
        this.comparator = m.comparator();
        initialize();
        buildFromSorted(m);
}
```

每种构造方法都调用了 void initialize()方法对 ConcurrentSkipListMap 进行初始化, void initialize()方法的代码如下。

```
private void initialize() {
    keySet = null;
    entrySet = null;
    values = null;
    descendingMap = null;
    head = new HeadIndex<K,V>(new Node<K,V>(null, BASE_HEADER, null),
                              null, null, 1);
}
```

void initialize()方法的代码对一些成员变量进行赋值,然后将head指针指向新创建的头索引节点。

6.2.4 添加数据

ConcurrentSkipListMap 主要提供了如下方法添加数据。

```
//向ConcurrentSkipListMap中添加键-值对
V put(K key, V value)
//向ConcurrentSkipListMap中添加键-值对
//ConcurrentSkipListMap中不存在key,才添加键-值对
V putIfAbsent(K key, V value)
```

从代码角度看,无论是 V put(K key, V value)方法还是 V putIfAbsent(K key, V value)方法,

最终都会调用 V doPut(K key, V value, boolean onlyIfAbsent)方法。这里以 V putIfAbsent(K key, V value)方法为例进行说明。由于 V putIfAbsent(K key, V value)方法的代码比较长，这里分为几部分进行说明。

（1）在单一链表上处理各节点的关系，代码如下。

```
//代码位于 java.util.concurrent.ConcurrentSkipListMap#doPut 方法中
//指向待添加的 Node 节点
Node<K,V> z;
//key 为 null，直接抛出 NullPointerException 异常
if (key == null)
    throw new NullPointerException();
Comparator<? super K> cmp = comparator;
outer: for (;;) {
    //遍历的节点 b 可能是小于并且最接近 key 的 Node 节点，也可能是底层链表的头节点
    //n 指向 b 的下一个节点
    for (Node<K,V> b = findPredecessor(key, cmp), n = b.next;;) {
        if (n != null) {
            Object v; int c;
            //f 指向 n 的下一个节点，b、n、f 三者的关系为 b->n->f
            Node<K,V> f = n.next;
            //开始 n 指向 b 的下一个节点，此时如果 n 未指向 b 的下一个节点
            //则说明有其他线程更新了节点关系，跳出循环重试
            if (n != b.next)
                break;
            //n 节点被删除
            if ((v = n.value) == null) {
                n.helpDelete(b, f);
                break;
            }
            //b 节点被删除
            if (b.value == null || v == n)
                break;
            //向后找到第 1 个大于 key 的节点
            if ((c = cpr(cmp, key, n.key)) > 0) {
                b = n;
                n = f;
                continue;
            }
            if (c == 0) {
                //存在 key 相同的节点
                if (onlyIfAbsent || n.casValue(v, value)) {
                    @SuppressWarnings("unchecked") V vv = (V)v;
                    return vv;
                }
                //n.casValue()更新失败，退出循环重试
                break;
            }
        }
```

```
        //创建z节点并插入z节点
        z = new Node<K,V>(key, value, n);
        //z插入成功后，b->z->n
        if (!b.casNext(n, z))
            break;
        //跳出最外层循环
        break outer;
    }
}
```

（2）增加链表的层级，构建各层级之间的关系，代码如下。

```
//代码位于java.util.concurrent.ConcurrentSkipListMap#doPut方法中
//创建随机数因子
int rnd = ThreadLocalRandom.nextSecondarySeed();
//增加链表的层级
if ((rnd & 0x80000001) == 0) {
    //层级
    int level = 1, max;
    while (((rnd >>>= 1) & 1) != 0)
        ++level;
    Index<K,V> idx = null;
    HeadIndex<K,V> h = head;
    //level没有超过最大层级
    if (level <= (max = h.level)) {
        for (int i = 1; i <= level; ++i)
            //使用头插法创建Index节点
            idx = new Index<K,V>(z, idx, null);
    }
    //level超过最大层级
    else {
        level = max + 1;
        //生成Index节点数组，数组的第1个位置不会被使用
        Index<K,V>[] idxs = (Index<K,V>[])new Index<?,?>[level+1];
        for (int i = 1; i <= level; ++i)
            idxs[i] = idx = new Index<K,V>(z, idx, null);
        //生成HeadIndex头索引节点
        for (;;) {
            h = head;
            //获取原来的最大层级
            int oldLevel = h.level;
            if (level <= oldLevel)
                break;
            HeadIndex<K,V> newh = h;
            //指向底层链表的头节点
            Node<K,V> oldbase = h.node;
            for (int j = oldLevel+1; j <= level; ++j)
                newh = new HeadIndex<K,V>(oldbase, newh, idxs[j], j);
            if (casHead(h, newh)) {
                h = newh;
                idx = idxs[level = oldLevel];
```

```
            break;
        }
    }
}
```

（3）构建新层级的 HeadIndex 头索引节点和 Index 索引节点，代码如下。

```
//代码位于 java.util.concurrent.ConcurrentSkipListMap#doPut 方法中
//构建新层级的 HeadIndex 头索引节点和 Index 索引节点
splice: for (int insertionLevel = level;;) {
    int j = h.level;
    for (Index<K,V> q = h, r = q.right, t = idx;;) {
        if (q == null || t == null)
            break splice;
        if (r != null) {
            Node<K,V> n = r.node;
            // compare before deletion check avoids needing recheck
            int c = cpr(cmp, key, n.key);
            if (n.value == null) {
                if (!q.unlink(r))
                    break;
                r = q.right;
                continue;
            }
            if (c > 0) {
                q = r;
                r = r.right;
                continue;
            }
        }

        if (j == insertionLevel) {
            if (!q.link(r, t))
                break; // restart
            if (t.node.value == null) {
                findNode(key);
                break splice;
            }
            if (--insertionLevel == 0)
                break splice;
        }

        if (--j >= insertionLevel && j < level)
            t = t.down;
        q = q.down;
        r = q.right;
    }
}
}
return null;
```

V doPut(K key, V value, boolean onlyIfAbsent)方法的主要逻辑是首先处理单一链表上各节点之间的关系，然后处理链表层级与层级之间的关系，最后构建新层级的 HeadIndex 头索引节点和 Index 索引节点。

6.2.5 读取数据

ConcurrentSkipListMap 主要提供了如下方法读取数据。

```
//获取 key 对应的 value 的值
V get(Object key)

//获取 key 对应的 value 的值
//如果获取的 value 的值为 null，则返回传入的 defaultvalue 的值
V getOrDefault(Object key, V defaultValue)
```

无论是 V get(Object key)方法还是 V getOrDefault(Object key, V defaultValue)方法，最终都会调用 V doGet(Object key)方法。V doGet(Object key)方法的代码如下。

```
private V doGet(Object key) {
    if (key == null)
        throw new NullPointerException();
    Comparator<? super K> cmp = comparator;
    outer: for (;;) {
        //获取 key 前面的节点 b
        for (Node<K,V> b = findPredecessor(key, cmp), n = b.next;;) {
            Object v; int c;
            //n 为 null，说明 key 对应的 Node 节点不存在
            //退出外层循环返回 null
            if (n == null)
                break outer;
            Node<K,V> f = n.next;
            //说明有其他线程已经修改了数据
            //退出循环重试
            if (n != b.next)
                break;
            //n 节点已经被删除，调用 helpDelete()方法
            //退出循环重试
            if ((v = n.value) == null) {
                n.helpDelete(b, f);
                break;
            }
            //b 节点已经被删除，退出循环重试
            if (b.value == null || v == n)
                break;
            //传入的 key 与 n 节点的 key 相等
            if ((c = cpr(cmp, key, n.key)) == 0) {
                @SuppressWarnings("unchecked")
                V vv = (V)v;
                //返回结果，可能为 null
```

```
            return vv;
        }
        if (c < 0)
            break outer;
        b = n;
        n = f;
    }
}
return null;
}
```

V doGet(Object key)方法的总体逻辑是先查找传入的 key 的前继节点，再通过判断 b 节点、b 的后继节点和 key 之间的关系来查找 key 对应的节点，并获取对应的 value。

6.2.6 修改数据

ConcurrentSkipListMap 主要提供了如下方法修改数据。

```
//将 ConcurrentSkipListMap 中 key 对应的值修改为 value
V replace(K key, V value)
//将 ConcurrentSkipListMap 中键等于 key、值等于 oldValue 的值修改为 newValue
boolean replace(K key, V oldValue, V newValue)
//按照一定的规则修改 ConcurrentSkipListMap 中的所有值
void replaceAll(BiFunction<? super K, ? super V, ? extends V> function)
```

这里，以 V replace(K key, V value)方法为例进行说明，V replace(K key, V value)方法的代码如下。

```
//将 ConcurrentSkipListMap 中 key 的值修改为 value
public V replace(K key, V value) {
    //如果 key 或者 value 为 null，则直接抛出 NullPointerException
    if (key == null || value == null)
        throw new NullPointerException();
    for (;;) {
        Node<K,V> n; Object v;
        //查找 key 对应的 Node 节点，将其赋值给 n
        //如果未找到节点，则直接返回 null
        if ((n = findNode(key)) == null)
            return null;
        //找到 key 对应的节点，如果节点的 value 的值不为 null
        //则使用 CAS 将节点对应的 value 的值修改成参数传进来的 value 的值
        //最终返回原来的值
        if ((v = n.value) != null && n.casValue(v, value)) {
            @SuppressWarnings("unchecked") V vv = (V)v;
            return vv;
        }
    }
}
```

V replace(K key, V value)方法的代码比较简单，根据传入的 key 找到对应的 Node 节点，如果未找到对应的 Node 节点，则直接返回 null；如果找到对应的 Node 节点，并且 Node 节点的 value 的值不为 null，则使用 CAS 更新 Node 的节点值；否则，继续下一次循环重试。

6.2.7 删除数据

ConcurrentSkipListMap 主要提供了如下方法删除数据。

```
//删除ConcurrentSkipListMap中key对应的数据
V remove(Object key)
//删除ConcurrentSkipListMap中键与key相等、值与value相等的数据
boolean remove(Object key, Object value)
//清空ConcurrentSkipListMap中所有的数据
void clear()
```

这里，以 V remove(Object key)方法为例进行说明，代码如下。

```
public V remove(Object key) {
    return doRemove(key, null);
}
```

V remove(Object key)方法直接调用了 V doRemove(Object key, Object value)方法删除数据，V doRemove(Object key, Object value)方法的代码如下。

```
final V doRemove(Object key, Object value) {
    if (key == null)
        throw new NullPointerException();
    Comparator<? super K> cmp = comparator;
    outer: for (;;) {
        //获取key节点对应的前继节点b，并将b的后继节点赋值给n
        for (Node<K,V> b = findPredecessor(key, cmp), n = b.next;;) {
            Object v; int c;
            //节点n为null，直接退出外层循环
            if (n == null)
                break outer;
            Node<K,V> f = n.next;
            //n不再是b的后继节点，说明有其他线程修改了数据
            //退出循环，重试
            if (n != b.next)
                break;
            //节点n被删除，调用helpDelete()方法处理数据
            //退出循环，重试
            if ((v = n.value) == null) {
                n.helpDelete(b, f);
                break;
            }
            //b节点被删除，退出循环，重试
            if (b.value == null || v == n)
                break;
            //key对应的节点不存在，直接退出外层循环
            if ((c = cpr(cmp, key, n.key)) < 0)
                break outer;
```

```
            //如果其他线程调用了 findPredecessor() 方法,又在 b 节点后增加新的节点
            //则需要向后遍历链表
            if (c > 0) {
                b = n;
                n = f;
                continue;
            }
            //传入的 value 的值与节点的 value 的值不相等
            //直接退出外层循环
            if (value != null && !value.equals(v))
                break outer;

            //使用 CAS 将节点 n 的 value 修改为 null
            //修改不成功则退出循环,重试
            if (!n.casValue(v, null))
                break;

            //通过比较并设置 b 节点的 next 节点的方式来实现删除操作
            if (!n.appendMarker(f) || !b.casNext(n, f))
                findNode(key);
            else {
                findPredecessor(key, cmp);
                //如果 head 的 right 引用为 null
                if (head.right == null)
                    tryReduceLevel();
            }
            @SuppressWarnings("unchecked")
            V vv = (V)v;
            //则返回删除的 value 的值
            return vv;
        }
    }
    return null;
}
```

该代码的逻辑总体上是将 ConcurrentSkipListMap 中所有层级上键为 key 的节点都删除。

6.2.8 遍历数据

ConcurrentSkipListMap 主要提供了如下方法遍历数据。

```
//获取所有的 key 并返回 NavigableSet 集合
NavigableSet<K> keySet()

//获取所有的 key 并返回 NavigableSet 集合
NavigableSet<K> navigableKeySet()

//获取所有的 value 并返回 Collection 集合
Collection<V> values()
```

```
//获取所有的键-值对并返回 Set 集合
Set<Map.Entry<K,V>> entrySet()

//获取所有的 key 并按照降序返回 NavigableSet 集合
NavigableSet<K> descendingKeySet()

//获取所有 key 的 Iterator 迭代器
Iterator<K> keyIterator()

//获取所有 value 的 Iterator 迭代器
Iterator<V> valueIterator()

//获取所有键-值对的 Iterator 迭代器
Iterator<Map.Entry<K,V>> entryIterator()

//按照一定的规则遍历
void forEach(BiConsumer<? super K, ? super V> action)
```

这里以 Iterator<K> keyIterator()方法为例介绍，代码如下。

```
Iterator<K> keyIterator() {
    return new KeyIterator();
}
```

在 Iterator<K> keyIterator()方法中，直接创建一个 KeyIterator 对象并返回，查看 KeyIterator 内部类的代码，如下所示。

```
final class KeyIterator extends Iter<K> {
    public K next() {
        Node<K,V> n = next;
        advance();
        return n.key;
    }
}
```

KeyIterator 内部类继承自 Iter 类，在 KeyIterator 内部类中，只有一个 next()方法，在 next() 方法中，先将 next 赋值给 Node 节点 n，在调用 void advance()方法后，返回节点 n 的 key。其中 void advance()方法在 KeyIterator 类的父类 Iter 类中，代码如下。

```
final void advance() {
    if (next == null)
        throw new NoSuchElementException();
    lastReturned = next;
    while ((next = next.next) != null) {
        Object x = next.value;
        if (x != null && x != next) {
            @SuppressWarnings("unchecked")
            V vv = (V)x;
            nextValue = vv;
            break;
        }
    }
}
```

    }
}
```

void advance()方法的主要逻辑是遍历 Node 节点，查找下一个节点，如果下一个节点的值 x 不为 null 且不等于下一个节点本身，则将下一个节点的值 x 的类型强转为 ConcurrentSkipListMap 的值的泛型类型 V，并赋值给 nextValue，退出循环。

### 6.2.9 使用案例

案例期望：证明 ConcurrentSkipListMap 是线程安全的，并且存储的数据是按照 key 进行排序的。实现的具体逻辑为循环 1000 次，每次都向线程池提交一个任务，线程池固定有 3 个线程处理任务，并且每 3 个任务作为一个批次同时执行，在各线程的内部，会将当前线程的名称作为 key，将 AtomicInteger 的递增值作为 value，存储到 ConcurrentSkipListMap 中。在 3 个线程处理完数据后，输出 ConcurrentSkipListMap 中的数据，输出结果中的 key 按照线程的名称排序，各线程输出的 AtomicInteger 值的累加结果为 1000。

案例程序代码如下。

```
public class ConcurrentSkipListMapTest {
    //同时并发执行的线程数
    private static final int THREAD_COUNT = 3;
    //执行的总次数
    private static final int TOTAL_COUNT = 1000;
    //创建 ConcurrentSkipListMap 实例
    private static final Map<String, AtomicInteger> MAP = new ConcurrentSkipListMap<>();

    public static void main(String[] args) throws InterruptedException {
        initMap();
        printMap();
    }
    //输出 map 中的数据
    private static void printMap() {
        for(Map.Entry<String, AtomicInteger> entry : MAP.entrySet()){
            System.out.println(entry.getKey() + " ======>>> " + entry.getValue());
        }
    }
    //向 map 中填充数据
    private static void initMap() throws InterruptedException {
        //创建一个线程池
        ExecutorService threadPool = Executors.newFixedThreadPool(THREAD_COUNT);
        final Semaphore semaphore = new Semaphore(THREAD_COUNT);
        final CountDownLatch countDownLatch = new CountDownLatch(TOTAL_COUNT);
        for (int i = 0; i < TOTAL_COUNT; i++) {
            threadPool.execute(() -> {
                try {
                    semaphore.acquire();//是否允许被执行
```
```

```java
            String key = Thread.currentThread().getName();
            AtomicInteger value = MAP.get(key);
            if (value == null){
                AtomicInteger zeroValue = new AtomicInteger(0);
                value = MAP.putIfAbsent(key, zeroValue);
                if (value == null){
                    value = zeroValue;
                }
            }
            value.incrementAndGet();
            semaphore.release();
        } catch (Exception e) {
        }
        countDownLatch.countDown();
    });
}
countDownLatch.await();
threadPool.shutdown();
```

首先定义并发执行的线程数 THREAD_COUNT 和执行的总次数 TOTAL_COUNT 两个常量，并创建一个 ConcurrentSkipListMap 对象。然后在 init()方法中创建一个固定有 3 个线程的线程池，在线程池中处理提交的任务。以当前线程的名称为 key，以 AtomicInteger 的递增值为 value，将各线程的键-值对存储到 ConcurrentSkipListMap 中，并使用 CountDownLatch 和 Semaphore 控制程序的并发，使得每次同时处理 3 个任务。接下来在 printMap()方法中迭代循环 ConcurrentSkipListMap 键-值对，输出 ConcurrentSkipListMap 中的所有键-值对信息。最后在 main()方法中依次调用 init()方法和 printMap()方法。

运行案例程序，输出结果如下。

```
pool-1-thread-1 ======>>> 511
pool-1-thread-2 ======>>> 273
pool-1-thread-3 ======>>> 216
```

从输出结果可以看出，ConcurrentSkipListMap 中的 key 是有序的，并且输出的所有 value 的值的和为 1000，符合预期。

注意：多次运行上述程序输出的结果可能不同，但线程名称都是从小到大输出的，并且所有线程输出的 value 的值的和都等于 1000，符合预期。另外，案例中使用的 Semaphore 和 CountDownLatch 会在本书的并发工具篇进行详细的介绍。

第 7 章

并发阻塞队列

在多线程高并发环境下，使用阻塞队列能够以线程安全的方式实现数据在多个线程之间的共享。大名鼎鼎的生产者—消费者模式就可以使用并发阻塞队列实现。

> 注意：为了让读者更好地理解并发阻塞队列，同时考虑全书的篇幅，在介绍并发阻塞队列的 ArrayBlockingQueue 时，会详细介绍添加、删除和获取数据的每一种方法，而在介绍其他并发阻塞队列时，针对添加、删除和获取队列中的数据，则分别以一个典型的方法进行介绍。有关并发阻塞队列的更多信息，读者可以关注"冰河技术"微信公众号阅读相关的技术文章。
>
> 另外，本章中的添加数据特指入队操作，删除数据特指出队操作。

7.1 并发阻塞队列简介

Java 中的并发阻塞队列都实现了 BlockingQueue 接口，并且可以为队列的容量设置一定的界限，当为并发阻塞队列设置了有界容量时，读写数据在一定条件下会发生阻塞。

7.1.1 概述

并发阻塞队列是线程安全的队列。在使用并发阻塞队列时，可以为队列设置一个有界容量，此时向队列中写数据，或者从队列中读数据，可能阻塞挂起线程，大致情况如下。

（1）在向队列中添加数据时，如果队列中数据的数量达到了设置的最大值，继续向队列中添加数据，线程就会被阻塞挂起，直到有其他线程从队列中读取数据，线程才会被唤醒。

（2）在从队列中读取数据时，如果队列为空，读取队列数据的线程就会被阻塞挂起，直到有其他线程向队列中添加数据，线程才会被唤醒。

并发阻塞队列可以分为单端阻塞队列和双端阻塞队列。单端阻塞队列的示意图如图 7-1 所示。

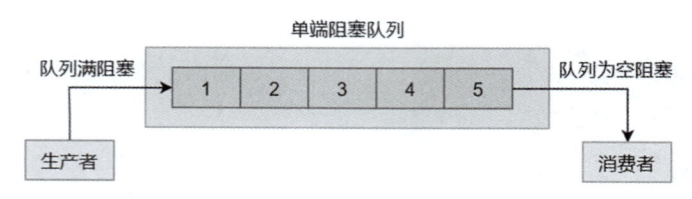

图 7-1 单端阻塞队列的示意图

单端阻塞队列只能向队列的一端添加数据,并且只能从队列的另一端消费数据。当生产者向单端阻塞队列添加数据时,如果队列已满,生产者就会阻塞挂起。当消费者从单端阻塞队列消费数据时,如果队列为空,消费者就会阻塞挂起。

双端阻塞队列的示意图如图 7-2 所示。

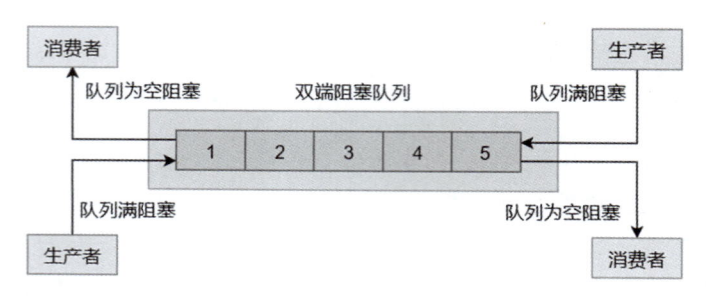

图 7-2 双端阻塞队列的示意图

双端阻塞队列可以分别在队列的两端添加数据和消费数据。当生产者向双端阻塞队列添加数据时,如果队列已满,生产者就会阻塞挂起。当消费者从双端阻塞队列消费数据时,如果队列为空,消费者就会阻塞挂起。

并发阻塞队列经常被用于生产者—消费者模式中,在生产者—消费者模式中,生产者负责向队列中添加数据,消费者负责从队列中读取并消费数据,而阻塞队列就是生产者用来存储数据的队列,也是消费者用来读取和消费数据的队列。

另外,在 Java 实现的线程池中,大量使用了并发阻塞队列。

注意:并发阻塞队列中也提供了非阻塞的方法向队列中添加数据、删除队列中的数据和获取队列中的数据。

7.1.2 类继承关系

Java 中的并发阻塞队列基本上都实现了 BlockingQueue,类继承关系如图 7-3 所示。

由图 7-3 可以看出,Java 中的并发阻塞队列包括 ArrayBlockingQueue、LinkedBlockingQueue、PriorityBlockingQueue、DelayQueue、SynchronousQueue、LinkedTransferQueue 和 LinkedBlockingDeque,这些队列都直接或间接地实现了 BlockingQueue 接口,说明这些队列都

是阻塞队列。其中，LinkedTransferQueue 直接实现了 TransferQueue 接口，而 LinkedBlockingDeque 直接实现了 BlockingDeque 接口。

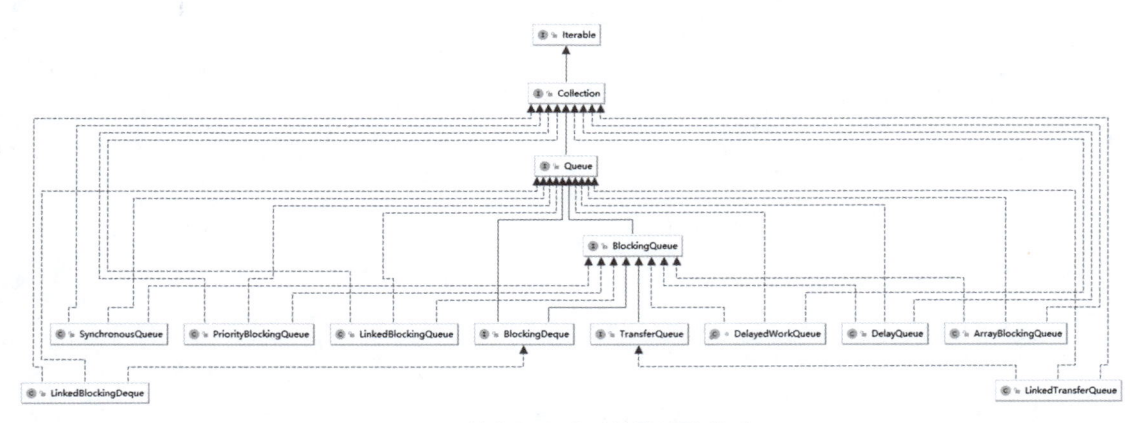

图 7-3 并发阻塞队列的类继承关系

7.1.3 常用方法

并发阻塞队列在 BlockingQueue 接口中定义了操作队列常用的一些方法，BlockingQueue 接口的代码如下。

```
public interface BlockingQueue<E> extends Queue<E> {
    //向队列尾部添加元素，如果队列已满，则抛出 IllegalStateException 异常
    boolean add(E e);

    //非阻塞添加，向队列尾部添加元素，如果队列已满，则立即返回 false
    boolean offer(E e);

    //阻塞添加，向队列尾部添加元素，如果队列已满，则一直阻塞等待
    void put(E e) throws InterruptedException;

    //限时阻塞添加，向队列尾部添加元素，如果在 timeout 时间内，线程就阻塞挂起
    //如果在 timeout 时间内添加成功或超出 timeout 时间，则返回
    boolean offer(E e, long timeout, TimeUnit unit) throws InterruptedException;

    //阻塞删除，移除队列头部的元素，并返回
    //如果队列为空，则阻塞挂起线程，直到队列中存在元素
    E take() throws InterruptedException;

    //非阻塞删除，移除队列头部的元素，并返回删除的元素
    //如果队列为空，则返回 null
    E poll();

    //限时阻塞删除，移除队列头部的元素并返回
    //如果队列为空，则在 timeout 时间内，阻塞挂起线程
```

```
//在timeout时间内队列中添加了数据，则线程被唤醒
//读取数据后返回，如果超出timeout时间队列中数据仍为空，则直接返回
E poll(long timeout, TimeUnit unit)
    throws InterruptedException;

//获取队列的剩余容量
int remainingCapacity();

//删除队列头部的元素，如果队列为空，则抛出NoSuchElementException异常
E remove();

//从队列中移除指定的元素
boolean remove(Object o);

//检测队列中是否包含某个元素
public boolean contains(Object o);

//获取队列头部的元素，如果队列为空，则抛出NoSuchElementException异常
E element();

//获取队列头部的元素，如果队列为空，则返回null
E peek()
//##############其他方法省略##################
}
```

可以将 BlockingQueue 接口提供的操作队列中数据的方法归为 3 类，分别为添加数据的方法、删除数据的方法和获取数据的方法，如表 7-1 所示。

表 7-1 BlockingQueue 接口提供的方法

	抛出异常	返回特殊值	阻塞	限时阻塞
添加数据	add(e)	offer(e)	put(e)	offer(e,timeout,unit)
删除数据	remove()	poll()	take()	poll(time, unit)
获取数据	element()	peek()	无	无

具体说明如下。

（1）添加数据的方法

- add(e)：向队列尾部添加数据成功返回 true，如果队列已满，则抛出 IllegalStateException 异常。
- offer(e)：向队列尾部添加数据成功返回 true，如果队列已满，则返回 false。
- put(e)：向队列尾部添加数据，如果队列已满，则阻塞挂起。
- offer(e,timeout,unit)：向队列尾部添加数据，如果在 timeout 时间内添加成功，则返回 true，否则返回 false。

（2）删除数据的方法

- remove()：获取并删除队列头部的数据，删除成功则返回删除的数据，如果队列为空，则抛出 NoSuchElementException 异常。
- remove(o)：删除队列中指定的数据，删除成功则返回 true，删除失败返回 false。
- poll()：获取并删除队列头部的数据，删除成功则返回删除的数据，如果队列为空，则返回 null。
- take()：获取并删除队列头部的数据，删除成功则返回删除的数据，如果队列为空，则一直阻塞。
- poll(time, unit)：获取并删除队列头部的数据，如在 timeout 时间内删除成功则返回删除的数据，否则返回 null。

（3）获取数据的方法

- element()：获取但不删除队列头部的数据，如果队列为空，则抛出 NoSuchElementException 异常。
- peek()：获取但不删除队列头部的数据，如果队列为空，则返回 null。

在表 7-1 中，对于数据的添加、删除和获取有 4 种不同的处理方式，分别为抛出异常、返回特殊值、阻塞和限时阻塞，具体说明如下。

（1）抛出异常：在添加、删除和获取数据时，如果不能立即执行，则直接抛出异常。

（2）返回特殊值：在添加、删除和获取数据时，如果不能立即执行，则返回一个特殊值（true、false 或 null）。

（3）阻塞：在添加和删除数据时，如果不能立即执行则阻塞挂起，直到可以执行。

（4）限时阻塞：在添加和删除数据时，如果不能立即执行，则在 timeout 时间内阻塞挂起，直到可以执行或者超时。

7.2 ArrayBlockingQueue

ArrayBlockingQueue 是一个常用的阻塞队列，其内部是基于数组实现的。

7.2.1 概述

ArrayBlockingQueue 是一个基于数组实现的有界阻塞队列，遵循先进先出（FIFO）的原则。ArrayBlockingQueue 支持以线程公平和非公平两种方式访问队列中的数据。

线程公平指先阻塞等待的线程先访问队列。线程非公平指先阻塞等待的线程不一定先访问队列，当队列可用时，被唤醒的线程都可以竞争访问队列的资格，先阻塞的线程可能会在后面访问队列。

ArrayBlockingQueue 是基于数组实现的线程安全的队列，适用于高并发环境。

7.2.2 核心成员变量

在 ArrayBlockingQueue 中，定义了一些处理队列的核心成员变量，代码如下。

```
final Object[] items;
int takeIndex;
int putIndex;
int count;
final ReentrantLock lock;
private final Condition notEmpty;
private final Condition notFull;
```

其中，各成员变量的含义如下。

- items：队列底层存储数据的数组。
- takeIndex：读取或者删除数据的索引。
- putIndex：添加数据的索引。
- count：队列中元素的数量。
- lock：可重入锁，用其实现线程阻塞和线程公平与非公平的访问队列。
- notEmpty：删除数据的等待条件。
- notFull：添加数据的等待条件。

7.2.3 初始化

ArrayBlockingQueue 提供了 3 个构造方法用于 ArrayBlockingQueue 类的初始化，如下所示。

```
//构造方法①
public ArrayBlockingQueue(int capacity) {
    this(capacity, false);
}
//构造方法②
public ArrayBlockingQueue(int capacity, boolean fair) {
    if (capacity <= 0)
        throw new IllegalArgumentException();
    this.items = new Object[capacity];
    lock = new ReentrantLock(fair);
    notEmpty = lock.newCondition();
    notFull =  lock.newCondition();
}
//构造方法③
public ArrayBlockingQueue(int capacity, boolean fair,
                    Collection<? extends E> c) {
    this(capacity, fair);
    final ReentrantLock lock = this.lock;
```

```
        lock.lock();
        try {
            int i = 0;
            try {
                for (E e : c) {
                    checkNotNull(e);
                    items[i++] = e;
                }
            } catch (ArrayIndexOutOfBoundsException ex) {
                throw new IllegalArgumentException();
            }
            count = i;
            putIndex = (i == capacity) ? 0 : i;
        } finally {
            lock.unlock();
        }
    }
```

（1）通过构造方法①传入队列的容量创建 ArrayBlockingQueue 对象，内部会直接调用构造方法②，传入 false 实例化非公平锁。

（2）通过构造方法②传入队列的容量和是否是公平锁的标识来创建 ArrayBlockingQueue 对象，在构造方法②中为 items 数组、lock 锁、notEmpty 等待条件和 notFull 等待条件进行实例化。

（3）通过构造方法③传入队列的容量、是否是公平锁的标识和 Collection 集合 c 创建 ArrayBlockingQueue 对象，内部会调用构造方法②创建 ArrayBlockingQueue 对象，对成员变量进行实例化，最终会将集合 c 中的数据添加到队列中。在添加的过程中，如果数组下标越界，则抛出 IllegalArgumentException 异常。

7.2.4 添加数据

ArrayBlockingQueue 中实现了如下添加数据的方法。

```
boolean add(E e)
boolean offer(E e)
void put(E e)
offer(E e, long timeout, TimeUnit unit)
```

1. boolean add(E e)方法

boolean add(E e)方法向队列添加数据失败会抛出 IllegalStateException 异常，代码如下。

```
public boolean add(E e) {
    return super.add(e);
}
```

ArrayBlockingQueue 的 boolean add(E e)方法直接调用了父类 AbstractQueue 的 boolean add(E e)方法，AbstractQueue 的 boolean add(E e)方法代码如下。

```
public boolean add(E e) {
    if (offer(e))
        return true;
    else
        throw new IllegalStateException("Queue full");
}
```

AbstractQueue 的 boolean add(E e)方法会调用 ArrayBlockingQueue 的 boolean offer(E e)方法，向队列添加数据成功则返回 true，失败则抛出 IllegalStateException 异常。

2. boolean offer(E e)方法

boolean offer(E e)方法向队列添加数据失败会返回 false，代码如下。

```
public boolean offer(E e) {
    checkNotNull(e);
    final ReentrantLock lock = this.lock;
    lock.lock();
    try {
        if (count == items.length)
            return false;
        else {
            enqueue(e);
            return true;
        }
    } finally {
        lock.unlock();
    }
}
```

boolean offer(E e)方法的代码比较简单，首先检测传入的参数 e 是否为空，如果为空则抛出 NullPointerException 异常。然后获取 ReentrantLock 锁，在获取锁成功后，判断队列中元素的数量是否等于数组的长度，如果等于数组的长度，则说明队列已满，直接返回 false，否则调用 void enqueue(E x)方法向队列中添加数据，添加成功则返回 true。最后释放 ReentrantLock 锁。

void enqueue(E x)方法的代码如下。

```
private void enqueue(E x) {
    final Object[] items = this.items;
    items[putIndex] = x;
    if (++putIndex == items.length)
        putIndex = 0;
    count++;
    notEmpty.signal();
}
```

可以看到 void enqueue(E x)方法的逻辑主要是向 items 数组中添加数据，当添加完数组中最后一个位置的元素时，又从数组的第 1 个位置开始添加，最后唤醒在 notEmpty 条件队列中等待的线程。

3. void put(E e)方法

void put(E e)方法向队列添加数据失败会一直阻塞，在阻塞期间如果有其他线程中断阻塞线程，则会抛出 InterruptedException 异常，代码如下。

```
public void put(E e) throws InterruptedException {
    checkNotNull(e);
    final ReentrantLock lock = this.lock;
    lock.lockInterruptibly();
    try {
        while (count == items.length)
            notFull.await();
        enqueue(e);
    } finally {
        lock.unlock();
    }
}
```

在 void put(E e)方法中，首先检测参数 e 是否为空，如果为空则抛出 NullPointerException 异常。随后获取 ReentrantLock 的可中断锁，获取锁成功后在 while 循环中判断队列中的元素数量与 items 数组长度是否相等。如果相等，则说明队列已满，调用 notFull.await()方法让线程在 notFull 条件上等待。如果队列中的元素数量与 items 数组的长度不相等，则说明队列未满，调用 void enqueue(E x)方法向队列中添加数据。最后释放 ReentrantLock 锁。

4. boolean offer(E e, long timeout, TimeUnit unit)

boolean offer(E e, long timeout, TimeUnit unit)方法如果在 timeout 时间内添加成功则返回 true，否则返回 false。在阻塞期间如果有其他线程中断阻塞线程，则会抛出 InterruptedException 异常。代码如下。

```
public boolean offer(E e, long timeout, TimeUnit unit) throws InterruptedException
{
    checkNotNull(e);
    long nanos = unit.toNanos(timeout);
    final ReentrantLock lock = this.lock;
    lock.lockInterruptibly();
    try {
        while (count == items.length) {
            if (nanos <= 0)
                return false;
            nanos = notFull.awaitNanos(nanos);
        }
        enqueue(e);
        return true;
    } finally {
        lock.unlock();
    }
}
```

在 boolean offer(E e, long timeout, TimeUnit unit)方法中，同样首先判断 e 是否为空，为空则抛出 NullPointerException 异常。随后将传入的 timeout 的单位转换成纳秒（ns），获取 ReentrantLock 可中断锁。接下来在 while 循环中判断队列中的元素数量与 items 数组长度是否相等。

如果相等则判断 nanos 是否小于或等于 0，如果成立则说明等待已经超时，直接返回 false；如果不成立则更新线程在 notFull 上的等待时间，并更新 nanos 的值。

如果队列中的元素数量与 items 数组长度不相等，则直接调用 void enqueue(E x)方法向队列中添加数据。最后释放 ReentrantLock 锁。

7.2.5 删除数据

ArrayBlockingQueue 主要提供了如下方法删除队列中的数据。

```
boolean remove(Object o)
E poll()
E take()
E poll(long timeout, TimeUnit unit)
```

1. boolean remove(Object o)方法

boolean remove(Object o)方法删除队列中指定的元素，删除成功则返回 true，否则返回 false，代码如下。

```
public boolean remove(Object o) {
    if (o == null) return false;
    final Object[] items = this.items;
    final ReentrantLock lock = this.lock;
    lock.lock();
    try {
        if (count > 0) {
            final int putIndex = this.putIndex;
            int i = takeIndex;
            do {
                if (o.equals(items[i])) {
                    removeAt(i);
                    return true;
                }
                if (++i == items.length)
                    i = 0;
            } while (i != putIndex);
        }
        return false;
    } finally {
        lock.unlock();
    }
}
```

在 boolean remove(Object o)方法中，当 o 为 null 时返回 false，随后获取 ReentrantLock 锁。获取锁成功后判断队列中的元素数量是否大于 0，如果大于 0 则说明队列中存在元素，接下来获取队列的 putIndex 和 takeIndex 索引，将 takeIndex 索引的值赋给变量 i。

判断 i 和 putIndex 是否相等，如果不相等，则一直执行 do-while 循环。在 do-while 循环中判断要删除的元素 o 是否等于队头元素，如果相等则调用 void removeAt(final int removeIndex)方法删除 removeIndex 处的数据并返回 true；如果不相等则继续判断 i+1 的值是否等于 items 数组的长度，当+1 的值等于 items 数组的长度时，将 i 的值重置为 0。

如果队列中没有要删除的数据，则返回 false。最终释放 ReentrantLock 锁。

2. E poll()方法

E poll()方法如果删除成功则返回删除的数据，否则返回 null，代码如下。

```
public E poll() {
    final ReentrantLock lock = this.lock;
    lock.lock();
    try {
        return (count == 0) ? null : dequeue();
    } finally {
        lock.unlock();
    }
}
```

E poll()方法的代码比较简单，首先获取 ReentrantLock 锁，获取锁成功后判断队列中的元素数量是否等于 0，如果等于 0 则直接返回 null，否则调用 E dequeue()方法删除队列头部的元素，最终释放 ReentrantLock 锁。

E dequeue()方法的代码如下所示。

```
private E dequeue() {
    final Object[] items = this.items;
    @SuppressWarnings("unchecked")
    E x = (E) items[takeIndex];
    items[takeIndex] = null;
    if (++takeIndex == items.length)
        takeIndex = 0;
    count--;
    if (itrs != null)
        itrs.elementDequeued();
    notFull.signal();
    return x;
}
```

E dequeue()方法的逻辑也比较简单，就是删除 items 数组中下标为 takeIndex 的数据，并调用 notFull.signal()方法唤醒阻塞在 notFull 上的线程。

3. E take()方法

E take()方法如果删除成功则返回删除的数据，否则一直阻塞。在阻塞期间如果有其他线程中断阻塞线程，则抛出 InterruptedException 异常。代码如下。

```
public E take() throws InterruptedException {
    final ReentrantLock lock = this.lock;
    lock.lockInterruptibly();
    try {
        while (count == 0)
            notEmpty.await();
        return dequeue();
    } finally {
        lock.unlock();
    }
}
```

在 E take()方法中，获取 ReentrantLock 可中断锁，使用 while 循环判断队列中的元素数量是否等于 0，如果等于，则调用 notEmpty 的 await()方法，让线程在 notEmpty 条件队列中阻塞等待；否则调用 E dequeue()方法删除队头数据。最终释放 ReentrantLock 锁。

4. E poll(long timeout, TimeUnit unit)方法

E poll(long timeout, TimeUnit unit)方法获取并删除队列头部的数据，如果在 timeout 时间内删除成功则返回删除的数据，否则返回 null。在阻塞期间如果有其他线程中断阻塞线程，则抛出 InterruptedException 异常。代码如下。

```
public E poll(long timeout, TimeUnit unit) throws InterruptedException {
    long nanos = unit.toNanos(timeout);
    final ReentrantLock lock = this.lock;
    lock.lockInterruptibly();
    try {
        while (count == 0) {
            if (nanos <= 0)
                return null;
            nanos = notEmpty.awaitNanos(nanos);
        }
        return dequeue();
    } finally {
        lock.unlock();
    }
}
```

在 E poll(long timeout, TimeUnit unit)方法中，首先将 timeout 的单位转换成纳秒。随后获取 ReentrantLock 可中断锁，获取锁成功后使用 while 循环判断队列中的元素数量是否等于 0，当队列中的元素数量等于 0 时，如果剩余的 timeout 小于或等于 0，则直接返回 null；如果剩余的 timeout 大于 0，则更新线程在 notEmpty 条件队列中阻塞等待的时间，并更新剩余的时间值。

当队列中的元素数量大于 0 时，直接调用 E dequeue()方法删除队头数据。最终释放 ReentrantLock 锁。

7.2.6 获取数据

ArrayBlockingQueue 主要提供了如下方法获取队列中的数据。

```
E element()
E peek()
```

1. E element()方法

E element()方法位于 ArrayBlockingQueue 类的父类 AbstractQueue 类中，表示获取但不删除队列头部的元素，如果队列为空，则抛出 NoSuchElementException 异常。代码如下。

```
public E element() {
   E x = peek();
   if (x != null)
      return x;
   else
      throw new NoSuchElementException();
}
```

在 E element()方法中，首先调用 ArrayBlockingQueue 类的 E peek()方法获取数据，如果获取的数据不为 null，则返回获取的数据，否则抛出 NoSuchElementException 异常。

2. E peek()方法

E peek()方法获取但不删除队列头部的元素，如果队列为空，则返回 null。代码如下。

```
public E peek() {
   final ReentrantLock lock = this.lock;
   lock.lock();
   try {
      return itemAt(takeIndex); // null when queue is empty
   } finally {
      lock.unlock();
   }
}
```

在 E peek()方法中，首先获取 ReentrantLock 锁，获取成功后调用 E itemAt(int i)方法返回数据，最终释放 ReentrantLock 锁。

E itemAt(int i)方法的代码如下。

```
final E itemAt(int i) {
   return (E) items[i];
}
```

E itemAt(int i)方法会直接返回 items 数组中下标为传入的参数 i 位置的元素。

7.3 LinkedBlockingQueue

LinkedBlockingQueue 是可选边界，其底层是基于链表实现的线程安全的队列，适用于高并发场景。

7.3.1 概述

LinkedBlockingQueue 主要通过构造方法决定，它提供了一个无参构造方法，传入的队列容量为 Integer 类型的最大值，可以看作无界队列。另外，它提供了传入队列容量的构造方法，此构造方法可以创建有界队列，也可以创建无界队列。

7.3.2 核心成员变量

LinkedBlockingQueue 定义了一些处理队列的核心成员变量，代码如下。

```
private final int capacity;
private final AtomicInteger count = new AtomicInteger();
transient Node<E> head;
private transient Node<E> last;
private final ReentrantLock takeLock = new ReentrantLock();
private final Condition notEmpty = takeLock.newCondition();
private final ReentrantLock putLock = new ReentrantLock();
private final Condition notFull = putLock.newCondition();
```

其中，各成员变量的含义如下。

- capacity：队列的容量。
- count：队列中当前元素的数量。
- head：队列的头节点。
- last：队列的尾节点。
- takeLock：take 锁，主要用于 take、poll 等方法。
- notEmpty：take 锁上的等待条件队列。
- putLock：put 锁，主要用于 put、offer 等方法。
- notFull：put 锁上的等待条件队列。

7.3.3 初始化

LinkedBlockingQueue 提供了 3 种构造方法用于初始化，代码如下。

```
//构造方法①
public LinkedBlockingQueue() {
    this(Integer.MAX_VALUE);
}
//构造方法②
public LinkedBlockingQueue(int capacity) {
```

```
        if (capacity <= 0) throw new IllegalArgumentException();
        this.capacity = capacity;
        last = head = new Node<E>(null);
    }
    //构造方法③
    public LinkedBlockingQueue(Collection<? extends E> c) {
        this(Integer.MAX_VALUE);
        final ReentrantLock putLock = this.putLock;
        putLock.lock(); // Never contended, but necessary for visibility
        try {
            int n = 0;
            for (E e : c) {
                if (e == null)
                    throw new NullPointerException();
                if (n == capacity)
                    throw new IllegalStateException("Queue full");
                enqueue(new Node<E>(e));
                ++n;
            }
            count.set(n);
        } finally {
            putLock.unlock();
        }
    }
```

（1）构造方法①为无参构造方法，直接调用构造方法②传入 Integer.MAX_VALUE 创建一个无界队列。

（2）构造方法②传入队列的容量创建队列，并初始化队列的头尾节点。

（3）构造方法③传入集合 c 创建队列，先调用调用构造方法②传入 Integer.MAX_VALUE 创建一个无界队列，再将集合 c 中的元素添加到队列中。在添加的过程中，如果集合 c 中的元素为空，则抛出 NullPointerException 异常，如果队列已满，则抛出 IllegalStateException 异常。

7.3.4 添加数据

这里以 void put(E e)方法为例说明 LinkedBlockingQueue 添加数据的方法，代码如下。

```
public void put(E e) throws InterruptedException {
    //如果传入的参数为空，则抛出 NullPointerException 异常
    if (e == null) throw new NullPointerException();
    int c = -1;
    Node<E> node = new Node<E>(e);
    final ReentrantLock putLock = this.putLock;
    //获取队列中元素的数量
    final AtomicInteger count = this.count;
    //获取可中断锁
    putLock.lockInterruptibly();
    try {
```

```
        //如果队列已满，则阻塞线程
        while (count.get() == capacity) {
            notFull.await();
        }
        //向队列中添加元素
        enqueue(node);
        //更新队列中元素的数量
        c = count.getAndIncrement();
        //如果队列未满，则继续向队列中添加元素
        if (c + 1 < capacity)
            notFull.signal();
    } finally {
        //释放锁
        putLock.unlock();
    }
    //如果队列不为空，则唤醒在notEmpty条件队列中等待的线程
    if (c == 0)
        signalNotEmpty();
}
```

在 void put(E e)方法中，首先判断传入的参数是否为空，如果是则抛出 InterruptedException 异常。然后根据传入的参数 e 创建一个 Node 节点，获取队列中当前元素的数量，并获取 ReentrantLock 可中断锁，获取锁成功后使用 while 循环判断队列中当前元素的数量，如果等于队列的容量，则说明队列已满，调用 notFull.await()使线程在 notFull 条件队列中阻塞等待。

如果当前元素数量不等于队列容量，说明队列未满，则调用 void enqueue(Node<E> node) 方法向队列中添加数据，并更新队列中元素的数量。

如果添加元素后的队列仍然未满，则唤醒在 notFull 条件队列中阻塞等待的线程继续向队列中添加数据。

最后，如果队列不为空，则唤醒在 notEmpty 条件队列中阻塞等待的线程消费队列中的数据。

注意：在 void put(E e)方法中，如果有如下代码表示队列不为空，则唤醒在 notEmpty 条件队列中等待的线程。

```
if (c == 0)
    signalNotEmpty();
```

这是因为，在方法中执行了如下代码更新队列中元素的数量。

```
c = count.getAndIncrement();
```

count.getAndIncrement()方法首先获取 count 的值，然后对 count 的值进行加 1 操作，如果返回的结果为 0，则说明实际的 count 值为 1，队列中已经存在一个元素，可以通知线程从队列中消费数据。

7.3.5 删除数据

这里以 take() 方法为例说明 LinkedBlockingQueue 删除数据的方法，代码如下。

```java
public E take() throws InterruptedException {
    E x;
    int c = -1;
    //获取队列中当前元素的数量
    final AtomicInteger count = this.count;
    final ReentrantLock takeLock = this.takeLock;
    //获取可中断锁
    takeLock.lockInterruptibly();
    try {
        //如果队列为空，则线程在 notEmpty 条件队列中阻塞等待
        while (count.get() == 0) {
            notEmpty.await();
        }
        //删除数据并返回删除的数据
        x = dequeue();
        //更新队列中元素的数量
        c = count.getAndDecrement();
        //如果队列不为空，则继续唤醒在 notEmpty 条件队列中阻塞等待的线程消费数据
        if (c > 1)
            notEmpty.signal();
    } finally {
        //释放锁
        takeLock.unlock();
    }
    //如果队列不满，则唤醒在 notFull 条件队列中阻塞等待的线程，向队列中添加数据
    if (c == capacity)
        signalNotFull();
    return x;
}
```

在 take() 方法中，首先获取队列中当前元素的数量，然后获取 takeLock 的可中断锁。获取成功后，使用 while 循环判断，如果当前队列中元素的数量等于 0，则线程在 notEmpty 条件队列中阻塞等待。否则，删除队头数据并返回删除的数据，同时更新队列中元素的数量。

删除队头数据后如果队列不为空，则继续唤醒在 notEmpty 条件队列中阻塞等待的线程消费队列中的数据，释放锁。

最终，如果队列不满，则唤醒在 notFull 条件队列中阻塞等待的线程，继续向队列中添加数据。

注意： 在 take() 方法中，如果有如下代码表示队列不满，则唤醒在 notFull 条件队列中阻塞等待的线程，继续向队列中添加数据。

```java
if (c == capacity)
    signalNotFull();
```

这是因为在 take()方法中，存在如下代码更新队列中元素的数量。
```
c = count.getAndDecrement();
```
上述代码先返回 count 的值，再对 count 的值进行减 1 操作。如果返回值为队列的容量，那么实际上队列中的元素数量为队列容量减 1，此时的队列未满，可以通知线程向队列中添加数据。

7.3.6 获取数据

这里以 E peek()方法为例说明 LinkedBlockingQueue 获取数据的方法，代码如下。

```java
public E peek() {
    //如果队列为空，则返回 null
    if (count.get() == 0)
        return null;
    final ReentrantLock takeLock = this.takeLock;
    //获取锁
    takeLock.lock();
    try {
        //获取队头节点
        Node<E> first = head.next;
        //如果队头节点为空则返回 null
        if (first == null)
            return null;
        else //如果队头节点不为空则返回队头节点的 item 元素
            return first.item;
    } finally {
        //释放锁
        takeLock.unlock();
    }
}
```

在 E peek()方法中，首先判断队列是否为空，如果队列为空则返回 null。然后获取 ReentrantLock 锁，成功后获取 head 头节点指向的队头节点。如果获取的队头节点为空，则直接返回 null。否则返回队头节点的 item 元素。最终释放 ReentrantLock 锁。

7.4 PriorityBlockingQueue

PriorityBlockingQueue 是一个基于平衡二叉树堆实现的带优先级的无界阻塞队列，每次消费者从队列中消费的都是优先级最高的数据或优先级最低的数据。

7.4.1 概述

当直接遍历 PriorityBlockingQueue 中的数据时，遍历出的数据可能不是有序的，添加到 PriorityBlockingQueue 中的数据默认会使用 compareTo()方法进行比较，也可以在构造方法中传入 Comparator 实现自定义的比较规则。

PriorityBlockingQueue 是一个线程安全的队列，在 PriorityBlockingQueue 中，不存在阻塞添加数据的方法，适用于高并发场景。

注意：在使用 PriorityBlockingQueue 时，要特别注意生产者生产数据的速度不能快于消费者消费数据的速度，否则可能耗尽堆内存。

7.4.2 核心成员变量

在 PriorityBlockingQueue 中，定义了一些处理队列的核心成员变量，代码如下。

```
private transient Object[] queue;
private transient int size;
private transient Comparator<? super E> comparator;
private final ReentrantLock lock;
private final Condition notEmpty;
private transient volatile int allocationSpinLock;
private PriorityQueue<E> q;
```

其中各成员变量的含义如下。

- queue：存放队列元素的数组。
- size：队列中元素的数量。
- comparator：元素大小比较器。
- lock：用来控制同一时间只能有一个线程进行入队和出队操作。
- notEmpty：在进行出队操作时，如果队列为空，就会将线程添加到 notEmpty 条件队列阻塞挂起。notEmpty 成员变量主要是实现 take()方法和 poll()方法的阻塞模式。与 ArrayBlockingQueue 和 LinkedBlockingQueue 不同的是，PriorityBlockingQueue 中没有 notFull 条件等待队列，这是因为在 PriorityBlockingQueue 中，入队操作是非阻塞的，并且 PriorityBlockingQueue 是无界队列。
- allocationSpinLock：自旋锁，使用 CAS 操作保证同一时刻只有一个线程进行扩容。
- q：数组实现的最小堆，在 writeObject()方法和 readObject()方法中用到。为了兼容之前的版本，只有在序列化和反序列化时才非空。

另外，在 PriorityBlockingQueue 内部定义了一个数组的默认初始容量为 11。

7.4.3 初始化

PriorityBlockingQueue 提供了如下构造方法进行 PriorityBlockingQueue 的初始化。

```
//构造方法①
public PriorityBlockingQueue() {
    this(DEFAULT_INITIAL_CAPACITY, null);
```

```java
}
//构造方法②
public PriorityBlockingQueue(int initialCapacity) {
    this(initialCapacity, null);
}
//构造方法③
public PriorityBlockingQueue(int initialCapacity,
                             Comparator<? super E> comparator) {
    if (initialCapacity < 1)
        throw new IllegalArgumentException();
    this.lock = new ReentrantLock();
    this.notEmpty = lock.newCondition();
    this.comparator = comparator;
    this.queue = new Object[initialCapacity];
}
//构造方法④
public PriorityBlockingQueue(Collection<? extends E> c) {
    this.lock = new ReentrantLock();
    this.notEmpty = lock.newCondition();
    boolean heapify = true;
    boolean screen = true;
    if (c instanceof SortedSet<?>) {
        SortedSet<? extends E> ss = (SortedSet<? extends E>) c;
        this.comparator = (Comparator<? super E>) ss.comparator();
        heapify = false;
    }
    else if (c instanceof PriorityBlockingQueue<?>) {
        PriorityBlockingQueue<? extends E> pq =
            (PriorityBlockingQueue<? extends E>) c;
        this.comparator = (Comparator<? super E>) pq.comparator();
        screen = false;
        if (pq.getClass() == PriorityBlockingQueue.class)
            heapify = false;
    }
    Object[] a = c.toArray();
    int n = a.length;
    if (a.getClass() != Object[].class)
        a = Arrays.copyOf(a, n, Object[].class);
    if (screen && (n == 1 || this.comparator != null)) {
        for (int i = 0; i < n; ++i)
            if (a[i] == null)
                throw new NullPointerException();
    }
    this.queue = a;
    this.size = n;
    if (heapify)
        heapify();
}
```

（1）构造方法①是默认的无参构造方法，会直接调用构造方法③传入默认的数组初始化容量，并且传入空的比较器来创建 PriorityBlockingQueue 对象。

（2）构造方法②传入队列中数组的初始容量，直接调用构造方法③传入数组初始容量和空比较器来创建 PriorityBlockingQueue 对象。

（3）构造方法③传入队列中数组的初始容量、自定义比较器创建 PriorityBlockingQueue 对象，并初始化 ReentrantLock 锁、notEmpty 条件等待队列和 queue 数组，同时为比较器 comparator 赋值。

（4）构造方法④传入 Collection 集合创建 PriorityBlockingQueue 对象，并将集合中的元素添加到队列中。在添加的过程中，如果元素为空则抛出 NullPointerException 异常。

在创建 PriorityBlockingQueue 时，如果没有指定 Comparator，并且向队列中添加的数据没有实现 Comparable 接口，就会抛出类型转换异常。

注意：初始化 PriorityBlockingQueue 传入的数组容量与初始化 ArrayBlockingQueue 传入的数组容量不同，ArrayBlockingQueue 传入的是数组的最大容量，而 PriorityBlockingQueue 传入的是数组的初始容量，在 PriorityBlockingQueue 中，数组的实际容量可能大于初始容量。

7.4.4 添加数据

这里以 void put(E e) 方法为例说明 PriorityBlockingQueue 添加数据的方法，代码如下。

```
public void put(E e) {
    offer(e);
}
```

在 void put(E e) 方法中，直接调用了 boolean offer(E e) 方法，boolean offer(E e) 方法的代码如下。

```
public boolean offer(E e) {
    if (e == null)
        throw new NullPointerException();
    final ReentrantLock lock = this.lock;
    //获取锁
    lock.lock();
    int n, cap;
    Object[] array;
    //如果当前元素的数量大于或等于队列容量，则进行扩容
    while ((n = size) >= (cap = (array = queue).length))
        tryGrow(array, cap);
    try {
        Comparator<? super E> cmp = comparator;
        //如果比较器为空，则按照元素的自然顺序比较并调整堆
        if (cmp == null)
            siftUpComparable(n, e, array);
        //如果比较器不为空，则按照比较器的规则比较元素并调整堆
        else
```

```
            siftUpUsingComparator(n, e, array, cmp);
        //添加元素成功后为队列中的元素数量加1
        size = n + 1;
        //唤醒在notEmpty条件队列中阻塞的线程消费队列数据
        notEmpty.signal();
    } finally {
        //释放锁
        lock.unlock();
    }
    return true;
}
```

boolean offer(E e)方法的逻辑相对比较清晰，在boolean offer(E e)方法中，首先判断传入的元素e，如果为空，则抛出NullPointerException。随后获取ReentrantLock锁，获取成功后，使用while循环进行判断，如果当前元素数量大于或等于队列容量，则对队列进行扩容。

接下来，如果当前元素的数量小于队列容量，则判断比较器是否为空，如果比较器为空，则按照元素的自然顺序比较元素并调整堆，否则按照比较器的规则比较元素并调整堆。

向队列中添加元素成功后将队列中的元素数量加1，唤醒在notEmpty条件队列中阻塞的线程消费队列中的数据，最后释放锁并返回true。

boolean offer(E e)方法的扩容操作调用了void tryGrow(Object[] array, int oldCap)方法，void tryGrow(Object[] array, int oldCap)方法的代码如下。

```
private void tryGrow(Object[] array, int oldCap) {
    lock.unlock();
    Object[] newArray = null;
    if (allocationSpinLock == 0 &&
        UNSAFE.compareAndSwapInt(this, allocationSpinLockOffset,
                                 0, 1)) {
        try {
            int newCap = oldCap + ((oldCap < 64) ?
                                   (oldCap + 2) :
                                   (oldCap >> 1));
            if (newCap - MAX_ARRAY_SIZE > 0) {
                int minCap = oldCap + 1;
                if (minCap < 0 || minCap > MAX_ARRAY_SIZE)
                    throw new OutOfMemoryError();
                newCap = MAX_ARRAY_SIZE;
            }
            if (newCap > oldCap && queue == array)
                newArray = new Object[newCap];
        } finally {
            allocationSpinLock = 0;
        }
    }
    if (newArray == null)
        Thread.yield();
```

```
    lock.lock();
    if (newArray != null && queue == array) {
        queue = newArray;
        System.arraycopy(array, 0, newArray, 0, oldCap);
    }
}
```

在 void tryGrow(Object[] array, int oldCap)方法中，首先释放锁。

allocationSpinLock 锁使用 CAS 操作保证同一时刻只有一个线程进行扩容操作，如果 CAS 操作成功，就会进入 try 语句块。如果数组原来的容量小于 64，则容量扩大 1 倍再加 2，否则直接将容量扩大 2 倍。如果超出了容量范围，则抛出 OutOfMemoryError 错误。并在 finally 代码块中释放 CAS 自旋锁。

接下来，如果数组为空，则调用 Thread.yield()方法让出 CPU 使用权，这是因为在 void tryGrow(Object[] array, int oldCap)方法中，使用 CAS 保证了同一时刻只有一个线程能够执行扩容操作。此时其他线程会一直进行 CAS 自旋，当 newArray 数组为空时，调用 Thread.yield()方法让出 CPU 使用权。

最后，扩容的线程执行完毕，而没有执行扩容的线程发现不满足扩容条件，就会跳过扩容逻辑继续向下执行。最后获取 ReentrantLock 锁并复制元素。

7.4.5　删除数据

这里以 take()方法为例说明 PriorityBlockingQueue 删除元素的方法，代码如下。

```
public E take() throws InterruptedException {
    final ReentrantLock lock = this.lock;
    //获取可中断锁
    lock.lockInterruptibly();
    E result;
    try {
        //在 while 循环中进行判断，如果出队元素为空则一直在 notEmpty 条件队列中阻塞等待
        while ( (result = dequeue()) == null)
            notEmpty.await();
    } finally {
        //释放锁
        lock.unlock();
    }
    //返回出队元素
    return result;
}
```

在 take()方法中，首先获取 ReentrantLock 可中断锁，成功后在 while 循环中进行判断，如果出队元素如果为空，则一直在 notEmpty 条件队列中阻塞等待，如果出队元素不为空，则释放锁并返回出队元素。

在 take()方法中调用了 E dequeue()方法进行出队操作，E dequeue()方法的代码如下。

```
private E dequeue() {
    int n = size - 1;
    if (n < 0)
        return null;
    else {
        Object[] array = queue;
        E result = (E) array[0];
        E x = (E) array[n];
        array[n] = null;
        Comparator<? super E> cmp = comparator;
        if (cmp == null)
            siftDownComparable(0, x, array, n);
        else
            siftDownUsingComparator(0, x, array, n, cmp);
        size = n;
        return result;
    }
}
```

E dequeue()方法执行的逻辑也比较简单，先取出最大堆或者最小堆的堆顶元素，再向下调整堆结构，直到整个堆结构稳定。

7.4.6 获取数据

这里以 E peek()方法为例说明 PriorityBlockingQueue 获取数据的方法，代码如下。

```
public E peek() {
    final ReentrantLock lock = this.lock;
    lock.lock();
    try {
        return (size == 0) ? null : (E) queue[0];
    } finally {
        lock.unlock();
    }
}
```

E peek()方法会查看堆顶的元素，如果堆顶元素为空则返回 null，否则返回堆顶元素。

7.5 DelayQueue

DelayQueue 是一个支持延时获取数据的无界阻塞队列，在 DelayQueue 中，只有元素指定的延迟时间到了，才能被获取。

7.5.1 概述

添加到 DelayQueue 中的数据不能被立刻获取，队列中的每个元素都被设置了一个过期时间，只有"过期"的元素才能被获取。

DelayQueue 中存储的元素不允许为 null，这些元素会根据过期时间进行排序。可以将最近过期的元素排到队列头部，将最晚过期的元素排到队列尾部。

DelayQueue 内部使用 PriorityQueue 存储数据，并使用 Leader/Followers 模式最小化不必要的等待时间，使用 ReentrantLock 锁实现线程的同步操作。存入 DelayQueue 中的元素需要实现 Delayed 接口和重写 Delayed 接口的 long getDelay(TimeUnit unit) 方法，并在 long getDelay(TimeUnit unit)方法中计算当前元素距离过期的时间。如果在读取 DelayQueue 中的元素时，发现没有过期的元素，那么可以立即返回 null，也可以使获取元素的线程阻塞。

由于 DelayQueue 是一个无界队列，所以添加数据不会阻塞。

使用 DelayQueue 可以实现精准定时任务和缓存超时机制等。

7.5.2 核心成员变量

DelayQueue 中定义了一些处理队列的核心成员变量，代码如下。

```
private final transient ReentrantLock lock = new ReentrantLock();
private final PriorityQueue<E> q = new PriorityQueue<E>();
private Thread leader = null;
private final Condition available = lock.newCondition();
```

各成员变量的含义如下。

- lock：ReentrantLock 可重入锁。
- q：存储数据的优先级队列。
- leader：leader 线程，成为领导者的线程，只会等待队列头部元素的 delay 时间，其他线程会一直等待，直到被唤醒。随着一个个队列头部元素出队，后面的线程相继成为 leader 线程，同一时刻只有一个线程作为 leader 线程等待队列头部元素的 delay 时间。
- available：条件等待队列，阻塞的线程会被添加到 available 条件等待队列中等待，直到被其他线程唤醒。

7.5.3 初始化

DelayQueue 提供了如下构造方法进行初始化。

```
//构造方法①
public DelayQueue() {}
//构造方法②
public DelayQueue(Collection <? extends E> c) {
    this.addAll(c);
}
```

(1)构造方法①是默认的无参构造方法,内部会初始化一个空的存储队列数据的 PriorityQueue 队列。

(2)构造方法②会传入一个 Collection 集合,在创建 DelayQueue 对象后,会将 Collection 集合中的元素添加到内部的 PriorityQueue 队列中。

7.5.4 添加数据

这里以 void put(E e)方法为例说明 DelayQueue 中添加元素的方法,代码如下。

```
public void put(E e) {
   offer(e);
}
```

在 void put(E e)方法中直接调用了 boolean offer(E e)方法,boolean offer(E e)方法的代码如下。

```
public boolean offer(E e) {
   final ReentrantLock lock = this.lock;
   lock.lock();
   try {
      //向内部队列添加元素 e
      q.offer(e);
      //如果原来的内部队列为空
      if (q.peek() == e) {
         //则重置 leader 线程
         leader = null;
         //唤醒 available 条件队列中的线程
         available.signal();
      }
      return true;
   } finally {
      lock.unlock();
   }
}
```

在 boolean offer(E e)方法中,首先获取 ReentrantLock 锁,获取成功后调用内部优先级队列 q 的 boolean offer(E e)方法将参数 e 添加到内部优先级队列 q 中,然后判断队列 q 的队头元素是否是 e。

如果队列 q 的队头元素是 e,则说明在添加元素 e 之前队列 q 为空,此时需要将 leader 线程设置为 null,并唤醒 available 条件队列中的线程。

最后返回 true 并释放 ReentrantLock 锁。

7.5.5 删除数据

这里以 take()方法为例说明 DelayQueue 中删除元素的方法,代码如下。

```java
public E take() throws InterruptedException {
    final ReentrantLock lock = this.lock;
    lock.lockInterruptibly();
    try {
        for (;;) {
            //获取队列头部的元素
            E first = q.peek();
            //获取的队头元素为空
            if (first == null)
                //线程被添加到 available 条件等待队列中阻塞等待
                available.await();
            else {
                //计算距离队列到期所剩的时间
                long delay = first.getDelay(NANOSECONDS);
                //如果剩余时间小于或等于 0,则说明已到期
                //调用内部队列 q 的 poll 方法弹出元素
                if (delay <= 0)
                    return q.poll();
                first = null;
                //如果 leader 线程不为空
                //则将当前线程加入 available 条件等待队列中进行阻塞等待
                if (leader != null)
                    available.await();
                else {
                    //如果 leader 线程为空,则设置当前线程为 leader 线程
                    Thread thisThread = Thread.currentThread();
                    leader = thisThread;
                    try {
                        //尝试等待元素过期
                        available.awaitNanos(delay);
                    } finally {
                        //在 leader 线程等待 available 条件时
                        //可能有其他线程获取到 ReentrantLock 锁更新了 leader 线程
                        //所以要判断 leader 线程是否和当前线程相等
                        //如果相等则将 leader 线程设置为空,以便进行下次循环
                        if (leader == thisThread)
                            leader = null;
                    }
                }
            }
        }
    } finally {
        //如果 leader 线程为空,则说明没有其他线程在等待
        //如果此时队列不为空,则唤醒 available 条件队列中的线程
        if (leader == null && q.peek() != null)
            available.signal();
        //释放锁
        lock.unlock();
    }
}
```

首先在 take() 方法中获取 ReentrantLock 可中断锁，获取成功后进入一个 try 代码块，try 代码块的逻辑会放在一个 for 循环中。

然后在 for 循环中获取队头元素，如果获取的队头元素为空，则说明队列为空，此时线程会被添加到 available 条件等待队列中阻塞等待。

如果获取的队头元素不为空，则计算距离队头元素过期所剩的时间，如果剩余的时间小于或等于 0，则调用内部优先级队列 q 的 poll() 方法弹出队头元素。

随后判断 leader 线程是否为空，如果不为空，说明已经有线程在等待队头元素的超时时间，则当前线程会被添加到 available 条件队列中进行阻塞等待。

如果 leader 线程为空，则首先将当前线程设置为 leader 线程。接下来，在内层的 try 代码块中尝试等待元素过期，在内层的 finally 代码块中判断当前线程是否是 leader 线程，如果当前线程是 leader 线程，则将 leader 线程设置为 null，以便进行下一次循环。之所以要判断 leader 线程是否是当前线程，是因为 leader 线程在等待 available 条件时，可能有其他线程获取到 ReentrantLock 锁并更新了 leader 线程。最后，在外层的 finally 代码块中，判断 leader 线程是否为空，如果 leader 线程为空，则说明没有其他线程在 available 条件队列中阻塞等待，如果此时内部队列 q 不为空，则唤醒 available 条件队列的线程消费队列中的数据。最终释放 ReentrantLock 锁。

另外，在上述代码中，当距离元素过期的时间小于或等于 0 时，调用了内部优先级队列 q 的 poll() 方法弹出元素，此时的 poll() 方法位于 PriorityQueue 类中，PriorityQueue 类的 E poll() 方法代码如下。

```
public E poll() {
    if (size == 0)
        return null;
    int s = --size;
    modCount++;
    E result = (E) queue[0];
    E x = (E) queue[s];
    queue[s] = null;
    if (s != 0)
        siftDown(0, x);
    return result;
}
```

E poll() 方法先获取并移除 queue 数组中第 1 个元素，如果 queue 数组中还有其他元素，则对其他元素进行重新排序，再返回获取的元素。

7.5.6 获取数据

这里，以 E peek() 方法为例说明 DelayQueue 中获取元素的方法，代码如下。

```
public E peek() {
   final ReentrantLock lock = this.lock;
   lock.lock();
   try {
      return q.peek();
   } finally {
      lock.unlock();
   }
}
```

在 E peek()方法中，先获取 ReentrantLock 锁，获取锁成功后直接调用内部优先级队列 q 的 peek()方法读取数据并返回结果，再释放 ReentrantLock 锁。

E peek()方法中调用了内部优先级队列 q 的 peek()方法，也就是 PriorityQueue 类的 E peek()方法，PriorityQueue 类的 E peek()方法的代码如下。

```
public E peek() {
   return (size == 0) ? null : (E) queue[0];
}
```

PriorityQueue 类的 E peek()方法先判断 queue 数组的长度是否为 0，如果是则直接返回 null，否则返回 queue 数组的第 1 个元素。

7.6 SynchronousQueue

SynchronousQueue 队列是一个阻塞队列，内部不存储数据，每次向队列中添加数据都必须等待其他线程从队列中移除数据。

7.6.1 概述

SynchronousQueue 是一个不存储任何元素的阻塞队列，每次向队列中添加数据都需要等待其他线程从队列中移除数据。

在创建 SynchronousQueue 时，可以指定公平模式或非公平模式。公平模式采用公平锁和一个先进先出的队列来对其他的生产者和消费者进行阻塞，利用队列的先进先出特性确保公平性。非公平模式采用非公平锁和一个先进后出的栈来对其他的生产者和消费者进行阻塞。

SynchronousQueue 不像 ArrayBlockingQueue 底层那样依赖 AQS 实现并发操作，而是使用 CAS 实现并发访问，SynchronousQueue 的吞吐量比 ArrayBlockingQueue 和 LinkedBlockingQueue 都要高。

注意：在使用 SynchronousQueue 的非公平模式时，如果生产者生产数据的速度比消费者消费数据的速度快，则可能导致某些生产者和消费者长时间无法处理数据。

7.6.2 重要常量与内部类

在 SynchronousQueue 内部定义了一些非常重要的常量与内部类，还定义了一个 Transferer 类型的成员变量 transfer，核心代码如下。

```
static final int NCPUS = Runtime.getRuntime().availableProcessors();
static final int maxTimedSpins = (NCPUS < 2) ? 0 : 32;
static final int maxUntimedSpins = maxTimedSpins * 16;
static final long spinForTimeoutThreshold = 1000L;
static final class TransferStack<E> extends Transferer<E> {
    //省略代码
}
static final class TransferQueue<E> extends Transferer<E> {
    //省略代码
}
private transient volatile Transferer<E> transferer;
```

各常量与内部类以及成员变量的含义如下。

- NCPUS：CPU 的核心数。
- maxTimedSpins：限时等待阻塞前自旋的次数，在单核 CPU 下为 0，在多核 CPU 下为 32。
- maxUntimedSpins：不限时等待阻塞前自旋的次数，数值为 maxTimedSpins * 16。
- spinForTimeoutThreshold：纳秒数，主要是为了更快地进行自旋操作。
- TransferStack：TransferStack 类继承 Transferer，按照先进后出的顺序存储元素，用于非公平模式。
- TransferQueue：TransferQueue 类继承 Transferer，按照先进先出的顺序存储元素，用于公平模式。
- transferer：Transferer 对象，在公平模式下是 TransferQueue 对象，在非公平模式下是 TransferStack 对象。

在 SynchronousQueue 的实现中，TransferStack 栈和 TransferQueue 队列都继承了 Transferer 类，而在 Transferer 类中只定义了一个 E transfer(E e, boolean timed, long nanos)抽象方法，TransferStack 栈和 TransferQueue 队列都要实现 E transfer(E e, boolean timed, long nanos)方法。

E transfer(E e, boolean timed, long nanos)方法既支持添加操作也支持获取操作，这是因为方法中设计的双重队列数据结构对于添加和获取操作是对称的，单独执行 E transfer(E e, boolean timed, long nanos)方法添加元素或者获取元素都会被阻塞，只有添加元素和获取元素两个操作匹配时，对应的线程才会同时返回。

7.6.3 初始化

SynchronousQueue 提供了如下构造方法对 SynchronousQueue 进行初始化。

```
//构造方法①
public SynchronousQueue() {
   this(false);
}
//构造方法②
public SynchronousQueue(boolean fair) {
   transferer = fair ? new TransferQueue<E>() : new TransferStack<E>();
}
```

（1）构造方法①是默认的无参构造方法，直接调用构造方法②，传入 false 以非公平的模式创建 SynchronousQueue 对象。

（2）构造方法②会传入一个表示是否是公平模式的标识，如果是公平模式则创建一个 TransferQueue 对象赋给 transferer 成员变量。如果是非公平模式则创建一个 TransferStack 对象赋给 transferer 成员变量。

其中，TransferQueue 是一个基于链表实现的先进先出的队列，而 TransferStack 是一个基于链表实现的先进后出的栈。

7.6.4 添加数据

这里以 void put(E e)方法为例说明 SynchronousQueue 添加数据的方法，代码如下。

```
public void put(E e) throws InterruptedException {
   if (e == null) throw new NullPointerException();
   if (transferer.transfer(e, false, 0) == null) {
      Thread.interrupted();
      throw new InterruptedException();
   }
}
```

判断传入的参数 e 是否为空，如果是则直接抛出 NullPointerException 异常，然后调用 transferer 的 transfer()方法，并传入参数 e 来传输数据，实际就是向 SynchronousQueue 队列中添加一个数据。如果调用 transferer 的 transfer()方法返回的结果为 null，则中断当前线程，并抛出 InterruptedException 异常。

7.6.5 删除数据

这里以 take()方法为例说明 SynchronousQueue 删除数据的方法，代码如下。

```
public E take() throws InterruptedException {
   E e = transferer.transfer(null, false, 0);
   if (e != null)
      return e;
```

```
        Thread.interrupted();
        throw new InterruptedException();
}
```

在 take()方法中,同样调用 transferer 的 transfer()方法传输数据,只不过传输数据时传入的数据为 null,接收 transfer()方法的返回值 e,如果返回值 e 不为空则直接返回 e,如果返回值 e 为空则中断当前线程,并且抛出 InterruptedException 异常。

7.6.6 不支持的方法

SynchronousQueue 不支持的方法有很多,例如 isEmpty()方法永远返回 true,size()方法永远返回 0 等。列举部分方法代码如下。

```
//判断队列是否为空,永远返回 true
public boolean isEmpty() {
    return true;
}
//获取队列中的元素数量,永远返回 0
public int size() {
    return 0;
}
//获取队列剩余容量,永远返回 0
public int remainingCapacity() {
    return 0;
}
//清空队列的元素,方法体为空
public void clear() {
}
//判断队列中是否包含元素 o,永远返回 false
public boolean contains(Object o) {
    return false;
}
//移除队列中指定的元素 o,永远返回 false
public boolean remove(Object o) {
    return false;
}
//判断队列中是否包含集合 c 中所有的元素
//返回传入的集合 c 是否为空
public boolean containsAll(Collection<?> c) {
    return c.isEmpty();
}
//移除队列中所有包含在集合 c 中的元素
//永远返回 false
public boolean removeAll(Collection<?> c) {
    return false;
}
//永远返回 false
public boolean retainAll(Collection<?> c) {
    return false;
```

```
}
//获取但不删除队头元素,永远返回null
public E peek() {
    return null;
}
//获取队列的迭代器,永远返回空迭代器
public Iterator<E> iterator() {
    return Collections.emptyIterator();
}
//返回队列的分割迭代器,永远返回空迭代器
public Spliterator<E> spliterator() {
    return Spliterators.emptySpliterator();
}
//将队列转换为数组,永远返回空数组
public Object[] toArray() {
    return new Object[0];
}
//将队列转换为数组,永远返回空数组
public <T> T[] toArray(T[] a) {
    if (a.length > 0)
        a[0] = null;
    return a;
}
```

在使用时,一定要注意上述方法是 SynchronousQueue 不支持的。

7.7 LinkedTransferQueue

LinkedTransferQueue 是一个由链表结构组成的无界阻塞队列,LinkedTransferQueue 会比一般的 Blocking 多几种方法。

7.7.1 概述

LinkedTransferQueue 并没有直接实现 BlockingQueue 接口,而是直接实现了 TransferQueue 接口,而 TransferQueue 接口继承了 BlockingQueue 接口。LinkedTransferQueue 相比其他阻塞队列多了 tryTransfer()和 transfer()方法。

LinkedTransferQueue 采用预占模式读写数据,在消费者线程从 LinkedTransferQueue 中获取数据时,如果 LinkedTransferQueue 中存在数据,则直接获取数据并返回。如果 LinkedTransferQueue 为空,就会生成一个元素为 null 的节点添加到 LinkedTransferQueue 中,接下来,消费者线程会在这个节点上阻塞等待。在后续生产者线程调用 transfer()方法时,不会将数据添加到 LinkedTransferQueue 中,而是直接将数据传递给消费者线程。

如果生产者线程在调用 transfer()方法时,未发现有在 LinkedTransferQueue 节点上等待的消费者线程,就会将数据添加到 LinkedTransferQueue 中,然后阻塞等待,直到有其他消费者线程

获取添加的元素。

在 LinkedTransferQueue 中存取数据使用的是同一个队列，队列的节点有两种类型，一种是数据节点，另一种是非数据节点，向队列中存取数据时有如下两种情况。

（1）向队列中添加数据时，先与队列头节点对比，如果头节点是数据节点，则生成一个数据节点添加到队列尾端；如果头节点是非数据节点，则进行匹配。当匹配不成功时，继续匹配链表的下一个节点，直到匹配到节点或者遍历完链表的所有节点。

（2）从队列中获取数据时，先与队列头节点对比，如果头节点是非数据节点，则生成一个非数据节点添加到队列尾端；如果头节点是数据节点，则进行匹配操作。当匹配不成功时，继续匹配链表的下一个节点，直到匹配到节点或者遍历完链表的所有节点。

LinkedTransferQueue 是一个线程安全的队列，适用于高并发场景。

7.7.2 重要常量与成员变量

LinkedTransferQueue 中定义了一些处理队列的常量与成员变量，代码如下。

```
private static final boolean MP = Runtime.getRuntime().availableProcessors() > 1;
private static final int FRONT_SPINS   = 1 << 7;
private static final int CHAINED_SPINS = FRONT_SPINS >>> 1;
static final int SWEEP_THRESHOLD = 32;
transient volatile Node head;
private transient volatile Node tail;
private transient volatile int sweepVotes;
private static final int NOW   = 0;
private static final int ASYNC = 1;
private static final int SYNC  = 2;
private static final int TIMED = 3;
```

其中，各常量与成员变量的具体含义如下。

- MP：是否是多核处理器。
- FRONT_SPINS：当节点是队列的第 1 个等待节点时，阻塞前的自旋次数。
- CHAINED_SPINS：处理前驱节点时，当前节点需要自旋的次数。
- SWEEP_THRESHOLD：在扫描队列之前，允许的最大删除失败数。
- head：队列的头节点。
- tail：队列的尾部指针，可能不是最后一个节点。
- sweepVotes：删除失败的次数。
- NOW：存取数据时立即返回，主要用于非超时的 poll() 和 tryTransfer() 方法。
- ASYNC：以异步非阻塞的方式存放数据，主要用于 offer() 方法、put() 方法和 add() 方法。
- SYNC：同步阻塞模式，如果调用时未匹配到则一直阻塞，直到匹配到，主要用于 transfer() 方法和 take() 方法。

- TIMED：超时模式，主要用于带超时调用的 poll()方法和 tryTransfer()方法。

7.7.3 重要内部类

LinkedTransferQueue 中有一个非常重要的内部类 Node，它在整体上实现了链表的结构，核心代码如下。

```
static final class Node {
   final boolean isData;
   volatile Object item;
   volatile Node next;
   volatile Thread waiter;
   Node(Object item, boolean isData) {
      UNSAFE.putObject(this, itemOffset, item);
      this.isData = isData;
   }
//##############其他代码省略################
}
```

内部类 Node 总体实现了单向链表的数据结构，内部存在如下几个比较重要的字段。

- isData：表示当前节点是否是数据节点，也就是区分生产者与消费者。
- item：存储的具体对象。
- next：指向链表的下一个节点。
- waiter：在节点上等待的线程，也就是持有元素的线程。

内部类 Node 会将存储的数据串联成一个单向链表的结构。

7.7.4 初始化

LinkedTransferQueue 提供了如下构造方法创建 LinkedTransferQueue 对象并进行初始化。

```
//构造方法①
public LinkedTransferQueue() {
}
//构造方法②
public LinkedTransferQueue(Collection<? extends E> c) {
   this();
   addAll(c);
}
```

（1）构造方法①是默认的无参构造方法，创建 LinkedTransferQueue 对象后不做后续处理。

（2）构造方法②传入集合 c，首先调用构造方法①创建 LinkedTransferQueue 对象，然后将集合 c 中的元素添加到队列中。

7.7.5 添加数据

这里以 void put(E e)方法为例说明 LinkedTransferQueue 中添加元素的方法，代码如下。

```
public void put(E e) {
    xfer(e, true, ASYNC, 0);
}
```

在 void put(E e)方法中直接调用了 E xfer(E e, boolean haveData, int how, long nanos)方法，E xfer(E e, boolean haveData, int how, long nanos)方法的代码如下。

```
private E xfer(E e, boolean haveData, int how, long nanos) {
    //如果存入的元素为 null, 则直接抛出 NullPointerException
    if (haveData && (e == null))
        throw new NullPointerException();
    Node s = null;
    //外层自旋循环, 内部处理失败后会进行重试
    retry:
    for (;;) {
        //循环中在同一时刻只会存储一种类型的节点
        //从链表的头节点开始进行匹配, 如果头节点已经被其他线程匹配
        //就会匹配下一个节点, 直到遍历完链表的所有节点或者已经匹配到元素
        for (Node h = head, p = h; p != null;) {
            //获取 p 节点是否是数据节点
            boolean isData = p.isData;
            //p 节点中存储的数据
            Object item = p.item;
            //如果 p 节点没有被匹配
            if (item != p && (item != null) == isData) {
                //且 p 节点的模式与传入的模式相同, 则不再匹配
                //退出内层循环尝试将节点添加到链表中
                if (isData == haveData)
                    break;
                //如果 p 节点的模式与传入的模式不同, 则尝试匹配
                //使用 CAS 将 p 节点的 item 值设置为 e
                //如果获取元素, 则 e 的值为 null
                //如果添加元素, 则 e 是具体的值
                if (p.casItem(item, e)) {
                    //这里的 for 循环主要控制多线程同时存取元素时出现的竞争
                    for (Node q = p; q != h;) {
                        Node n = q.next;  // update by 2 unless singleton
                        if (head == h && casHead(h, n == null ? q : n)) {
                            h.forgetNext();
                            break;
                        }                  // advance and retry
                        if ((h = head)   == null ||
                            (q = h.next) == null || !q.isMatched())
                            break;         // unless slack < 2
                    }
                    LockSupport.unpark(p.waiter);
                    return LinkedTransferQueue.<E>cast(item);
                }
            }
            //到这里, 可能 p 已经匹配完成, 或者尝试匹配失败
```

```
            //此时p的next引用可能还没修改,也可能p的next引用指向了自己
            Node n = p.next;
            //如果p的next引用还没修改,则获取p的next节点进行重试
            //否则获取head节点进行重试
            p = (p != n) ? n : (h = head);
        }
        //如果不立即返回
        if (how != NOW) {
            //则新建s节点
            if (s == null)
                s = new Node(e, haveData);
            //尝试将s节点添加到队列中
            Node pred = tryAppend(s, haveData);
            //入队失败进行重试
            if (pred == null)
                continue retry;
            //如果不是异步操作,则等待进行匹配
            if (how != ASYNC)
                return awaitMatch(s, pred, e, (how == TIMED), nanos);
        }
        return e;
    }
}
```

E xfer(E e, boolean haveData, int how, long nanos)方法的代码看起来比较复杂,核心逻辑如下。

(1)找到 head 节点,从 head 节点向后遍历,跳过已经匹配过的节点,当找到模式相反的节点时,进行匹配并返回。这里可能根据需要修改 head 指向的节点。

(2)如果没有找到匹配的节点,则判断是否需要立即返回(是否等于 NOW),如果需要则立即返回。主要用于非超时的 poll()和 tryTransfer()方法。

(3)如果不需要立即返回,就构建一个节点,添加到队列中,如果是异步操作(是否等于 ASYNC)则直接返回,否则阻塞等待匹配操作。主要用于 offer()方法、put()方法和 add()方法。

在 E xfer(E e, boolean haveData, int how, long nanos)方法中调用了 Node tryAppend(Node s, boolean haveData)方法,代码如下。

```
private Node tryAppend(Node s, boolean haveData) {
    for (Node t = tail, p = t;;) {
        Node n, u;
        //队头节点未初始化
        if (p == null && (p = head) == null) {
            //使用 CAS 将 s 节点设置为队头节点
            if (casHead(null, s))
                return s;
        }
        //检测到异常,直接返回 null
```

```
        else if (p.cannotPrecede(haveData))
            return null;
        //遍历节点,找到 tail 的位置
        else if ((n = p.next) != null)
            p = p != t && t != (u = tail) ? (t = u) :
                (p != n) ? n : null;
        //尝试将 s 节点插入队列尾部
        //如果插入失败,则说明已经有其他线程插入了节点
        //此时 p 节点后移并重试
        else if (!p.casNext(null, s))
            p = p.next;
        else {
            if (p != t) {
                //此时 s 节点已经成功添加到队列中
                //如果 p 不等于 t 就更新 tail 的引用
                //将 tail 引用设置为 s 节点
                while ((tail != t || !casTail(t, s)) &&
                    (t = tail)   != null &&
                    (s = t.next) != null &&
                    (s = s.next) != null && s != t);
            }
            //返回 s 的前一个元素
            return p;
        }
    }
}
```

上述代码的总体逻辑如下。

(1) 当队列为空时,刚刚添加到队列的 s 节点是队列中的唯一节点,会返回 s 节点本身。

(2) 当队列不为空时存在两种情况,如下所示。

- 成功将节点 s 添加到队列中最后一个节点 p 的后面,此时返回 s 的上一个节点 p。
- 队列中存在与 s 节点互补的未匹配的节点,返回 null。

另外,在 E xfer(E e, boolean haveData, int how, long nanos)方法中,也调用了 E awaitMatch(Node s, Node pred, E e, boolean timed, long nanos)方法,代码如下。

```
private E awaitMatch(Node s, Node pred, E e, boolean timed, long nanos) {
    //如果存在超时逻辑,则计算超时时间
    final long deadline = timed ? System.nanoTime() + nanos : 0L;
    //获取当前线程
    Thread w = Thread.currentThread();
    //记录自旋的次数
    int spins = -1;
    //随机数,主要用于随机让一些自旋的线程让出 CPU 使用权
    ThreadLocalRandom randomYields = null;
    for (;;) {
        Object item = s.item;
```

```java
        //节点 s 存储的数据不等于 e
        if (item != e) {
            //主要是把 s 中的 item 更新为 s 本身,并将 s 中的 waiter 属性设置为 null
            s.forgetContents();
            //返回在队列中匹配的元素
            return LinkedTransferQueue.<E>cast(item);
        }
        //如果当前线程被中断,或者超时到期
        //就通过 CAS 操作更新 s 中 item 属性的值,将其设置为 s 本身
        if ((w.isInterrupted() || (timed && nanos <= 0)) &&
                s.casItem(e, s)) {
            //尝试删除 s 节点
            unsplice(pred, s);
            //没匹配到,返回元素值
            return e;
        }
        //如果自旋次数小于 0
        if (spins < 0) {
            //则计算自旋次数
            if ((spins = spinsFor(pred, s.isData)) > 0)
                //初始化随机数
                randomYields = ThreadLocalRandom.current();
        }
        else if (spins > 0) {
            //自旋次数减 1
            --spins;
            //随机让一个线程让出 CPU 使用权
            if (randomYields.nextInt(CHAINED_SPINS) == 0)
                Thread.yield();
        }
        //如果 s 的 waiter 属性为空
        else if (s.waiter == null) {
            //则将 s 的 waiter 属性设置为当前线程
            s.waiter = w;
        }
        //如果是超时逻辑
        else if (timed) {
            //则计算超时时间,并超时阻塞
            nanos = deadline - System.nanoTime();
            if (nanos > 0L)
                LockSupport.parkNanos(this, nanos);
        }
        else {
            //否则直接阻塞,等待被其他线程唤醒
            LockSupport.park(this);
        }
    }
}
```

在 E awaitMatch(Node s, Node pred, E e, boolean timed, long nanos)方法中，线程有自旋、让出 CPU 使用权和阻塞 3 种情况，直到 s 节点被匹配或者取消。另外，如果一个节点可能被优先匹配，这个节点就优先自旋，而不是直接阻塞，在达到一定的自旋次数时才会阻塞。这样的设计主要是避免线程的阻塞和唤醒消耗过多的资源。

7.7.6 删除数据

这里以 E take()方法为例说明 LinkedTransferQueue 删除数据的方法，代码如下。

```
public E take() throws InterruptedException {
    E e = xfer(null, false, SYNC, 0);
    if (e != null)
        return e;
    Thread.interrupted();
    throw new InterruptedException();
}
```

在 E take()方法中，同样调用了 E xfer(E e, boolean haveData, int how, long nanos)方法从队列中获取并移除数据，如果获取的数据不为空，则直接返回获取的数据。否则，中断当前线程并抛出 InterruptedException 异常。

7.7.7 获取数据

这里以 E peek()方法为例说明 LinkedTransferQueue 获取数据的方法，代码如下。

```
public E peek() {
    return firstDataItem();
}
```

E peek()方法中直接调用了 E firstDataItem()方法并返回数据，E firstDataItem()方法的代码如下。

```
private E firstDataItem() {
    for (Node p = head; p != null; p = succ(p)) {
        Object item = p.item;
        if (p.isData) {
            if (item != null && item != p)
                return LinkedTransferQueue.<E>cast(item);
        }
        else if (item == null)
            return null;
    }
    return null;
}
```

E firstDataItem()方法的主要逻辑就是查找并返回第 1 个不匹配的数据节点的数据。

7.7.8　新增方法

LinkedTransferQueue 直接实现了 TransferQueue 接口,比其他阻塞队列多了移交数据的方法和消费者相关的方法,这些方法主要在 TransferQueue 接口中,代码如下。

```
public interface TransferQueue<E> extends BlockingQueue<E> {
    boolean tryTransfer(E e);
    void transfer(E e) throws InterruptedException;
    boolean tryTransfer(E e, long timeout, TimeUnit unit)
        throws InterruptedException;
    boolean hasWaitingConsumer();
    int getWaitingConsumerCount();
}
```

每种方法的含义如下。

- tryTransfer(e):尝试将数据转移给等待的消费者线程,此方法不会阻塞线程。如果此时没有其他线程消费数据,则立即返回失败(false),数据也不会被添加到队列的尾部。
- transfer(e):将数据转移给消费者后进入阻塞状态,直到有其他线程对数据进行 take 或 poll 操作,否则,当前线程会将数据添加到队列尾部,并且等待其他线程消费数据。
- tryTransfer(e, timeout, unit):与 tryTransfer(e)方法的具体功能类似,只是设置了超时时间。
- hasWaitingConsumer():判断是否有消费者线程正在等待消费队列中的数据,如果至少有一个消费者线程正在等待消费队列中的数据,则返回 true。
- getWaitingConsumerCount():返回队列中正在等待消费数据的消费者数量。

7.8　LinkedBlockingDeque

LinkedBlockingDeque 是一个基于双向链表实现的双向阻塞队列,能够分别从队列的两端添加和移除数据。

7.8.1　概述

LinkedBlockingDeque 同时支持先进先出(FIFO)和先进后出(FILO)。LinkedBlockingDeque 并没有直接实现 BlockingQueue 接口,而是直接实现了 BlockingDeque 接口,而 BlockingDeque 接口继承了 BlockingQueue 接口,所以 LinkedBlockingDeque 也是一个阻塞队列。

在创建 LinkedBlockingDeque 时,可以手动指定队列的容量,如果不指定,则默认队列的容量是 Integer.MAX_VALUE。

由于 LinkedBlockingDeque 是双向阻塞队列,所以与其他阻塞队列相比,除了实现了 BlockingQueue 接口的方法,还提供了 xxxFirst()方法和 xxxLast()方法,xxxFirst()方法表示从队列的头部操作数据,xxxLast()方法表示从队列的尾部操作数据。

LinkedBlockingDeque 的 xxxFirst()方法总结如表 7-2 所示。

表 7-2　LinkedBlockingDeque 的 xxxFirst()方法

	抛出异常	返回特殊值	阻塞	限时阻塞
添加数据	addFirst(o)	offerFirst(o)	putFirst(o)	offerFirst(o,timeout,timeunit)
删除数据	removeFirst(o)	pollFirst(o)	takeFirst(o)	pollFirst(timeout, timeunit)
获取数据	getFirst(o)	peekFirst(o)		

LinkedBlockingDeque 的 xxxLast()方法总结如表 7-3 所示。

表 7-3　LinkedBlockingDeque 的 xxxLast()方法

	抛出异常	返回特殊值	阻塞	限时阻塞
添加数据	addLast(o)	offerLast(o)	putLast(o)	offerLast(o,timeout,timeunit)
删除数据	removeLast(o)	pollLast(o)	takeLast(o)	pollLast(timeout, timeunit)
获取数据	getLast(o)	peekLast(o)		

从表 7-2 和表 7-3 可以看出，xxxFirst()方法和 xxxLast()方法对于数据的添加、删除和获取都有 4 种处理方式，分别为抛出异常、返回特殊值、阻塞和限时阻塞。这 4 种处理方式的说明见 7.1.3 节。

LinkedBlockingDeque 是一个线程安全的队列，适用于高并发场景。当同一个线程既是队列的生产者也是队列的消费者时，可以使用 LinkedBlockingDeque。如果在某种场景下，需要生产者支持在队列的两端添加数据，消费者支持在队列的两端消费数据，那么也可以使用 LinkedBlockingDeque 实现。

7.8.2　核心成员变量

LinkedBlockingDeque 中定义了一些处理队列的核心成员变量，代码如下。

```
transient Node<E> first;
transient Node<E> last;
private transient int count;
private final int capacity;
final ReentrantLock lock = new ReentrantLock();
private final Condition notEmpty = lock.newCondition();
private final Condition notFull = lock.newCondition();
```

各成员变量的具体含义如下。

- first：队列的头节点。
- last：队列的尾节点。
- count：队列中当前元素的数量。

- capacity：队列的容量，也就是能够容纳的元素数量。
- lock：全局独占锁。
- notEmpty：当消费者线程从队列中移除数据时，如果队列为空，则线程会在 notEmpty 条件队列中阻塞等待，直到超时或者有生产者线程向队列中添加了数据。
- notFull：当生产者线程向队列中添加数据时，如果队列已满，则线程会在 notFull 条件队列中阻塞等待，直到超时或者有消费者线程从队列中移除了数据。

7.8.3　重要内部类

LinkedBlockingDeque 类中有一个重要的内部类 Node，它实现了一个双向链表结构，代码如下。

```
static final class Node<E> {
    //链表中存储的数据
    E item;
    //指向当前节点的前驱节点
    Node<E> prev;
    指向当前节点的后继节点
    Node<E> next;
    //构造方法
    Node(E x) {
        item = x;
    }
}
```

Node 类中定义了一个存储具体数据的 item 成员变量，一个指向当前节点的前驱节点的 prev 成员变量和一个指向当前节点的后继节点的 next 成员变量。

7.8.4　初始化

LinkedBlockingDeque 提供了如下构造方法对 LinkedBlockingDeque 进行初始化。

```
//构造方法①
public LinkedBlockingDeque() {
    this(Integer.MAX_VALUE);
}
//构造方法②
public LinkedBlockingDeque(int capacity) {
    if (capacity <= 0) throw new IllegalArgumentException();
    this.capacity = capacity;
}
//构造方法③
public LinkedBlockingDeque(Collection<? extends E> c) {
    this(Integer.MAX_VALUE);
    final ReentrantLock lock = this.lock;
    lock.lock();
    try {
```

```
        for (E e : c) {
            if (e == null)
                throw new NullPointerException();
            if (!linkLast(new Node<E>(e)))
                throw new IllegalStateException("Deque full");
        }
    } finally {
       lock.unlock();
    }
}
```

（1）构造方法①，默认无参构造方法，直接调用构造方法②，传入 Integer.MAX_VALUE 作为队列容量来创建 LinkedBlockingDeque 对象。

（2）构造方法②，传入队列容量创建 LinkedBlockingDeque 对象，如果传入的队列容量小于或等于 0，则直接抛出 IllegalArgumentException 异常。

（3）构造方法③，传入集合 c 创建 LinkedBlockingDeque，调用构造方法②，传入 Integer.MAX_VALUE 作为队列容量来创建 LinkedBlockingDeque 对象，并将集合 c 中的元素添加到队列中。如果添加的元素为空，则抛出 NullPointerException 异常，如果在添加的过程中队列已满，则抛出 IllegalStateException 异常。

7.8.5 添加数据

这里以 void putFirst(E e)方法为例说明 LinkedBlockingDeque 添加数据的方法，代码如下。

```
public void putFirst(E e) throws InterruptedException {
    if (e == null) throw new NullPointerException();
    Node<E> node = new Node<E>(e);
    final ReentrantLock lock = this.lock;
    lock.lock();
    try {
        while (!linkFirst(node))
            notFull.await();
    } finally {
        lock.unlock();
    }
}
```

在 void putFirst(E e)方法中，首先判断添加的元素是否为空，如果是则直接抛出 NullPointerException 异常。然后将传入的元素 e 封装成一个 Node 对象，获取 ReentrantLock 锁。获取成功后使用 while 循环调用 boolean linkFirst(Node<E> node)方法将 Node 对象添加到队列的头部，如果添加失败，则将当前线程添加到 notFull 条件队列中阻塞等待，如果添加成功，则释放 ReentrantLock 锁。

在 void putFirst(E e)方法中调用了 boolean linkFirst(Node<E> node)方法，boolean linkFirst(Node<E> node)方法的代码如下。

```
private boolean linkFirst(Node<E> node) {
    if (count >= capacity)
        return false;
    Node<E> f = first;
    node.next = f;
    first = node;
    if (last == null)
        last = node;
    else
        f.prev = node;
    ++count;
    notEmpty.signal();
    return true;
}
```

在 boolean linkFirst(Node<E> node)方法中，首先判断队列中当前元素的数量是否大于或等于队列的容量，如果是则直接返回 false，将传入的 node 节点添加到队列的头部，并使队列的头节点 first 指向新添加的 node 节点。

如果队列的尾节点 last 为空，说明在添加 node 节点之前队列为空，则让 last 指向 node 节点。如果尾节点 last 不为空，则将 f 节点的 prev 引用指向 node 节点，然后将队列的元素数量加 1，并唤醒在 notEmpty 条件队列中阻塞等待的线程消费队列中的数据，最终返回 true。

7.8.6 删除数据

这里以 E takeFirst()方法为例说明 LinkedBlockingDeque 删除数据的方法，代码如下。

```
public E takeFirst() throws InterruptedException {
    final ReentrantLock lock = this.lock;
    lock.lock();
    try {
        E x;
        while ( (x = unlinkFirst()) == null)
            notEmpty.await();
        return x;
    } finally {
        lock.unlock();
    }
}
```

在 E takeFirst()方法中，首先获取 ReentrantLock 锁，获取成功后使用 while 循环调用 E unlinkFirst()方法移除队列中头部的元素，如果调用 E unlinkFirst()方法返回的元素为 null，说明移除元素失败，则将当前线程放入 notEmpty 条件队列中阻塞等待。如果调用 E unlinkFirst()方法返回的元素不为 null，则直接返回对应的元素数据，并释放 ReentrantLock 锁。

在 E takeFirst()方法中调用了 E unlinkFirst()方法，代码如下。

```
private E unlinkFirst() {
    Node<E> f = first;
    if (f == null)
        return null;
    Node<E> n = f.next;
    E item = f.item;
    f.item = null;
    f.next = f; // help GC
    first = n;
    if (n == null)
        last = null;
    else
        n.prev = null;
    --count;
    notFull.signal();
    return item;
}
```

在 E unlinkFirst()方法中，首先获取队列的头节点，将其赋值给 f，如果头节点为空，则直接返回 null。然后移除队列的头节点，并返回移除的元素。如果移除后的队列为空，则将 last 尾节点设置为 null，否则，将移除的头节点的下一个节点的 prev 引用设置为 null。随后将队列中元素的数量减 1，唤醒在 notFull 条件队列中阻塞等待的线程向队列中添加数据。最后返回移除的数据。

7.8.7 获取数据

这里以 E peekFirst()方法为例说明 LinkedBlockingDeque 获取数据的方法，代码如下。

```
public E peekFirst() {
    final ReentrantLock lock = this.lock;
    lock.lock();
    try {
        return (first == null) ? null : first.item;
    } finally {
        lock.unlock();
    }
}
```

在 E peekFirst()方法中，首先获取 ReentrantLock 锁，然后判断队列的头节点是否为空。如果为空则直接返回 null，否则，返回队列中头节点的 item 值。最后，释放 ReentrantLock 锁。

注意：LinkedBlockingDeque 提供了比较多的添加数据、删除数据和获取数据的方法，限于篇幅这里只简单分析了 void putFirst(E e)方法、E takeFirst()方法和 E peekFirst()方法，读者可自行分析其他方法。

7.9 并发阻塞队列案例

并发阻塞队列在实际环境中有非常多的应用场景，例如，在 JDK 提供的线程池中，就大量使用了并发阻塞队列。本节基于并发阻塞队列实现两个具有代表性的案例，一个是生产者与消费者模型，另一个是按周期执行定时任务。

7.9.1 生产者与消费者模型

生产者与消费者模型是并发阻塞队列中一个非常典型且重要的应用场景。本节基于 ArrayBlockingQueue 实现生产者与消费者模型。

1. 案例需求

预期效果：生产者线程向 ArrayBlockingQueue 队列中添加数据，消费者线程从 ArrayBlockingQueue 队列中消费数据。如果 ArrayBlockingQueue 队列已满，则生产者线程会被阻塞，如果 ArrayBlockingQueue 队列为空，则消费者线程会被阻塞。

2. 案例实现

案例的实现步骤如下。

（1）在 mykit-concurrent-chapter07 子工程下的 io.binghe.concurrent.chapter07 包下创建 ProducerAndConsumerTest 类。并在 ProducerAndConsumerTest 类中定义 3 个静态变量。

```
//定义一个生产者线程池
private static ThreadPoolExecutor producerThreadPool = new ThreadPoolExecutor(20,
20, 500,
 TimeUnit.MILLISECONDS, new ArrayBlockingQueue<>(1000));
//定义一个消费者线程池
private static ThreadPoolExecutor consumerThreadPool = new ThreadPoolExecutor(20,
20, 500,
 TimeUnit.MILLISECONDS, new ArrayBlockingQueue<>(1000));
//定义一个阻塞队列
private static BlockingQueue<Long> queue = new ArrayBlockingQueue<>(5);
//发送数据的计数器,生产者每向队列中添加一个数据,值就+1
private static AtomicLong atomicLong = new AtomicLong(0);
```

各静态变量的含义如下。

- producerThreadPool：生产者线程池，用于生产者向队列中添加数据。
- consumerThreadPool：消费者线程池，用于消费者消费队列中的数据。
- queue：阻塞队列，生产者向队列中添加数据，消费者从队列中消费数据，定义的队列最大容量为 5，也就是队列中最多能容纳 5 个元素。
- atomicLong：发送数据的计数器，生产者每向队列中添加 1 个数据，其值就加 1。

（2）实现生产者生产数据的方法，代码如下。

```java
//生产者生产数据
private static void produceData(){
    //向线程池中提交10个任务
    IntStream.range(0, 10).forEach((i) -> {
        producerThreadPool.execute(() -> {
            try {
                Thread.sleep(200);
                //获取线程的名称
                String threadName = Thread.currentThread().getName();
                //生产的数据
                Long data = atomicLong.incrementAndGet();
                //将数据写入队列
                queue.put(data);
                System.out.println("生产者线程: " + threadName +
                            " 向队列添加的数据为: " + data);
            } catch (InterruptedException e) {
                logger.error("生产者生产数据异常: {}" ,e);
            }
        });
    });
}
```

生产者生产数据的逻辑比较简单,向生产者线程池中连续提交10个任务,处理每个任务时线程池都会创建一个线程,获取线程的名称,将计数器新增的数据作为生产者生产的数据添加到队列中。队列的最大容量为5,而向队列中提交的数据为10个,如果没有消费者线程消费队列中的数据,在生产者向队列中添加5个数据后,后续添加数据的线程就会被阻塞。

(3)实现消费者消费数据方法的代码如下。

```java
//消费者消费数据
private static void consumerData(){
    //向线程池中提交10个任务
    IntStream.range(0, 10).forEach((i) -> {
        consumerThreadPool.execute(()->{
            try {
                Thread.sleep(200);
                //获取线程的名称
                String threadName = Thread.currentThread().getName();
                //从队列中获取数据
                Long data = queue.take();
                System.out.println("消费者线程:" + threadName + " 获取的数据为:"
                            + data);
            } catch (InterruptedException e) {
                logger.error("消费者消费数据异常: {}" ,e);
            }
        });
    });
}
```

消费者消费数据的逻辑也比较简单,向消费者线程池提交10个任务,处理每个任务时线程池都会创建一个线程,每个线程都会获取当前线程的名称,从队列中获取数据并打印。如果消费者从队列中消费数据时队列为空,消费者线程就会被阻塞。

至此,生产者生产数据、消费者消费数据的主体逻辑已经实现。

3. 案例测试

(1)在main()方法中只调用生产者生产数据的方法produceData(),同时关闭线程池,代码如下。

```
public static void main(String[] args){
    produceData();
    producerThreadPool.shutdown();
    consumerThreadPool.shutdown();
}
```

运行上述程序,输出结果如下。

```
生产者线程: pool-1-thread-5 向队列添加的数据为: 5
生产者线程: pool-1-thread-2 向队列添加的数据为: 2
生产者线程: pool-1-thread-4 向队列添加的数据为: 1
生产者线程: pool-1-thread-1 向队列添加的数据为: 3
生产者线程: pool-1-thread-3 向队列添加的数据为: 4
```

从输出结果可以看出,当生产者向线程池提交10个生产数据的任务时,由于队列的最大容量为5,所以生产者线程只生产并向队列中添加了5个数据就被阻塞,同时,程序未退出。说明如果ArrayBlockingQueue队列已满,生产者线程就会被阻塞。

(2)在main()方法中只调用消费者消费数据的方法consumerData(),同时关闭线程池,代码如下。

```
public static void main(String[] args){
    consumerData();
    producerThreadPool.shutdown();
    consumerThreadPool.shutdown();
}
```

运行上述程序,不会输出任何结果,也不会退出。说明如果ArrayBlockingQueue队列为空,消费者线程就会被阻塞。

(3)在main()方法中同时调用生产者生产数据的方法produceData()和消费者消费数据的方法consumerData(),并关闭线程池,代码如下。

```
public static void main(String[] args){
    produceData();
    consumerData();
    producerThreadPool.shutdown();
    consumerThreadPool.shutdown();
}
```

运行上述程序,输出结果如下。

```
生产者线程: pool-1-thread-6 向队列添加的数据为: 4
生产者线程: pool-1-thread-3 向队列添加的数据为: 10
生产者线程: pool-1-thread-5 向队列添加的数据为: 6
生产者线程: pool-1-thread-9 向队列添加的数据为: 5
生产者线程: pool-1-thread-1 向队列添加的数据为: 7
消费者线程: pool-2-thread-2 获取的数据为: 4
生产者线程: pool-1-thread-2 向队列添加的数据为: 9
生产者线程: pool-1-thread-8 向队列添加的数据为: 2
消费者线程: pool-2-thread-1 获取的数据为: 10
消费者线程: pool-2-thread-7 获取的数据为: 6
消费者线程: pool-2-thread-4 获取的数据为: 2
消费者线程: pool-2-thread-8 获取的数据为: 3
消费者线程: pool-2-thread-5 获取的数据为: 8
生产者线程: pool-1-thread-4 向队列添加的数据为: 8
生产者线程: pool-1-thread-10 向队列添加的数据为: 3
消费者线程: pool-2-thread-3 获取的数据为: 9
消费者线程: pool-2-thread-10 获取的数据为: 7
消费者线程: pool-2-thread-6 获取的数据为: 1
消费者线程: pool-2-thread-9 获取的数据为: 5
生产者线程: pool-1-thread-7 向队列添加的数据为: 1
```

结果中正常输出了生产者向队列中添加的数据，同时，消费者也正常消费了数据。说明生产者线程向 ArrayBlockingQueue 队列中添加数据正常，消费者线程从 ArrayBlockingQueue 队列中消费数据正常。

综上所述，实现的案例程序符合预期。

注意：在输出结果中，部分消费者消费数据的日志在生产者生产数据的日志之前输出，这是因为在生产者线程执行 queue.put(data) 代码向队列中添加完数据、准备输出日志时，CPU 时间片已使用完，随即将当前生产者线程挂起，转而执行消费者线程。消费者从队列中消费数据并输出日志，之后生产者线程被唤醒，CPU 转而执行生产者线程，输出生产者的日志。这时，消费者消费数据的日志就会先于生产者生产数据的日志输出。

更多有关 CPU 时间片与线程切换的知识，读者可以参考《深入理解高并发编程：核心原理与案例实战》一书。

7.9.2 按周期执行的定时任务

使用并发阻塞队列中的 DelayQueue 可以非常方便地实现按周期执行的定时任务。

1. 案例需求

基于 DelayQueue 实现按 1s 的周期执行定时任务。

2. 案例实现

（1）由于添加到 DelayQueue 中的元素必须实现 Delayed 接口，所以首先在

io.binghe.concurrent.chapter07 包下创建一个 TimerTask 类，使其实现 Delayed 接口。又由于需要将 TimerTask 类放入单独的线程中执行，所以让 TimerTask 类实现 Runnable 接口，代码如下。

```java
public class TimerTask implements Delayed, Runnable {
    //延迟时长，出队时判断数据在队列中是否达到了interval时长
    //如果未达到interval时长，则继续等待
    //如果已达到interval时长，则出队
private long interval;
    //数据放入队列的时间
    //结合系统当前时间判断是否已经达到延迟时间
private long queueTime;
    //延迟队列
private DelayQueue<TimerTask> delayQueue;
    //执行的任务
    private Runnable task;
    public TimerTask(DelayQueue<TimerTask> delayQueue, Runnable task,  long interval) {
        this.interval = interval;
        this.delayQueue = delayQueue;
        this.task = task;
    }
    @Override
    public void run() {
        //执行具体的任务
        task.run();
        //每次执行任务时都更新time时间
        queueTime = System.currentTimeMillis();
        delayQueue.put(this);
    }
    @Override
    public long getDelay(TimeUnit unit) {
        return unit.convert((this.queueTime + this.interval) -
          System.currentTimeMillis(), TimeUnit.MILLISECONDS);
    }
    @Override
    public int compareTo(Delayed o) {
        return (int)(this.queueTime - ((TimerTask)o).getQueueTime());
    }
    public long getQueueTime() {
        return queueTime;
    }
}
```

TimerTask 类的逻辑比较简单，首先定义了 4 个成员变量，分别为延迟的时长 interval、数据添加到 DelayQueue 队列中的时间 queueTime、延迟队列 DelayQueue 对象 delayQueue 和执行的具体任务 task。其中，延迟的时长 interval、延迟队列 DelayQueue 对象 delayQueue 和执行的具体任务 task 会通过 TimerTask 的构造方法传递进来。

在run()方法中，先调用task的run()方法执行具体的任务，再将queueTime赋值为当前系统的时间，并将当前TimerTask对象添加到delayQueue队列中。

在long getDelay(TimeUnit unit)方法中，计算距离执行下次任务所剩的时间，在int compareTo(Delayed o)中实现任务的排序规则。

（2）在io.binghe.concurrent.chapter07包下创建一个Timer类，代码如下。

```java
//定时任务
public class Timer {
    private static final Logger logger = LoggerFactory.getLogger(Timer.class);
    //延迟队列
    private static DelayQueue<TimerTask> delayQueue = new DelayQueue<>();
    //创建线程池，执行任务
    private static ExecutorService threadPool = Executors.newFixedThreadPool(2);
    //定时执行任务的周期频率
    private static final Long INTERVAL = 1000L;
    public void schedule(Runnable task, long interval){
        threadPool.execute(new TimerTask(delayQueue, task, interval));
        threadPool.execute(() -> {
            while (true){
                try {
                    TimerTask timerTask = delayQueue.take();
                    threadPool.execute(timerTask);
                } catch (InterruptedException e) {
                    logger.error("执行定时任务抛出了异常");
                }
            }
        });
    }
    public static void main(String[] args){
        new Timer().schedule(()-> {
            System.out.println(Thread.currentThread().getName() + "
                            线程执行任务的当前时间为: " + new Date());
        },INTERVAL);
    }
}
```

在Timer类中，先定义了一个泛型类型为TimerTask的DelayQueue延迟队列对象delayQueue，创建一个固定有两个线程的线程池threadPool，因为向DelayQueue队列中添加任务和获取任务都需要在线程池中执行，所以这里创建的是固定有两个线程的线程池。接下来定义了一个执行周期为1s的常量INTERVAL。

在Timer类的实例方法void schedule(Runnable task, long interval)中实现定时任务的核心逻辑，首先，调用线程池threadPool的execute()方法向线程池提交一个TimerTask任务，此时线程池中会执行TimerTask的run()方法，将TimerTask添加到DelayQueue队列中。

然后继续调用线程池threadPool的execute()方法，向线程池提交一个Runable任务，在

Runnable 任务的 run()方法中，以 while(true)循环的形式不断从 DelayQueue 队列中尝试获取 TimerTask 任务，如果获取到任务，则将任务再次提交到线程池，执行任务的 run()方法，将任务再次添加到 DelayQueue 队列。以此往复执行。

最后，在 main()方法中创建 Timer 类的对象，并执行 void schedule(Runnable task, long interval)方法，传入要具体执行的任务和周期性任务的间隔时间。

至此，案例程序的核心代码实现完毕。

3. 案例测试

运行上述程序，输出结果如下。

```
pool-1-thread-1 线程执行任务的当前时间为：Sun Sep 11 04:23:57 CST 2022
pool-1-thread-1 线程执行任务的当前时间为：Sun Sep 11 04:23:58 CST 2022
pool-1-thread-1 线程执行任务的当前时间为：Sun Sep 11 04:23:59 CST 2022
pool-1-thread-1 线程执行任务的当前时间为：Sun Sep 11 04:24:00 CST 2022
pool-1-thread-1 线程执行任务的当前时间为：Sun Sep 11 04:24:01 CST 2022
```

从输出结果可以看出，每隔 1s 会执行一次具体的任务，符合预期。

第 8 章

并发非阻塞队列

JDK 提供的并发阻塞队列一般都是基于 ReentrantLock 锁实现线程安全的,在高并发环境下会涉及锁竞争的问题,在一定程度上影响队列的性能。除了并发阻塞队列,JDK 还提供了并发非阻塞队列,与并发阻塞队列不同的是,并发非阻塞队列并不是基于 ReentrantLock 锁实现线程安全的,而是基于自旋和 CAS 算法实现的。

> 注意:本章的添加数据特指入队操作,删除数据特指出队操作。

8.1 并发非阻塞队列简介

JDK 的并发非阻塞队列都直接或间接实现了 Queue 接口,并且都是无界队列,在并发非阻塞队列上读写数据时,线程不会阻塞。

8.1.1 概述

并发非阻塞队列是线程安全的队列,在多线程高并发环境下无须担心线程安全问题。同时,并发非阻塞队列是基于链表实现的无界队列,在向并发非阻塞队列添加数据时,线程不会阻塞;在从并发非阻塞队列中移除或者获取数据时,线程也不会阻塞。

在某种角度上,并发非阻塞队列可以分为单端非阻塞队列和双端非阻塞队列。

单端非阻塞队列的示意图如图 8-1 所示。单端非阻塞队列只支持向队列的一端添加数据,并且只支持从队列的另一端消费数据。

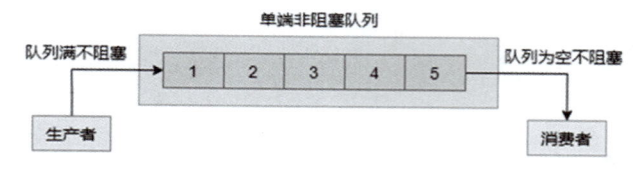

图 8-1 单端非阻塞队列示意图

双端非阻塞队列的示意图如图 8-2 所示，双端非阻塞队列支持从队列的两端添加和移除数据。

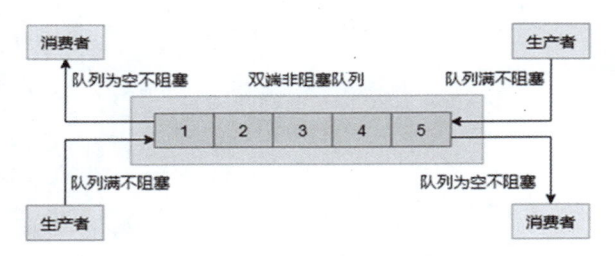

图 8-2　双端非阻塞队列示意图

8.1.2　类继承关系

JDK 提供的并发非阻塞队列容器类都直接或间接地实现了 Queue 接口，类继承关系如图 8-3 所示。

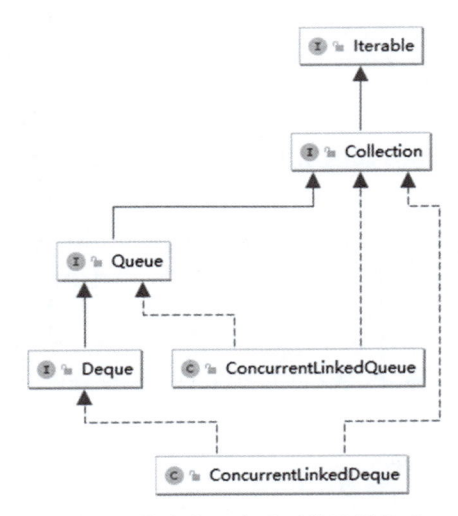

图 8-3　并发非阻塞队列类继承关系

Java 提供的并发非阻塞队列包括 ConcurrentLinkedQueue 和 ConcurrentLinkedDeque。其中，ConcurrentLinkedQueue 直接实现了 Queue 接口。而 ConcurrentLinkedDeque 直接实现了 Deque 接口，Deque 接口继承了 Queue 接口，所以 ConcurrentLinkedDeque 间接实现了 Queue 接口。

8.1.3　常用方法

ConcurrentLinkedQueue 和 ConcurrentLinkedDeque 都实现了 Queue 接口，所以它们都具有先进先出（FIFO）的特性。

Queue 接口中定义了 ConcurrentLinkedQueue 和 ConcurrentLinkedDeque 的一些常用方法，代码如下。

```
public interface Queue<E> extends Collection<E> {
    boolean add(E e);
    boolean offer(E e);
    E remove();
    E poll();
    E element();
    E peek();
}
```

Queue 接口提供的方法的具体说明如下。

- boolean add(E e)：向队列尾部添加数据，如果添加的数据为空，则抛出 NullPointerException 异常，如果数据添加失败，则抛出 IllegalStateException 异常。
- boolean offer(E e)：向队列尾部添加数据，如果添加成功则返回 true，否则返回 false。如果添加的数据为空，则抛出 NullPointerException 异常。
- E remove()：移除队列头部的数据，并且返回移除的数据，如果队列为空，则抛出 NoSuchElementException 异常。
- E poll()：移除队列头部的数据，并且返回移除的数据，如果队列为空，则返回 null。
- E element()：获取队列头部的数据，但不会删除数据，如果队列为空，则抛出 NoSuchElementException 异常。
- E peek()：获取队列头部的数据，但不会删除数据，如果队列为空，则返回 null。

ConcurrentLinkedDeque 除了间接实现了 Queue 接口，还直接实现了 Deque 接口，所以，ConcurrentLinkedDeque 支持在队列的两端分别添加数据和移除数据。也就是说，ConcurrentLinkedDeque 不仅实现了先进先出（FIFO）的特性，也实现了先进后出（FILO）的特性。

Deque 接口的代码提供的方法包括双端队列方法、单端队列方法、栈方法和集合方法 4 种，代码如下。

（1）双端队列方法。

```
//方法代码位于java.util.Deque接口中
void addFirst(E e);
void addLast(E e);
boolean offerFirst(E e);
boolean offerLast(E e);
E removeFirst();
E removeLast();
E pollFirst();
E pollLast();
E getFirst();
```

```
E getLast();
E peekFirst();
E peekLast();
boolean removeFirstOccurrence(Object o);
boolean removeLastOccurrence(Object o);
```

Deque 接口中双端队列方法的具体含义如下。

- void addLast(E e)：向队列尾部添加数据，如果添加失败，则抛出 IllegalStateException 异常，如果添加的数据为空，则抛出 NullPointerException 异常。
- boolean offerFirst(E e)：向队列头部添加数据，添加成功返回 true，添加失败返回 false。如果添加的元素为空，则抛出 NullPointerException 异常。
- boolean offerLast(E e)：向队列尾部添加数据，添加成功返回 true，否则返回 false。如果添加的元素为空，则抛出 NullPointerException 异常。
- E removeFirst()：移除并返回队列头部的节点，如果队列为空，则抛出 NoSuchElementException 异常。
- E removeLast()：移除并返回队列尾部的节点，如果队列为空，则抛出 NoSuchElementException 异常。
- E pollFirst()：移除并返回队列头部的节点，如果队列为空，则返回 null。
- E pollLast()：移除并返回队列尾部的节点，如果队列为空，则返回 null。
- E getFirst()：获取队列头部的数据，但不会删除数据。如果队列为空，则抛出 NoSuchElementException 异常。
- E getLast()：获取队列尾部的数据，但不会删除数据。如果队列为空，则抛出 NoSuchElementException 异常。
- E peekFirst()：获取队列头部的数据，但不会删除数据。如果队列为空，则返回 null。
- E peekLast()：获取队列尾部的数据，但不会删除数据。如果队列为空，则返回 null。
- boolean removeFirstOccurrence(Object o)：从队列头节点开始，移除第一次出现的元素 o，如果移除成功，则返回 true。如果传入的元素 o 为空，并且队列不允许元素为空，则抛出 NullPointerException 异常。
- boolean removeLastOccurrence(Object o)：从队列尾节点开始，移除第一次出现的元素 o。如果移除成功，则返回 true。如果传入的元素 o 为空，并且队列不允许元素为空，则抛出 NullPointerException 异常。

（2）单端队列方法。

```
//如果数据添加失败则抛出 IllegalStateException 异常
boolean add(E e);
boolean offer(E e);
E remove();
E poll();
```

```
E element();
E peek();
```

Deque 接口中单端队列方法的具体含义如下。

- boolean add(E e)：向队列尾部添加数据，如果添加的数据为空，则抛出 NullPointerException 异常，如果数据添加失败，则抛出 IllegalStateException 异常。
- boolean offer(E e)：向队列尾部添加数据，添加成功则返回 true，否则返回 false。如果添加的数据为空，则抛出 NullPointerException 异常。
- E remove()：移除队列头部的数据，并且返回移除的数据。如果队列为空，则抛出 NoSuchElementException 异常。
- E poll()：移除队列头部的数据，并且返回移除的数据。如果队列为空，则返回 null。
- E element()：获取队列头部的数据，但不会删除数据。如果队列为空，则抛出 NoSuchElementException 异常。
- E peek()：获取队列头部的数据，但不会删除数据。如果队列为空，则返回 null。

（3）栈方法。

```
//方法代码位于 java.util.Deque 接口中
void push(E e);
E pop();
```

栈方法的具体含义如下。

- void push(E e)：添加数据到队列头部，如果传入的元素 e 为空，并且队列不允许元素为空，则抛出 NullPointerException 异常。
- E pop()：移除并返回队列头部的节点。如果添加的数据为空，则抛出 NoSuchElementException 异常。

（4）集合方法。

```
boolean remove(Object o);
boolean contains(Object o);
public int size();
Iterator<E> iterator();
Iterator<E> descendingIterator();
```

集合方法的具体含义如下。

- boolean remove(Object o)：移除队列中指定的元素 o，如果成功则返回 true，如果传入的元素 o 为空，并且队列不允许元素为空，则抛出 NullPointerException 异常。
- boolean contains(Object o)：检测队列中是否包含元素 o，如果队列中包含元素 o，则返回 true。如果传入的元素 o 为空，并且队列不允许元素为空，则抛出 NullPointerException 异常。
- int size()：获取队列中元素的数量。

- Iterator<E> iterator()：获取从头节点到尾节点遍历队列的迭代器。
- Iterator<E> descendingIterator()：获取从尾节点到头节点遍历队列的迭代器。

8.1.4 并发非阻塞队列与并发阻塞队列的区别

并发非阻塞队列在数据入队和出队时不会阻塞，而并发阻塞队列在数据入队和出队时可能阻塞。除了这一点显著的区别外，二者还有以下不同点。

（1）并发阻塞队列内部基于 ReentrantLock 锁实现，并发非阻塞队列内部基于自旋和 CAS 实现。

（2）并发阻塞队列都直接或间接地实现了 BlockingQueue 接口，并发非阻塞队列都直接或间接地实现了 Queue 接口。

（3）并发阻塞队列中的 ArrayBlockingQueue 和 LinkedBlockingQueue 支持有界，其他都是无界队列。而并发非阻塞队列都是无界队列。

（4）当多个线程同时操作一个并发阻塞队列时，不需要额外的同步操作，队列会自动平衡负载情况，阻塞处理太快的，减少处理速度上的差异。

（5）当多个线程需要同时访问一个队列时，选择并发非阻塞队列更优。

（6）并发阻塞队列更多的用于任务队列，并发非阻塞队列更多的用于消息队列。

（7）大部分并发阻塞队列更适用于单生产者—单消费者和多生产者—单消费者的场景，并发非阻塞队列更适用于单生产者—多消费者和多生产者—多消费者的场景。

8.2 ConcurrentLinkedQueue

ConcurrentLinkedQueue 是 Java 从 JDK 1.5 版本开始提供的一个基于链表实现的高并发队列，也是一个单向非阻塞队列。

8.2.1 概述

ConcurrentLinkedQueue 实现了 Queue 接口，其内部没有采用锁机制，而是采用循环和 CAS 的方式实现线程安全。采用 CAS 原子性更新数据可以省去加锁时间，也保证了数据的一致性。

ConcurrentLinkedQueue 内部采用先进先出（FIFO）的规则对节点进行排序，向队列的尾部添加数据，从队列的头部获取数据。

ConcurrentLinkedQueue 中定义了 head 引用和 tail 引用，head 引用指向队列的头节点，tail 引用在某个时刻不一定指向队列的尾节点。每个节点的内部都由一个 item 元素和指向下一个节点的 next 引用组成，前后节点之间通过 next 引用进行关联。

ConcurrentLinkedQueue 适用于多线程高并发的场景。

注意：在使用 ConcurrentLinkedQueue 时，需要注意以下几点。

（1）ConcurrentLinkedQueue 内部是基于链表实现的，但是并未单独记录内部元素的数量，所以每次获取 ConcurrentLinkedQueue 中元素的数量时，都需要遍历整个链表。

（2）ConcurrentLinkedQueue 内部采用无锁 CAS 算法保证线程安全，当调用 ConcurrentLinkedQueue 的 size()方法获取元素数量时，可能有其他线程正在修改（添加或移除）队列中的数据，造成调用 ConcurrentLinkedQueue 的 size()方法获取的元素数量发生变化。

（3）在使用 ConcurrentLinkedQueue 的 remove()方法移除元素时，需要注意内存泄漏问题，具体查看随书源码的 README.md 文件。

8.2.2 核心成员变量

ConcurrentLinkedQueue 中定义了一些处理队列的核心成员变量，代码如下。

```
//head引用
private transient volatile Node<E> head;
//tail引用
private transient volatile Node<E> tail;
```

ConcurrentLinkedQueue 内部由 head 引用和 tail 引用组成，head 引用指向内部链表的头节点，tail 引用不一定指向内部链表的尾节点。

8.2.3 重要内部类

ConcurrentLinkedQueue 中有一个非常重要的内部类 Node，Node 类在整体上实现了链表结构，核心代码如下。

```
private static class Node<E> {
    volatile E item;
    volatile Node<E> next;
    Node(E item) {
        UNSAFE.putObject(this, itemOffset, item);
    }
    boolean casItem(E cmp, E val) {
        return UNSAFE.compareAndSwapObject(this, itemOffset, cmp, val);
    }
    void lazySetNext(Node<E> val) {
        UNSAFE.putOrderedObject(this, nextOffset, val);
    }
    boolean casNext(Node<E> cmp, Node<E> val) {
        return UNSAFE.compareAndSwapObject(this, nextOffset, cmp, val);
    }
//###############其他代码省略###############
}
```

在 Node 类中，定义了一个泛型类型的 item 成员变量，表示在 Node 节点中存储的具体元素信息。同时，定义了一个 Node 类型的 next 成员变量，表示指向下一个 Node 节点的引用。另外，在 Node 类中，还定义了通过 CAS 算法修改节点和元素的方法。

8.2.4　初始化

ConcurrentLinkedQueue 提供了如下构造方法来创建 ConcurrentLinkedQueue 对象并进行初始化。

```
//构造方法①
public ConcurrentLinkedQueue() {
    head = tail = new Node<E>(null);
}
//构造方法②
public ConcurrentLinkedQueue(Collection<? extends E> c) {
    Node<E> h = null, t = null;
    for (E e : c) {
        checkNotNull(e);
        Node<E> newNode = new Node<E>(e);
        if (h == null)
            h = t = newNode;
        else {
            t.lazySetNext(newNode);
            t = newNode;
        }
    }
    if (h == null)
        h = t = new Node<E>(null);
    head = h;
    tail = t;
}
```

（1）构造方法①为默认的无参构造方法，在默认的无参构造方法中，会默认将 head 引用和 tail 引用都指向 item 为空的节点，此时会创建一个空的 ConcurrentLinkedQueue 队列。

（2）构造方法②会传入一个 Collection 集合 c，在创建 ConcurrentLinkedQueue 时，会将集合 c 中的元素添加到 ConcurrentLinkedQueue 队列中。

8.2.5　添加数据

这里以 boolean add(E e)方法为例说明 ConcurrentLinkedQueue 添加数据的方法，代码如下。

```
public boolean add(E e) {
    return offer(e);
}
```

在 boolean add(E e)方法中直接调用了 boolean offer(E e)方法，代码如下。

```java
public boolean offer(E e) {
    //检测 e 是否为空
    //如果为空，则抛出 NullPointerException 异常
    checkNotNull(e);
    //创建一个 item 元素为 e 的节点
    final Node<E> newNode = new Node<E>(e);

    //遍历队列中的链表，定位 tail 节点
    for (Node<E> t = tail, p = t;;) {
        //此时 p 节点就是 tail 节点
        //q 节点是 p 节点的后继节点
        Node<E> q = p.next;
        //p 节点的后继节点 q 为空
        //也就是 p 节点是队列的尾节点
        if (q == null) {
            //设置 p 节点的下一个节点为新创建的节点
            if (p.casNext(null, newNode)) {
                //如果 p 节点不是 tail 节点
                //则将新创建的节点设置为 tail 节点
                if (p != t)
                    casTail(t, newNode);
                return true;
            }
        }
        //p 节点与 q 节点相等
        else if (p == q)
            //判断添加新节点前的尾节点是否与添加新节点后的尾节点相等
            //如果相等，则将 p 赋值为 head 节点
            //如果不相等，则将 p 赋值为添加新节点后的尾节点
            p = (t != (t = tail)) ? t : head;
        else
            //如果 p 节点不等于 t 节点，且添加新节点前的尾节点不等于添加新节点后的尾节点
            //则将 p 节点赋值为添加新节点后的尾节点
            //否则，将 p 节点赋值为 q 节点
            p = (p != t && t != (t = tail)) ? t : q;
    }
}
```

boolean offer(E e)方法的代码看起来比较简单，但背后还是有一定设计逻辑的。由于 ConcurrentLinkedQueue 支持多个线程并发向 ConcurrentLinkedQueue 队列中添加数据，所以其 tail 节点并不总是队列的尾节点。每次向队列添加数据时，都需要先通过 tail 节点定位队列的尾节点，队列的尾节点可能是 tail 节点，也可能是 tail 节点的下一个节点。

在 boolean offer(E e)方法的代码中，先通过循环的方式定位 tail 节点，并将其赋值给 p 节点，再获取 p 节点的后继节点 q，并判断 q 节点是否为空。

如果 q 节点为空，则 p 节点就是队列的尾节点。此时将 p 节点的下一个节点设置为新添加的节点，并且将 tail 节点指向新添加的节点，返回 true。

如果 q 节点不为空，则说明有其他线程更新了尾节点，此时需要重新获取当前队列的尾节点，并进行重试。

注意：在 boolean offer(E e)方法中处理 p 节点和 q 节点的关系时需要注意，当一个线程向队列中添加元素，另一个线程从队列中移除元素时，可能出现 p 节点与 q 节点相等的情况。

8.2.6 删除数据

这里以 E poll()方法为例说明 ConcurrentLinkedQueue 删除数据的方法，代码如下。

```
public E poll() {
    restartFromHead:
    for (;;) {
        //p节点指向队列的头节点
        for (Node<E> h = head, p = h, q;;) {
            //获取p节点的item元素
            E item = p.item;
            //如果p节点的item不为空，则通过cas将p节点的item元素设置为空
            if (item != null && p.casItem(item, null)) {
                //如果p节点不等于h节点
                if (p != h)
                    //则更新头节点
                    updateHead(h, ((q = p.next) != null) ? q : p);
                return item;
            }
            //如果头节点的元素为空，或者头节点被其他线程修改了
            //则获取p节点的下一个节点q
            //如果p节点的下一个节点q为空，则说明此时队列为空
            else if ((q = p.next) == null) {
                //更新头节点
                updateHead(h, p);
                return null;
            }
            //如果p节点与q节点相等，则重新开始循环
            else if (p == q)
                continue restartFromHead;
            //如果p的下一个节点q不为空，则将p指向节点q
            else
                p = q;
        }
    }
}
```

E poll()方法的主要逻辑是先获取队列的头节点，将其赋值给节点 p。判断 p 节点对应的元素是否为空，如果为空，则说明存在其他线程移除了队列头部的元素，否则使用 CAS 将头节点的引用设置为空。如果 CAS 操作成功，则直接返回头节点对应的 item 元素。如果 CAS 操作失败，则说明此时有其他线程移除了队列头部的元素，并更新了头节点，需要重新获取头节点，

并尝试重新移除队列头部的元素。

8.2.7 获取数据

这里以 E peek()方法为例说明 ConcurrentLinkedQueue 获取数据的方法，代码如下。

```
public E peek() {
    restartFromHead:
    for (;;) {
        //获取头节点，将其赋值给节点p
        for (Node<E> h = head, p = h, q;;) {
            //获取p节点的item元素
            E item = p.item;
            //如果p节点的item元素不为空，或者p节点的下一个节点q为空
            if (item != null || (q = p.next) == null) {
                //则更新头节点
                updateHead(h, p);
                //返回p节点对应的item元素
                return item;
            }
            //如果p节点与q节点相等，则重新获取队列头部的元素
            else if (p == q)
                continue restartFromHead;
            else
                //如果p节点对应的item元素为空，并且p节点的下一个节点q不为空
                //则将q节点赋值给p节点并重试
                p = q;
        }
    }
}
```

E peek()方法的主要逻辑是首先获取队列头节点的元素，判断头节点的元素是否为空，如果为空，则表示存在其他线程已经移除队列头部的元素，如果不为空，则使用 CAS 将头节点的引用设置为空。如果 CAS 操作成功，则返回头节点的 item 元素的值，如果 CAS 操作不成功，则表示存在其他线程已经移除队列头部的元素并且更新了头节点，此时需要重新获取头节点并重试。

8.2.8 性能对比案例

ConcurrentLinkedQueue 与 LinkedBlockingQueue 都是基于链表实现的线程安全的高并发队列，LinkedBlockingQueue 是阻塞队列，ConcurrentLinkedQueue 是非阻塞队列。

1. 案例需求

使用 JMH 对 ConcurrentLinkedQueue 和 LinkedBlockingQueue 的性能进行简单的测试与对比。

2. 案例实现

（1）在 mykit-concurrent-chapter08 工程的 io.binghe.concurrent.chapter08 包下新建

QueuePerformanceTest 类，在 QueuePerformanceTest 类中实现 ConcurrentLinkedQueue 和 LinkedBlockingQueue 的性能对比。

（2）在 QueuePerformanceTest 类上添加与 JMH 基准测试相关的注解，代码如下。

```
@Warmup(iterations = 10)
@Measurement(iterations = 10)
@Fork(1)
@BenchmarkMode(Mode.AverageTime)
@OutputTimeUnit(TimeUnit.MICROSECONDS)
@State(Scope.Group)
public class QueuePerformanceTest {
```

上述注解表示在使用 JMH 进行基准测试时，预热 10 次，执行 10 次，并输出每个方法调用的平均响应时间。

（3）在 QueuePerformanceTest 类中，定义一个测试的数据元素常量 DATA，并定义类型分别为 LinkedBlockingQueue 和 ConcurrentLinkedQueue 的两个成员变量，在 setUp()方法中对其进行实例化操作，代码如下。

```
//参与测试的LinkedBlockingQueue
private LinkedBlockingQueue<String> linkedBlockingQueue;
//参与测试的ConcurrentLinkedQueue
private ConcurrentLinkedQueue<String> concurrentLinkedQueue;
@Setup(Level.Invocation)
public void setUp(){
    linkedBlockingQueue = new LinkedBlockingQueue<>();
    concurrentLinkedQueue = new ConcurrentLinkedQueue<>();
}
```

（4）在 QueuePerformanceTest 类中创建基准测试方法 linkedBlockingQueueAdd() 和 linkedBlockingQueuePoll()，分别对应 LinkedBlockingQueue 类中的 boolean add(E e)方法与 E poll()方法，代码如下。

```
@Benchmark
@GroupThreads(5)
@Group("linkedBlockingQueue")
public void linkedBlockingQueueAdd(){
    linkedBlockingQueue.add(DATA);
}
@Benchmark
@GroupThreads(5)
@Group("linkedBlockingQueue")
public void linkedBlockingQueuePoll(){
    linkedBlockingQueue.poll();
}
```

在 JMH 中参与基准测试的方法中需要添加@Benchmark 注解，方法会使用 5 个线程进行测试，linkedBlockingQueueAdd()方法和 linkedBlockingQueuePoll()方法在 JMH 中被归为同一组，组名为 linkedBlockingQueue。

（5）在 QueuePerformanceTest 类中创建基准测试方法 concurrentLinkedQueueAdd ()和 concurrentLinkedQueuePoll ()，分别对应 ConcurrentLinkedQueue 类中的 boolean add(E e)方法和 E poll()方法，代码如下。

```
@Benchmark
@GroupThreads(5)
@Group("concurrentLinkedQueue")
public void concurrentLinkedQueueAdd(){
    concurrentLinkedQueue.add(DATA);
}
@Benchmark
@GroupThreads(5)
@Group("concurrentLinkedQueue")
public void concurrentLinkedQueuePoll(){
    concurrentLinkedQueue.poll();
}
```

方法会使用 5 个线程进行测试，concurrentLinkedQueueAdd () 方法和 concurrentLinkedQueuePoll ()方法在 JMH 中被归为同一组，组名为 concurrentLinkedQueue。

（6）在 QueuePerformanceTest 类中创建 main()方法，运行 JMH 基准测试程序，代码如下。

```
public static void main(String[] args) throws RunnerException {
    final Options opt = new OptionsBuilder()
                    .include(QueuePerformanceTest.class.getName()).build();
    new Runner(opt).run();
}
```

至此，整个案例实现完毕。

3．案例测试

运行案例程序，输出结果如下。

```
Benchmark                                                   Mode  Cnt   Score   Error  Units
concurrentLinkedQueue                                       avgt   10   0.266 ± 0.012  us/op
concurrentLinkedQueue:concurrentLinkedQueueAdd              avgt   10   0.289 ± 0.018  us/op
concurrentLinkedQueue:concurrentLinkedQueuePoll             avgt   10   0.244 ± 0.019  us/op
linkedBlockingQueue                                         avgt   10   1.090 ± 0.119  us/op
linkedBlockingQueue:linkedBlockingQueueAdd                  avgt   10   1.306 ± 0.187  us/op
linkedBlockingQueue:linkedBlockingQueuePoll                 avgt   10   0.873 ± 0.110  us/op
```

从输出结果可以看出如下信息。

（1）执行 ConcurrentLinkedQueue 类的 boolean add(E e)方法平均耗时 0.289μs，误差±0.018μs。

（2）执行 ConcurrentLinkedQueue 类的 E poll()方法平均耗时 0.244μs，误差±0.019μs。

（3）执行 LinkedBlockingQueue 类的 boolean add(E e)方法平均耗时 1.306μs，误差±0.187μs。

（4）执行 LinkedBlockingQueue 类的 E poll()方法平均耗时 0.873μs，误差±0.110μs。

综上所述，无论是在执行 boolean add(E e)方法时，还是在执行 E poll()方法时，ConcurrentLinkedQueue 的性能都优于 LinkedBlockingQueue 的性能。

注意：在不同的计算机上执行上述基准测试程序，输出的结果可能略有不同，但是 ConcurrentLinkedQueue 的性能都是优于 LinkedBlockingQueue 的。

另外，关于 JMH 相关的知识，读者可以关注"冰河技术"微信公众号阅读相关文章。

8.3 ConcurrentLinkedDeque

ConcurrentLinkedDeque 是 Java 从 JDK 1.7 版本开始提供的一个基于链表实现的高并发队列，是一个双向非阻塞队列。

8.3.1 概述

ConcurrentLinkedDeque 是一个基于双向链表实现的线程安全的无界非阻塞队列，实现了 Deque 接口，其内部与 ConcurrentLinkedQueue 一样，不是采用锁机制，而是采用循环和 CAS 的方式实现线程安全，免去了加锁的时间，也保证了数据的一致性。

ConcurrentLinkedDeque 内部采用先进先出（FIFO）和先进后出（FILO）的规则对节点进行排序，既可以向队列的头部添加数据，也可以向队列的尾部添加数据。既可以从队列的头部删除数据，也可以从队列的尾部删除数据。

ConcurrentLinkedDeque 中除了定义了 head 引用和 tail 引用，还定义了前后终止节点的标志 PREV_TERMINATOR 和 NEXT_TERMINATOR，每个节点都由 item 元素、指向上一个节点的 prev 引用，以及指向下一个节点的 next 引用组成，前后节点之间通过 prev 引用和 next 引用关联。

ConcurrentLinkedDeque 的 head 引用在某个时刻不一定指向队列的头节点，头节点会接近 head 引用指向的节点，通过 head 引用可以快速找到队列的头节点。同样的，ConcurrentLinkedDeque 的 tail 引用在某个时刻不一定指向尾节点，尾节点会接近 tail 引用指向的节点，通过 tail 引用可以快速找到队列的尾节点。

ConcurrentLinkedDeque 适用于多线程高并发的场景。

8.3.2 核心成员变量

在 ConcurrentLinkedDeque 中，定义了一些处理队列的核心成员变量，代码如下。

```
//head引用，在某个时刻不一定指向头节点
private transient volatile Node<E> head;
```

```
//tail 引用,在某个时刻不一定指向尾节点
private transient volatile Node<E> tail;
//前后终止节点的标志
private static final Node<Object> PREV_TERMINATOR, NEXT_TERMINATOR;
```

ConcurrentLinkedDeque 中除了定义了 head 引用和 tail 引用,还定义了前后终止节点的标志 PREV_TERMINATOR 和 NEXT_TERMINATOR,head 引用不一定指向队列的头节点,tail 引用也不一定指向队列的尾节点。

8.3.3 重要内部类

ConcurrentLinkedDeque 中有一个非常重要的内部类 Node,Node 类在整体上实现了双向链表结构,Node 类的核心代码如下。

```
static final class Node<E> {
    volatile Node<E> prev;
    volatile E item;
    volatile Node<E> next;
    Node() {
    }
    Node(E item) {
        UNSAFE.putObject(this, itemOffset, item);
    }
    boolean casItem(E cmp, E val) {
        return UNSAFE.compareAndSwapObject(this, itemOffset, cmp, val);
    }
    void lazySetNext(Node<E> val) {
        UNSAFE.putOrderedObject(this, nextOffset, val);
    }
    boolean casNext(Node<E> cmp, Node<E> val) {
        return UNSAFE.compareAndSwapObject(this, nextOffset, cmp, val);
    }
    void lazySetPrev(Node<E> val) {
        UNSAFE.putOrderedObject(this, prevOffset, val);
    }
    boolean casPrev(Node<E> cmp, Node<E> val) {
        return UNSAFE.compareAndSwapObject(this, prevOffset, cmp, val);
    }
    //##############省略其他代码##############
}
```

Node 类封装的每个节点都由 item 元素、指向上一个节点的 prev 引用,以及指向下一个节点的 next 引用组成,前后节点之间通过 prev 引用和 next 引用关联。

另外,在 Node 类中,还定义了通过 CAS 算法修改节点和元素的方法。

8.3.4 初始化

ConcurrentLinkedDeque 提供了如下构造方法来创建 ConcurrentLinkedDeque 对象并进行初

始化。

```
//构造方法①
public ConcurrentLinkedDeque() {
    head = tail = new Node<E>(null);
}
//构造方法②
public ConcurrentLinkedDeque(Collection<? extends E> c) {
    Node<E> h = null, t = null;
    for (E e : c) {
        checkNotNull(e);
        Node<E> newNode = new Node<E>(e);
        if (h == null)
            h = t = newNode;
        else {
            t.lazySetNext(newNode);
            newNode.lazySetPrev(t);
            t = newNode;
        }
    }
    initHeadTail(h, t);
}
```

（1）构造方法①为默认的无参构造方法，默认将 head 引用和 tail 引用都指向 item 为空的节点，此时会创建一个空的 ConcurrentLinkedDeque 队列。

（2）构造方法②会传入一个 Collection 集合 c，在创建 ConcurrentLinkedDeque 时，会将集合 c 中的元素添加到 ConcurrentLinkedDeque 队列中。

8.3.5 添加数据

这里以 boolean add(E e)方法为例说明 ConcurrentLinkedDeque 添加数据的方法，代码如下。

```
public boolean add(E e) {
    return offerLast(e);
}
```

在 boolean add(E e)方法中，直接调用了 boolean offerLast(E e)方法，代码如下。

```
public boolean offerLast(E e) {
    linkLast(e);
    return true;
}
```

在 boolean offerLast(E e)方法中，调用 void linkLast(E e)方法将元素 e 添加到队列尾部，并返回 true，代码如下。

```
private void linkLast(E e) {
    checkNotNull(e);
    final Node<E> newNode = new Node<E>(e);
    restartFromTail:
```

```
    for (;;)
        for (Node<E> t = tail, p = t, q;;) {
            if ((q = p.next) != null &&
                (q = (p = q).next) != null)
                p = (t != (t = tail)) ? t : q;
            else if (p.prev == p)
                continue restartFromTail;
            else {
                newNode.lazySetPrev(p);
                if (p.casNext(null, newNode)) {
                    if (p != t)
                        casTail(t, newNode);
                    return;
                }
            }
        }
}
```

void linkLast(E e)方法的主要逻辑是找到队列的尾节点，将新创建的节点添加到尾节点的后面，并连接尾节点和新添加的节点。如果尾节点已经被其他线程移除，则重新获取尾节点并重试。

8.3.6 删除数据

这里以 E poll()方法为例说明 ConcurrentLinkedDeque 删除数据的方法，代码如下。

```
public E poll(){
    return pollFirst();
}
```

在 E poll()方法中，直接调用了 E pollFirst()方法，代码如下。

```
public E pollFirst() {
    for (Node<E> p = first(); p != null; p = succ(p)) {
        E item = p.item;
        if (item != null && p.casItem(item, null)) {
            unlink(p);
            return item;
        }
    }
    return null;
}
```

E pollFirst()方法的逻辑是找到队列的头节点，获取头节点的 item 元素，如果 item 元素不为空，并且通过 CAS 将头节点的 item 元素设置为 null 成功，则调用 void unlink(Node<E> x)方法移除节点，并返回 item 元素的值，否则返回 null。void unlink(Node<E> x)方法的代码如下。

```
void unlink(Node<E> x) {
    final Node<E> prev = x.prev;
    final Node<E> next = x.next;
    //如果 x 节点的前驱节点为空，则表示 x 节点是头节点
    if (prev == null) {
```

```
        unlinkFirst(x, next);
//如果 x 节点的后继节点为空，则表示 x 节点是尾节点
} else if (next == null) {
        unlinkLast(x, prev);
//x 既不是头节点也不是尾节点
} else {
    Node<E> activePred, activeSucc;
    boolean isFirst, isLast;
    //用来记录逻辑删除的节点数量
    int hops = 1;
    //遍历 x 的前驱节点
    for (Node<E> p = prev; ; ++hops) {
        if (p.item != null) {
            activePred = p;
            isFirst = false;
            break;
        }
        Node<E> q = p.prev;
        //如果 p 的前驱节点为 null，则 p 是头节点
        if (q == null) {
            //p 被其他线程从队列中移除
            if (p.next == p)
                return;
            activePred = p;
            isFirst = true;
            break;
        }
        //p 被其他线程从队列中移除
        else if (p == q)
            return;
        else
            //将 p 的前驱节点赋值给 p，继续重试
            p = q;
    }
    //遍历后继节点
    for (Node<E> p = next; ; ++hops) {
        if (p.item != null) {
            activeSucc = p;
            isLast = false;
            break;
        }
        Node<E> q = p.next;
        //如果 p 的后继节点为空，则 p 是尾节点
        if (q == null) {
            //p 被其他线程从队列中移除
            if (p.prev == p)
                return;
            activeSucc = p;
            isLast = true;
            break;
```

```
            }
            //p 被其他线程从队列中移除
            else if (p == q)
                return;
            else
                //将 p 的后继节点赋值给 p,继续重试
                p = q;
        }
        if (hops < HOPS
            && (isFirst | isLast))
            return;
        //移除有效前驱和后继节点之间的有效节点,包括 x,使得前驱节点和后继节点相连
        skipDeletedSuccessors(activePred);
        skipDeletedPredecessors(activeSucc);
        //如果有效前驱节点是队头节点或者后继节点是队尾节点,则尝试 gc-unlink
        if ((isFirst | isLast) &&
            //检查前驱节点和后继节点的状态,确保没有发生变化
            (activePred.next == activeSucc) &&
            (activeSucc.prev == activePred) &&
            (isFirst ? activePred.prev == null : activePred.item != null) &&
            (isLast  ? activeSucc.next == null : activeSucc.item != null)) {
            // 更新 head,确保 x 不可达
            updateHead();
            // 更新 tail,确保 x 不可达
            updateTail();
            //更新节点 x
            x.lazySetPrev(isFirst ? prevTerminator() : x);
            x.lazySetNext(isLast  ? nextTerminator() : x);
        }
    }
}
```

void unlink(Node<E> x)方法的代码看起来比较复杂,但实现的核心逻辑比较简单。首先,如果 x 节点是队列的头节点则移除。然后,判断 x 节点是否是队列的尾节点,如果是则移除。如果 x 节点既不是队列的头节点也不是队列的尾节点,则遍历 x 节点的有效前驱节点并记录,同时,遍历 x 节点的有效后继节点并记录。最后,将 x 节点的有效前驱节点的 next 引用指向 x 节点的有效后继节点,并且将 x 节点的有效后继节点的 prev 引用指向 x 节点的有效前驱节点。

8.3.7 获取数据

这里,以 E peek()方法为例说明 ConcurrentLinkedDeque 获取数据的方法,代码如下。

```
public E peek(){
    return peekFirst();
}
```

E peek()方法直接调用了 E peekFirst()方法,代码如下。

```
public E peekFirst() {
    for (Node<E> p = first(); p != null; p = succ(p)) {
```

```
            E item = p.item;
            if (item != null)
                return item;
        }
        return null;
}
```

E peekFirst()方法的主要逻辑就是找到队列的头节点，获取头节点的 item 元素，如果头节点的 item 元素不为空，则直接返回 item 元素，否则返回 null。

8.3.8 性能对比案例

ConcurrentLinkedDeque 与 LinkedBlockingDeque 都是基于双向链表实现的线程安全的高并发队列，LinkedBlockingDeque 是阻塞队列，ConcurrentLinkedDeque 是非阻塞队列。

1. 案例需求

使用 JMH 对 ConcurrentLinkedDeque 和 LinkedBlockingDeque 的性能进行简单的测试与对比。

2. 案例实现

本案例实现的步骤与 8.2.8 节中案例实现的步骤相同，笔者直接给出完整的代码。

```
@Warmup(iterations = 10)
@Measurement(iterations = 10)
@Fork(1)
@BenchmarkMode(Mode.AverageTime)
@OutputTimeUnit(TimeUnit.MICROSECONDS)
@State(Scope.Group)
public class DequePerformanceTest {
    //测试的数据
    private static final String DATA = "binghe";
    //参与测试的 LinkedBlockingDeque
    private LinkedBlockingDeque<String> linkedBlockingDeque;
    //参与测试的 ConcurrentLinkedDeque
    private ConcurrentLinkedDeque<String> concurrentLinkedDeque;
    @Setup(Level.Invocation)
    public void setUp(){
        linkedBlockingDeque = new LinkedBlockingDeque<>();
        concurrentLinkedDeque = new ConcurrentLinkedDeque<>();
    }
    @Benchmark
    @GroupThreads(5)
    @Group("linkedBlockingDeque")
    public void linkedBlockingDequeAdd(){
        linkedBlockingDeque.add(DATA);
    }
    @Benchmark
    @GroupThreads(5)
    @Group("linkedBlockingDeque")
    public void linkedBlockingDequePoll(){
```

```java
        linkedBlockingDeque.poll();
    }
    @Benchmark
    @GroupThreads(5)
    @Group("concurrentLinkedDeque")
    public void concurrentLinkedDequeAdd(){
        concurrentLinkedDeque.add(DATA);
    }
    @Benchmark
    @GroupThreads(5)
    @Group("concurrentLinkedDeque")
    public void concurrentLinkedDequePoll(){
        concurrentLinkedDeque.poll();
    }
    public static void main(String[] args) throws RunnerException {
        final Options opt = new OptionsBuilder()
                        .include(DequePerformanceTest.class.getName()).build();
        new Runner(opt).run();
    }
}
```

在案例程序中，主要对 ConcurrentLinkedDeque 和 LinkedBlockingDeque 的 boolean add(E e) 和 E poll() 方法进行了基准性能测试。

3. 案例测试

运行案例程序，输出结果如下。

```
Benchmark                                                Mode   Cnt   Score     Error    Units
concurrentLinkedDeque                                    avgt   10    0.283  ±  0.034    us/op
concurrentLinkedDeque:concurrentLinkedDequeAdd           avgt   10    0.272  ±  0.034    us/op
concurrentLinkedDeque:concurrentLinkedDequePoll          avgt   10    0.294  ±  0.058    us/op
linkedBlockingDeque                                      avgt   10    1.379  ±  0.127    us/op
linkedBlockingDeque:linkedBlockingDequeAdd               avgt   10    1.401  ±  0.098    us/op
linkedBlockingDeque:linkedBlockingDequePoll              avgt   10    1.357  ±  0.203    us/op
```

从输出结果可以看出如下信息。

（1）ConcurrentLinkedDeque 类的 boolean add(E e) 方法平均耗时 0.272μs，误差 ± 0.034μs。

（2）执行 ConcurrentLinkedDeque 类的 E poll() 方法平均耗时 0.294μs，误差 ± 0.058μs。

（3）执行 LinkedBlockingDeque 类的 boolean add(E e) 方法平均耗时 1.401μs，误差 ± 0.098μs。

（4）执行 LinkedBlockingDeque 类的 E poll() 方法平均耗时 1.357μs，误差 ± 0.203μs。

综上所述，无论执行 boolean add(E e) 方法还是 E poll() 方法，ConcurrentLinkedDeque 的性能都优于 LinkedBlockingDeque。

注意：在不同的计算机上执行上述基准测试程序时，输出结果可能略有不同，但是 ConcurrentLinkedDeque 的性能都是优于 LinkedBlockingDeque 的。

第 9 章

并发工具类

目前，几乎所有的互联网项目都涉及多线程并发编程，即使在实际的业务开发过程中不会涉及太多并发编程，其底层依赖的第三方库、运行程序代码使用的 Web 容器等，多多少少也会涉及并发编程。可以这么说，并发编程已经融入几乎所有的互联网项目的开发与运行中了。

然而，深入掌握并发编程技术并不是一件容易的事情，好在 Java 的 JDK 提供了一系列并发编程工具类，能够极大地简化并发编程。

9.1 CountDownLatch 工具类

在实际开发业务的过程中，经常遇到一种场景：某个线程需要等到其他所有线程运行结束后再运行。当明确知道创建的线程数量并且需要创建的线程较少时，可以使用 Thread 类创建线程，同时使用 Thread 类的 join()方法等待线程执行结束。不过，如果需要创建的线程比较多，或者并不明确具体要创建多少线程，或者需要使用线程池，这种方式就显得力不从心了。好在 JDK 提供了 CountDownLatch 工具类解决这个问题。

9.1.1 概述及重要方法说明

CountDownLatch 是 Java 从 JDK 1.5 版本开始提供的一个非常实用的多线程控制工具类，它能够实现一个或多个线程等待其他所有线程执行完毕，或者在其他线程完成某种操作之后再执行。

CountDownLatch 内部是基于 AQS 实现的，会维护一个记录未完成线程数量的计数器。在调用 CountDownLatch 的构造方法时为这个计数器赋一个大于 0 的值。随后每调用一次 CountDownLatch 的 countDown()方法，这个计数器的值都会减 1。而调用 CountDownLatch 的 await()方法的线程会被添加到 AQS 的阻塞队列中阻塞等待，当计数器的值减为 0 时，阻塞的线程被唤醒继续执行。

CountDownLatch 的计数器不能被重置，一旦计数器的值减为 0，CountDownLatch 就不能再使用了，如果需要再使用 CountDownLatch 实现相应的功能，就必须重新创建一个 CountDownLatch 对象。另外，调用 CountDownLatch 的 await()方法和 countDown()方法都无须加锁。

注意：CountDownLatch 的 countDown()方法除了可以用于不同线程触发计数器减 1，也可以用于同一线程触发计数器减 1。

另外，CountDownLatch 底层基于 AQS 实现，有关 AQS 的核心原理，读者可以参考《深入理解高并发编程：核心原理与案例实战》一书。

CountDownLatch 提供的比较重要的方法的代码和说明分别如下。

（1）构造方法的代码如下。

```
public CountDownLatch(int count) {
    if (count < 0) throw new IllegalArgumentException("count < 0");
    this.sync = new Sync(count);
}
```

CountDownLatch 的构造方法会在创建 CountDownLatch 对象时传入一个计数器的值，这个值就是 CountDownLatch 内部记录的等待工作线程的数量，或者同一个线程中操作步骤的数量。

（2）void await()方法的代码如下。

```
public void await() throws InterruptedException {
    sync.acquireSharedInterruptibly(1);
}
```

调用 CountDownLatch 的 await()方法的线程会被阻塞，当其他所有线程都调用了 countDown()方法，或者同一线程执行了所有步骤并调用了 countDown()方法后，CountDownLatch 的计数减为 0，会唤醒调用 await()方法被阻塞的线程。

另外，当其他线程调用了被阻塞线程的 interrupt()方法时，被阻塞的线程会被中断，抛出 InterruptedException 异常并返回。

（3）boolean await(long timeout, TimeUnit unit)方法的代码如下。

```
public boolean await(long timeout, TimeUnit unit)
    throws InterruptedException {
    return sync.tryAcquireSharedNanos(1, unit.toNanos(timeout));
}
```

调用 CountDownLatch 中 await(long timeout, TimeUnit unit)方法的线程会被阻塞，如果其他所有线程都调用了 countDown()方法，或者同一线程执行了所有步骤并调用了 countDown()方法，CountDownLatch 的计数就会减为 0，唤醒调用 await(long timeout, TimeUnit unit)方法被阻塞的线程。如果超过 timeout 时间，也会唤醒调用 await(long timeout, TimeUnit unit)方法被阻塞的线程。

另外，线程调用 await(long timeout, TimeUnit unit)方法会被阻塞，在阻塞期间，当其他线程调用了被阻塞线程的 interrupt()方法时，被阻塞的线程会被中断，抛出 InterruptedException 异常并返回。

（4）void countDown()方法的代码如下。

```
public void countDown() {
   sync.releaseShared(1);
}
```

当调用 countDown()方法时，计数器的值减 1，如果计数器的值减为 0，则唤醒因为调用 await()方法或 await(long timeout, TimeUnit unit)方法被阻塞的线程。

（5）long getCount()方法的代码如下。

```
public long getCount() {
   return sync.getCount();
}
```

调用 getCount()方法会返回当前计数器的值。

9.1.2 使用案例

1. 案例需求

假设张三、李四、王五一起去旅游，导游正在出发地等待他们的到来，只有张三、李四和王五都到达才能出发，使用 CountDownLatch 模拟这个过程。

2. 案例实现

导游正在等待张三、李四和王五的到来，可以将导游看作调用 CountDownLatch 的 await()方法被阻塞的线程，将张三、李四和王五看作 3 个不同的线程，每人到达出发地后都调用 CountDownLatch 的 countDown()方法将计数减 1，当 3 人都到达出发地时，计数就会减为 0，张三、李四和王五 3 个线程执行完毕，唤醒阻塞的导游线程。很容易分析出计数器的初始值为 3。

案例的实现过程如下。

（1）在 mykit-concurrent-chapter09 工程下的 io.binghe.concurrent.chapter09 包下新建 CountDownLatchTask 类模拟张三、李四和王五去出发地的过程，CountDownLatchTask 类的代码如下。

```
//模拟张三、李四、王五的线程
public class CountDownLatchTask implements Runnable {
   private String name;
   private CountDownLatch countDownLatch;
   public CountDownLatchTask(String name, CountDownLatch countDownLatch) {
      this.name = name;
      this.countDownLatch = countDownLatch;
   }
```

```
    @Override
    public void run() {
        try {
            Random random = new Random();
            Thread.sleep(random.nextInt(2000));
            System.out.println(this.name + "--到达出发地");
            countDownLatch.countDown();
        } catch (InterruptedException e) {
            e.printStackTrace();
        }
    }
}
```

CountDownLatchTask 类的代码比较简单，首先实现了 Runnable 接口，并定义了一个 String 类型的成员变量 name 和一个 CountDownLatch 类型的成员变量 countDownLatch，两个成员变量都会通过构造方法传递进来。

然后在 run()方法中使用 Random 让当前线程随机休眠 2s 以内。接下来，输出 xxx--到达出发地的日志。最后调用 CountDownLatch 的 countDown()方法使计数减 1。

（2）在 io.binghe.concurrent.chapter09 包下新建 CountDownLatchTest 类，用来模拟导游等待张三、李四和王五到达出发地，一起去目的地的过程，代码如下。

```
//模拟导游等待张三、李四和王五一起去旅游
public class CountDownLatchTest {
    public static void main(String[] args){
        //创建一个初始值为 3 的 CountDownLatch
        CountDownLatch countDownLatch = new CountDownLatch(3);
        //导游线程
        new Thread(()->{
            try {
                System.out.println("导游--等待张三、李四和王五到达出发地");
                countDownLatch.await();
                System.out.println("张三、李四和王五都已经到达出发地，一起跟随导游出发去目的地旅游");
            } catch (InterruptedException e) {
                e.printStackTrace();
            }
        }, "daoyou-thread").start();
        //张三线程
        new Thread(new CountDownLatchTask("张三", countDownLatch)).start();
        //李四线程
        new Thread(new CountDownLatchTask("李四", countDownLatch)).start();
        //王五线程
        new Thread(new CountDownLatchTask("王五", countDownLatch)).start();
    }
}
```

在 CountDownLatchTest 类的 main()方法中，首先创建一个初始值为 3 的 CountDownLatch。

然后创建一个导游线程，在新建的导游线程的 run()方法中，输出导游等待张三、李四和王五到达出发地的日志，随后调用 CountDownLatch 的 await()方法等待计数减为 0，当计数减为 0 时，输出"张三、李四和王五都已经到达出发地，一起跟随导游出发去目的地旅游"。

接下来，分别创建模拟张三、李四和王五到出发地的线程，在创建 CountDownLatchTask 对象时传递的名称分别为张三、李四和王五，并将 countDownLatch 传递到 CountDownLatchTask 类的构造方法中，分别启动 3 个线程。

至此，整个案例程序实现完毕。

3. 案例测试

运行 CountDownLatchTest 类的 main()方法，输出结果如下。

```
导游--等待张三、李四和王五到达出发地
李四--到达出发地
王五--到达出发地
张三--到达出发地
张三、李四和王五都已经到达出发地，一起跟随导游出发去目的地旅游
```

从输出结果可以看出，程序完整地模拟出导游等待张三、李四和王五到达出发地，一起去旅游的过程，符合预期。

9.2　CyclicBarrier 工具类

为了使计数器的值可以被重置并能够反复使用，JDK 的开发者开发了 CyclicBarrier。

9.2.1　概述及重要方法说明

CyclicBarrier 是 Java 从 JDK 1.5 版本开始提供的一个线程同步工具类，与 CountDownLatch 类似，可以实现某个线程等待其他线程的场景。不过 CyclicBarrier 实现的功能要比 CountDownLatch 强大得多。

CyclicBarrier 还能够实现多个线程之间的计数等待，也就是在多个线程之间互相等待对方的场景下，当所有的线程都达到某个屏障点时，继续执行后续的业务逻辑。

CyclicBarrier 也适用于某个串行化任务被拆分成多个并行执行的子任务的场景，当所有的子任务都执行结束后，再继续执行后续的逻辑。

CyclicBarrier 的计数器是可以重置的，也就是说，当 CyclicBarrier 的计数器的值减为 0 时，可以被重置为初始值。

在调用 CyclicBarrier 的构造方法创建 CyclicBarrier 对象时，CyclicBarrier 可以传入一个 Runnable 接口类型的对象，当计数器的值减为 0 时，会自动调用 Runnable 接口类型对象的 run() 方法。

注意：在创建对象时，一旦指定 CyclicBarrier 的线程计数器就不能手动修改。在调用 int await()方法、int await(long timeout, TimeUnit unit)方法时计数器会减 1，在调用 void reset()方法时，计数器的值重置为创建对象时的初始值。

CyclicBarrier 提供的比较重要的方法的代码和说明分别如下。

（1）构造方法的代码如下。

```
//构造方法①
public CyclicBarrier(int parties, Runnable barrierAction) {
    if (parties <= 0) throw new IllegalArgumentException();
    this.parties = parties;
    this.count = parties;
    this.barrierCommand = barrierAction;
}
//构造方法②
public CyclicBarrier(int parties) {
    this(parties, null);
}
```

两种构造方法的具体含义如下。

- 构造方法①会传入一个线程计数器 parties 和 Runnable 类型的 barrierAction 事件创建 CyclicBarrier 对象，当 parties 的值减为 0 时，会自动执行 barrierAction 的 run()方法。
- 构造方法②会传入一个线程计数器 parties 创建 CyclicBarrier 对象，构造方法②会调用构造方法①，传入线程计数器 parties 和一个 null 创建 CyclicBarrier 对象。

（2）int await()方法的代码如下。

```
public int await() throws InterruptedException, BrokenBarrierException {
    try {
        return dowait(false, 0L);
    } catch (TimeoutException toe) {
        throw new Error(toe);
    }
}
```

当线程调用 int await()方法时，当前线程会被阻塞，待其他线程执行 await()方法进入屏障点、线程计数器减为 0 后，阻塞的线程都会被唤醒。

（3）int await(long timeout, TimeUnit unit)方法的代码如下。

```
public int await(long timeout, TimeUnit unit)
    throws InterruptedException,
        BrokenBarrierException,
        TimeoutException {
    return dowait(true, unit.toNanos(timeout));
}
```

该方法与 int await()方法的功能总体相似，只是加入了超时机制。除 int await()方法的唤醒机制外，超过 timeout 的时长也能自动唤醒。

（4）void reset()方法的代码如下。

```
public void reset() {
    final ReentrantLock lock = this.lock;
    lock.lock();
    try {
        breakBarrier();
        nextGeneration();
    } finally {
        lock.unlock();
    }
}
```

void reset()方法主要用来将线程计数器重置为初始值。

（5）boolean isBroken()方法的代码如下。

```
public boolean isBroken() {
    final ReentrantLock lock = this.lock;
    lock.lock();
    try {
        return generation.broken;
    } finally {
        lock.unlock();
    }
}
```

boolean isBroken()方法主要用于返回 broken 状态，当某个线程执行了 int await()方法或者 int await(long timeout, TimeUnit unit)方法被阻塞时，如果有其他线程中断了该线程，则 boolean isBroken()方法会返回 true。

（6）int getParties()方法的代码如下。

```
public int getParties() {
    return parties;
}
```

获取创建 CyclicBarrier 对象时传入的 parties 值。

9.2.2　使用案例

1．案例需求

模拟 3 个任务，分别执行提交订单、扣减库存、生成物流单的操作，每个任务执行完毕都输出"当前提交订单、扣减库存、生成物流单完毕"。

2．案例实现

模拟 3 个不同的任务分别执行提交订单、扣减库存、生成物流单的操作，如果使用

CountDownLatch，则要为每个任务创建一个 CountDownLatch 实例。如果使用 CyclicBarrier，则在每个任务执行完毕后重置线程计数器的值，继续执行下一个任务。

案例的实现过程如下。

（1）在 mykit-concurrent-chapter09 工程下的 io.binghe.concurrent.chapter09 包下创建 CyclicBarrierTask 类，用于模拟不同任务执行提交订单、扣减库存、生成物流单的操作，CyclicBarrierTask 类的代码如下。

```java
//分别模拟提交订单、扣减库存、生成物流单的操作
public class CyclicBarrierTask implements Runnable{
    private String task;
    private CyclicBarrier cyclicBarrier;
    public CyclicBarrierTask(String task, CyclicBarrier cyclicBarrier) {
        this.task = task;
        this.cyclicBarrier = cyclicBarrier;
    }
    @Override
    public void run() {
        IntStream.rangeClosed(1, 3).forEach((i) -> {
            try {
                Random random = new Random();
                Thread.sleep(random.nextInt(2000));
                System.out.println("任务" + i + "--完成" + task + "的操作");
                cyclicBarrier.await();
            } catch (Exception e) {
                e.printStackTrace();
            }
            //重置计数器的值
            cyclicBarrier.reset();
        });
    }
}
```

CyclicBarrierTask 类实现了 Runnable 接口，在 CyclicBarrierTask 类中，首先定义了一个 String 类型的成员变量 task 和一个 CyclicBarrier 类型的成员变量 cyclicBarrier，两个成员变量会通过 CyclicBarrierTask 类的构造方法进行赋值。

接下来，在 run()方法中使用 IntStream 构造一个初始值为 1、终止值为 3 的循环，每次循环时，都将当前线程随机休眠 2s 以内，然后输出任务 x-完成 xxx 的操作日志，并调用 CyclicBarrier 的 await()方法使线程计数器减 1。随后重置线程计数器的值。

（2）在 io.binghe.concurrent.chapter09 包下创建 CyclicBarrierTest 类，代码如下。

```java
//模拟不同任务提交订单、扣减库存、生成物流单
public class CyclicBarrierTest {

    public static void main(String[] args){
```

```
    CyclicBarrier cyclicBarrier = new CyclicBarrier(3, () -> {
        System.out.println("当前提交订单、扣减库存、生成物流单完毕");
    });
    new Thread(new CyclicBarrierTask("提交订单", cyclicBarrier)).start();
    new Thread(new CyclicBarrierTask("扣减库存", cyclicBarrier)).start();
    new Thread(new CyclicBarrierTask("生成物流单", cyclicBarrier)).start();
    }
}
```

在 CyclicBarrierTest 类的 main()方法中创建 CyclicBarrier 对象，最后，创建提交订单、扣减库存和生成物流单 3 个线程并启动。

3．案例测试

运行 CyclicBarrierTest 类的代码，输出结果如下。

```
任务 1--完成扣减库存的操作
任务 1--完成提交订单的操作
任务 1--完成生成物流单的操作
当前提交订单、扣减库存、生成物流单完毕
任务 2--完成生成物流单的操作
任务 2--完成扣减库存的操作
任务 2--完成提交订单的操作
当前提交订单、扣减库存、生成物流单完毕
任务 3--完成提交订单的操作
任务 3--完成扣减库存的操作
任务 3--完成生成物流单的操作
当前提交订单、扣减库存、生成物流单完毕
```

从输出结果可以看出，任务 1、任务 2 和任务 3，分别执行提交订单、扣减库存和生成物流单的操作，每个任务执行完毕后，都会输出当前提交订单、扣减库存、生成物流单完毕的日志，符合预期。

9.3 Phaser 工具类

Phaser 是 Java 从 JDK 1.7 版本开始提供的多线程同步工具类，功能类似于 CountDownLatch 和 CyclicBarrier 的合集。

9.3.1 概述及重要方法说明

JDK1.7 版本提供的 Phaser 整合了 CountDownLatch 和 CyclicBarrier 的优点。

Phaser 适用于将一个大任务拆分为多个小任务，拆分后的每个小任务都可以多个线程并发执行，且需要完成上一个阶段的任务才可以执行下一个阶段的任务的场景。虽然这种场景使用 CountDownLatch 和 CyclicBarrier 也能实现，但是使用 Phaser 会更加灵活。

Phaser 提供的比较重要的方法的代码和说明分别如下。

（1）构造方法的代码如下。

```
//构造方法①
public Phaser() {
    this(null, 0);
}
//构造方法②
public Phaser(int parties) {
    this(null, parties);
}
//构造方法③
public Phaser(Phaser parent) {
    this(parent, 0);
}
//构造方法④
public Phaser(Phaser parent, int parties) {
    if (parties >>> PARTIES_SHIFT != 0)
        throw new IllegalArgumentException("Illegal number of parties");
    int phase = 0;
    this.parent = parent;
    if (parent != null) {
        final Phaser root = parent.root;
        this.root = root;
        this.evenQ = root.evenQ;
        this.oddQ = root.oddQ;
        if (parties != 0)
            phase = parent.doRegister(1);
    }
    else {
        this.root = this;
        this.evenQ = new AtomicReference<QNode>();
        this.oddQ = new AtomicReference<QNode>();
    }
    this.state = (parties == 0) ? (long)EMPTY :
        ((long)phase << PHASE_SHIFT) |
        ((long)parties << PARTIES_SHIFT) |
        ((long)parties);
}
```

Phaser 对外提供 4 种构造方法。

- 构造方法①，默认的无参构造方法，直接调用构造方法④创建 Phaser 对象。
- 构造方法②，传入参与者计数器，直接调用构造方法④创建 Phaser 对象。
- 构造方法③，传入 Phaser 对象，直接调用构造方法④创建 Phaser 对象。
- 构造方法④，核心构造方法，传入 Phaser 对象和参与者计数器创建 Phaser 对象。

通过构造方法也可以看出以下几点。

- 在通过构造方法创建 Phaser 对象时，parties 参数的值不能超过 65535。

- 如果在创建 Phaser 对象时传入了父级 Phaser 对象,则当前 Phaser 对象会共享父级 Phaser 对象的线程等待队列。
- 如果在创建 Phaser 对象时传入了父级 Phaser 对象,并且传入的参与者计数器的值大于 0,则将当前创建的 Phaser 对象注册到父级 Phaser 对象中,此时执行的阶段数 phase 也来自父级。所以,当前创建的 Phaser 对象的阶段数和终止标记始终与父级 Phaser 对象相同。

(2)动态注册方法的代码如下。

```
//方法①
//一次注册一个参与者
public int register() {
    return doRegister(1);
}
//方法②
//一次注册多个参与者
public int bulkRegister(int parties) {
    if (parties < 0)
        throw new IllegalArgumentException();
    if (parties == 0)
        return getPhase();
    return doRegister(parties);
}
```

在 Phaser 对外提供的动态注册方法中,int register()方法表示一次注册一个参与者,int bulkRegister(int parties)方法表示一次可以注册多个参与者,这也是两者的区别所在。

另外,int bulkRegister(int parties)方法会对参与者数量 parties 进行校验,如果 parties 小于 0,则抛出 IllegalArgumentException 异常,如果 parties 等于 0,则返回阶段数。两种方法都会调用 int doRegister(int registrations)方法进行注册。

(3)阻塞等待方法的代码如下。

```
//方法①
public int awaitAdvance(int phase) {
    final Phaser root = this.root;
    long s = (root == this) ? state : reconcileState();
    int p = (int)(s >>> PHASE_SHIFT);
    if (phase < 0)
        return phase;
    if (p == phase)
        return root.internalAwaitAdvance(phase, null);
    return p;
}

//方法②
public int awaitAdvanceInterruptibly(int phase)
    throws InterruptedException {
    final Phaser root = this.root;
```

```
        long s = (root == this) ? state : reconcileState();
        int p = (int)(s >>> PHASE_SHIFT);
        if (phase < 0)
            return phase;
        if (p == phase) {
            QNode node = new QNode(this, phase, true, false, 0L);
            p = root.internalAwaitAdvance(phase, node);
            if (node.wasInterrupted)
                throw new InterruptedException();
        }
        return p;
}
//方法③
public int awaitAdvanceInterruptibly(int phase, long timeout, TimeUnit unit)
    throws InterruptedException, TimeoutException {
        long nanos = unit.toNanos(timeout);
        final Phaser root = this.root;
        long s = (root == this) ? state : reconcileState();
        int p = (int)(s >>> PHASE_SHIFT);
        if (phase < 0)
            return phase;
        if (p == phase) {
            QNode node = new QNode(this, phase, true, true, nanos);
            p = root.internalAwaitAdvance(phase, node);
            if (node.wasInterrupted)
                throw new InterruptedException();
            else if (p == phase)
                throw new TimeoutException();
        }
        return p;
}
```

在 Phaser 对外提供的阻塞等待方法中，每种方法的前半部分的主体逻辑都相同：对传入的阶段数 phase 进行校验，如果当前的 Phaser 已经终止或者当前的阶段数与传入的阶段数不相等，则直接返回，否则，在方法①中阻塞当前线程直到阶段数增长。在方法②中阻塞当前线程时，可能抛出 InterruptedException 异常。在方法③中阻塞当前线程时，可能抛出 InterruptedException 和 TimeoutException。

在实际项目中使用 Phaser 工具类时，如果捕获到抛出的异常，那么通常需要调用 forceTermination 方法来强制终止程序。

（4）到达与撤销方法的代码如下。

```
//方法①
//到达
public int arrive() {
    return doArrive(ONE_ARRIVAL);
}
//方法②
```

```java
//到达并撤销
public int arriveAndDeregister() {
    return doArrive(ONE_DEREGISTER);
}
```

在 Phaser 对外提供的到达与撤销方法中，int arrive()方法表示到达，int arriveAndDeregister() 方法表示到达并撤销。两种方法都调用了 int doArrive(int adjust)方法。

int arrive()方法调用 int doArrive(int adjust)方法传递的是 ONE_ARRIVAL（值是 1），int arriveAndDeregister()方法调用 int doArrive(int adjust)方法传递的是 ONE_DEREGISTER（值是 65535）。本质上都是通过传递数值到 int doArrive (int adjust)方法中，直接作用到 Phaser 的成员变量 state 上来完成对应的操作。

值得一提的是，在 JDK 1.8 之前，调用 doArrive()方法传递的是 boolean 类型的值。

（5）到达与等待方法的代码如下。

```java
public int arriveAndAwaitAdvance() {
    final Phaser root = this.root;
    for (;;) {
        long s = (root == this) ? state : reconcileState();
        int phase = (int)(s >>> PHASE_SHIFT);
        //phase 的值小于 0，直接返回
        if (phase < 0)
            return phase;
        int counts = (int)s;
        int unarrived = (counts == EMPTY) ? 0 : (counts & UNARRIVED_MASK);
        if (unarrived <= 0)
            throw new IllegalStateException(badArrive(s));
        if (UNSAFE.compareAndSwapLong(this, stateOffset, s,
                                      s -= ONE_ARRIVAL)) {
            if (unarrived > 1)
                return root.internalAwaitAdvance(phase, null);
            if (root != this)
                return parent.arriveAndAwaitAdvance();
            long n = s & PARTIES_MASK;  // base of next state
            int nextUnarrived = (int)n >>> PARTIES_SHIFT;
            if (onAdvance(phase, nextUnarrived))
                n |= TERMINATION_BIT;
            else if (nextUnarrived == 0)
                n |= EMPTY;
            else
                n |= nextUnarrived;
            int nextPhase = (phase + 1) & MAX_PHASE;
            n |= (long)nextPhase << PHASE_SHIFT;
            if (!UNSAFE.compareAndSwapLong(this, stateOffset, s, n))
                return (int)(state >>> PHASE_SHIFT); // terminated
            releaseWaiters(phase);
            return nextPhase;
```

在 int arriveAndAwaitAdvance()中，如果当前参与者不是最后一个到达的参与者，则会通过执行 Phaser 的 int internalAwaitAdvance(int phase, QNode node)方法进行阻塞。当最后到达的参与者执行方法时，会依次调用父节点的 arriveAndAwaitAdvance()方法将阻塞的参与者交由父节点执行到达操作，当所有的参与者都到达后，会将更新 state 成员变量的操作交由根 Phaser 完成。

（6）终结方法的代码如下。

```
//动态终结方法
protected boolean onAdvance(int phase, int registeredParties) {
    return registeredParties == 0;
}
//强制终结方法
public void forceTermination() {
    final Phaser root = this.root;
    long s;
    while ((s = root.state) >= 0) {
        if (UNSAFE.compareAndSwapLong(root, stateOffset,
                                s, s | TERMINATION_BIT)) {
            // 唤醒所有的线程
            releaseWaiters(0);
            releaseWaiters(1);
            return;
        }
    }
}
```

Phaser 提供了动态终结和强制终结两种终结方法，其中 boolean onAdvance(int phase, int registeredParties)方法是动态终结方法，void forceTermination()方法是强制终结方法。

- boolean onAdvance(int phase, int registeredParties) 方法会被 int arrive()、int arriveAndDeregister() 和 int arriveAndAwaitAdvance()方法调用，默认当参与者数量减为 0 时，返回 true，导致整体被终结。可以通过在 Phaser 的子类中重写 boolean onAdvance(int phase, int registeredParties)方法实现在某个阶段返回 true 来完成终结操作。
- void forceTermination()方法会不断尝试将根 Phaser 的状态设置为终止状态，设置成功后会唤醒所有阻塞等待的线程。

（7）获取状态的相关方法如下。

```
public final int getPhase() {
    return (int)(root.state >>> PHASE_SHIFT);
}
public int getRegisteredParties() {
    return partiesOf(state);
}
```

```
public int getArrivedParties() {
    return arrivedOf(reconcileState());
}
public int getUnarrivedParties() {
    return unarrivedOf(reconcileState());
}
public boolean isTerminated() {
    return root.state < 0L;
}
```

各方法的具体含义如下。

- int getPhase()方法表示获取当前的阶段数，当阶段数达到最大值 Integer.MAX_VALUE 后会从 0 开始，如果当前 Phaser 已经被终止，则返回负数。如果返回的是负数，则可以将这个负数的值加上 Integer.MAX_VALUE 获得在终止前的阶段数。
- int getRegisteredParties()方法表示获取参与者的总数量。
- int getArrivedParties()方法表示获取已到达当前阶段的参与者数量。
- int getUnarrivedParties()方法表示获取未到达当前阶段的参与者数量。
- boolean isTerminated()方法表示如果 Phaser 已经被终止，则返回 true，否则返回 false。

9.3.2 使用案例

1. 案例需求

聚餐时通常有这样的场景：等所有人都到场后开始点餐，点完菜后，每人再点一份饮料，点完餐后开始用餐。使用 Phaser 模拟这个过程。

2. 案例实现

（1）在 io.binghe.concurrent.chapter09 包下新建 PhaserDinner 类，继承 Phaser，模拟聚餐的每个阶段，例如人已经到齐、菜点完了、饮料点完了 3 个阶段，代码如下。

```
//模拟聚餐的过程
public class PhaserDinner extends Phaser {
    @Override
    protected boolean onAdvance(int phase, int registeredParties) {
        switch (phase){
            case 0:
                return allArrive();
            case 1:
                return allOrdered();
            case 2:
                return allDrink();
            default:
                return true;
        }
    }
}
```

```
        private boolean allDrink(){
            System.out.println("饮料点完了");
            return false;
        }
        private boolean allOrdered(){
            System.out.println("菜点完了");
            return false;
        }
        private boolean allArrive(){
            System.out.println("所有人都到齐了，开始点餐");
            return false;
        }
}
```

在 PhaserDinner 类中，分别定义了人已经到齐的方法 allArrive()、菜已经点完的方法 allOrdered()和饮料已经点完的方法 allDrink()。然后在重写的 onAdvance()方法中根据 phase 的值调用不同的方法。

（2）在 io.binghe.concurrent.chapter09 包下新建 PhaserTask 类，实现聚餐中每个阶段的细节，代码如下。

```
//模拟聚餐的任务
public class PhaserTask implements Runnable {
    private String name;
    private Phaser phaser;
    public PhaserTask(String name, Phaser phaser) {
        this.name = name;
        this.phaser = phaser;
    }
    @Override
    public void run() {
        //到达聚餐地点
        System.out.println(name + "--到达聚餐地点");
        phaser.arriveAndAwaitAdvance();
        //点菜
        System.out.println(name + "--点了一份香辣牛肉");
        phaser.arriveAndAwaitAdvance();
        //点饮料
        System.out.println(name + "--点了一份常温豆奶");
        phaser.arriveAndAwaitAdvance();
    }
}
```

PhaserTask 实现了 Runnable 接口。在 PhaserTask 类中，定义了一个 String 类型的成员变量 name，表示参与聚餐的名称，一个 Phaser 型的成员变量 phaser。然后在 run()方法中分别输出 3 个阶段的细节信息。

（3）在 io.binghe.concurrent.chapter09 包下新建 PhaserTest 类，在 PhaserTest 类的 main()方法中新建 PhaserDinner 对象，然后分别以 3 个线程模拟张三、李四和王五的聚餐过程，在模拟每

个人时，都调用 PhaserDinner 的 register()方法进行注册。代码如下。

```java
public class PhaserTest {
    public static void main(String[] args){
        PhaserDinner phaserDinner = new PhaserDinner();
        new Thread(new PhaserTask("张三", phaserDinner)).start();
        phaserDinner.register();
        new Thread(new PhaserTask("李四", phaserDinner)).start();
        phaserDinner.register();
        new Thread(new PhaserTask("王五", phaserDinner)).start();
        phaserDinner.register();
    }
}
```

PhaserTest 类的实现比较简单。

3. 案例测试

运行 PhaserTest 类的 main()方法，输出结果如下。

```
张三--到达聚餐地点
王五--到达聚餐地点
李四--到达聚餐地点
所有人都到齐了，开始点餐
李四--点了一份香辣牛肉
张三--点了一份香辣牛肉
王五--点了一份香辣牛肉
菜点完了
王五--点了一份常温豆奶
张三--点了一份常温豆奶
李四--点了一份常温豆奶
饮料点完了
```

从输出结果可以看出，张三、李四和王五先后到达聚餐地点后开始点餐，随后，张三、李四和王五都点了一份香辣牛肉，菜点完后，张三、李四和王五都点了一份常温豆奶，饮料点完后开始吃饭。整个模拟结果符合预期。

9.4 Semaphore 工具类

Semaphore 是从 Java 1.5 版本开始提供的一个高并发编程工具类，它可以控制同时访问某一系统资源的线程数量。

9.4.1 概述及重要方法说明

可以将 Semaphore 看成一种信号量，在创建 Semaphore 对象时传入信号量许可的初始值。如果线程要访问共享资源，则可以申请一定数量的信号量许可，如果许可不够，则线程会被阻塞。线程执行完毕，会释放申请的信号量许可，以便后续线程继续使用。

Semaphore 相当于一个共享锁，它允许多个线程同时拥有一定数量的信号量许可，只要线程拥有的信号量许可数量不超过总的信号量许可数量，就允许多个线程同时访问系统资源。

Semaphore 提供了公平信号量与非公平信号量两种模式。Semaphore 可以用于流量控制等场景，例如最多允许有多少个请求访问接口，最多有多少个连接访问数据库等。

Semaphore 提供的比较重要的方法的代码和说明分别如下。

（1）构造方法的代码如下。

```
//构造方法①
public Semaphore(int permits) {
    sync = new NonfairSync(permits);
}
//构造方法②
public Semaphore(int permits, boolean fair) {
    sync = fair ? new FairSync(permits) : new NonfairSync(permits);
}
```

各构造方法的具体含义如下。

- 构造方法①会调用构造方法②传入初始信号量的值创建 Semaphore 对象，默认是非公平模式。
- 构造方法②传入初始信号量的值和是否是公平模式的标识创建 Semaphore 对象。

（2）获取信号量许可的方法代码如下。

```
public void acquire() throws InterruptedException {
    sync.acquireSharedInterruptibly(1);
}
public void acquireUninterruptibly() {
    sync.acquireShared(1);
}
public boolean tryAcquire() {
    return sync.nonfairTryAcquireShared(1) >= 0;
}
public boolean tryAcquire(long timeout, TimeUnit unit)
    throws InterruptedException {
    return sync.tryAcquireSharedNanos(1, unit.toNanos(timeout));
}
public void acquire(int permits) throws InterruptedException {
    if (permits < 0) throw new IllegalArgumentException();
    sync.acquireSharedInterruptibly(permits);
}
public void acquireUninterruptibly(int permits) {
    if (permits < 0) throw new IllegalArgumentException();
    sync.acquireShared(permits);
}
public boolean tryAcquire(int permits) {
    if (permits < 0) throw new IllegalArgumentException();
```

```
        return sync.nonfairTryAcquireShared(permits) >= 0;
}
public boolean tryAcquire(int permits, long timeout, TimeUnit unit)
    throws InterruptedException {
    if (permits < 0) throw new IllegalArgumentException();
    return sync.tryAcquireSharedNanos(permits, unit.toNanos(timeout));
}
```

各方法的具体含义如下。

- void acquire()方法：表示从当前信号量中获取一个许可，如果获取到则线程立即返回；否则当前线程一直阻塞，直到获取到可用的许可，或者被其他线程中断。
- void acquireUninterruptibly()方法：表示从当前信号量中获取一个许可，如果获取到，则线程立即返回。否则当前线程会一直阻塞，直到获取到可用的许可。
- boolean tryAcquire()方法：表示尝试从当前的信号量中获取一个许可，如果获取到许可则立即返回 true，否则返回 false。
- boolean tryAcquire(long timeout, TimeUnit unit)方法：表示尝试从当前的信号量中获取一个许可，如果获取到许可则立即返回 true；否则线程一直阻塞，直到获取到许可，或者被其他线程中断，或者超时。
- void acquire(int permits)方法：表示从当前信号量中获取指定数量的许可，如果能够获取到足够的许可则线程立刻返回；否则线程一直阻塞，直到获取到足够的许可，或者被其他线程中断。
- void acquireUninterruptibly(int permits)方法：表示从当前信号量中获取指定数量的许可。如果够获取到足够的许可则线程立刻返回；否则线程会一直阻塞，直到获取到足够的许可。
- boolean tryAcquire(int permits)方法：表示尝试从当前的信号量中获取指定数量的许可，如果获取到足够的许可则立即返回 true；否则返回 false。
- boolean tryAcquire(int permits, long timeout, TimeUnit unit)方法：表示尝试从当前的信号量中获取指定数量的许可，如果获取到足够的许可则立即返回 true；否则线程一直阻塞，直到获取到足够的许可，或者被其他线程中断，或者超时。

（3）释放信号量许可的方法代码如下。

```
public void release() {
    sync.releaseShared(1);
}
public void release(int permits) {
    if (permits < 0) throw new IllegalArgumentException();
    sync.releaseShared(permits);
}
```

各方法的具体含义如下。

- void release()方法表示释放一个许可,将其返还到信号量中。
- void release(int permits)方法表示释放指定数量的许可,将其返还到信号量中。

(4)获取信号量许可方法的代码如下。

```
public int availablePermits() {
   return sync.getPermits();
}
public int drainPermits() {
   return sync.drainPermits();
}
```

各方法的具体含义如下。

- int availablePermits():获取当前信号量中剩余的许可数量。
- int drainPermits():获取当前信号量中可用的全部许可。

9.4.2 使用案例

1. 案例需求

使用 Semaphore 模拟同一时刻最多只能有 2 个线程同时执行某种方法的逻辑。

2. 案例实现

使用 Semaphore 能够很轻松地实现案例的需求,具体实现步骤如下。

(1)在 io.binghe.concurrent.chapter09 包下创建 SemaphoreTask 类,实现 Runnable 接口,使用 Semaphore 来控制线程访问方法的最大数量,代码如下。

```
// Semaphore 案例程序
public class SemaphoreTask implements Runnable {
   private Semaphore semaphore;
   public SemaphoreTask(Semaphore semaphore) {
      this.semaphore = semaphore;
   }
   @Override
   public void run() {
      try {
         semaphore.acquire();
         Thread.sleep(1000);
         System.out.println(Thread.currentThread().getName() + "--执行方法--" + new Date());
         semaphore.release();
      } catch (InterruptedException e) {
         e.printStackTrace();
      }
   }
}
```

SemaphoreTask 类实现了 Runnable 接口，在 SemaphoreTask 类中，定义了一个 Semaphore 型的成员变量 semaphore，并通过 SemaphoreTask 类的构造方法进行赋值。接下来，在 run()方法中获取一个信号量许可后休眠 1s，输出某个线程--执行了方法--时间的日志。随后释放信号量许可。

（2）在 io.binghe.concurrent.chapter09 包下创建 SemaphoreTest 类，代码如下。

```java
public class SemaphoreTest {
    public static void main(String[] args){
        Semaphore semaphore = new Semaphore(2);
        ExecutorService threadPool = Executors.newFixedThreadPool(6);
        IntStream.range(0, 6).forEach((i) -> {
            threadPool.submit(new SemaphoreTask(semaphore));
        });
        threadPool.shutdown();
    }
}
```

在 SemaphoreTest 类的 main()方法中首先创建了一个信号量许可数量为 2 的 Semaphore 对象，然后创建一个固定有 6 个线程的线程池，随后循环 6 次，每次都向线程池提交一个 SemaphoreTask 任务,在创建 SemaphoreTask 对象时将创建的 Semaphore 对象传入 SemaphoreTask 类的构造方法中。最后关闭线程池。

3. 案例测试

运行 SemaphoreTest 案例的 main()方法，输出结果如下。

```
pool-1-thread-1--执行方法--Fri Sep 16 00:00:32 CST 2022
pool-1-thread-2--执行方法--Fri Sep 16 00:00:32 CST 2022
pool-1-thread-4--执行方法--Fri Sep 16 00:00:33 CST 2022
pool-1-thread-3--执行方法--Fri Sep 16 00:00:33 CST 2022
pool-1-thread-5--执行方法--Fri Sep 16 00:00:34 CST 2022
pool-1-thread-6--执行方法--Fri Sep 16 00:00:34 CST 2022
```

pool-1-thread-1 线程与 pool-1-thread-2 线程输出的时间相同，pool-1-thread-3 线程与 pool-1-thread-4 线程输出的时间相同，pool-1-thread-5 线程与 pool-1-thread-6 线程输出的时间相同。因此，在同一时刻，最多有两个线程执行了输出日志的逻辑，符合预期。

9.5　Exchanger 工具类

Exchanger 是 Java 从 JDK 1.5 版本开始提供的一种线程间协作的工具类，它支持泛型，能够实现线程之间的数据交换。

9.5.1　概述及重要方法说明

在使用 Exchanger 实现线程之间的数据交换时，如果一个线程调用了 Exchanger 的

exchange()方法，就会进入阻塞状态，只有在另一个线程也调用了exchange()方法后，当前线程才能继续执行后面的逻辑。此时，两个线程会交换数据。

所以，在使用Exchanger交换线程的数据时，需要有成对的线程调用exchange()方法。同时Exchanger的exchange()方法是可以重复使用的。也就是说，两个线程之间可以通过Exchanger不断交换数据。

注意：Exchanger仅用于两个线程之间交换数据，超过两个线程使用同一个Exchanger交换数据，结果是无法确定的。

Exchanger提供的比较重要的方法的代码和说明分别如下。

（1）构造方法的代码如下。

```
public Exchanger () {
   participant = new Participant();
}
```

在Exchanger中，只有一个默认的无参构造方法，并且在这种方法中会实例化一个Participant类型的对象，并赋值给Participant类型的成员变量participant。

（2）数据交换方法的代码如下。

```
//启动交换并等待另一个线程调用exchange()方法
public V exchange(V x) throws InterruptedException {
   Object v;
   Object item = (x == null) ? NULL_ITEM : x;
   if ((arena != null ||
        (v = slotExchange(item, false, 0L)) == null) &&
       ((Thread.interrupted() || // disambiguates null return
        (v = arenaExchange(item, false, 0L)) == null)))
      throw new InterruptedException();
   return (v == NULL_ITEM) ? null : (V)v;
}

//启动交换并在timeout时间内等待另一个线程调用exchange()方法
//如果等待时间超过timeout就会停止等待
public V exchange(V x, long timeout, TimeUnit unit)
   throws InterruptedException, TimeoutException {
   Object v;
   Object item = (x == null) ? NULL_ITEM : x;
   long ns = unit.toNanos(timeout);
   if ((arena != null ||
        (v = slotExchange(item, true, ns)) == null) &&
       ((Thread.interrupted() ||
        (v = arenaExchange(item, true, ns)) == null)))
      throw new InterruptedException();
   if (v == TIMED_OUT)
      throw new TimeoutException();
```

```
        return (v == NULL_ITEM) ? null : (V)v;
}
```

各方法的具体含义如下。

- V exchange(V x)方法：启动交换并等待另一个线程调用 exchange()方法。
- V exchange(V x, long timeout, TimeUnit unit)方法：启动交换并在 timeout 时间内等待另一个线程调用 exchange()方法，如果等待的时间超过 timeout 就会停止等待。

9.5.2 使用案例

1. 案例需求

假设有这样一个场景：张三去超市买商品，付完款后收银员才会将商品交给张三。使用 Exchanger 模拟这个过程。

2. 案例实现

张三和收银员好比两个不同的线程，张三线程给收银员线程付款信息，然后收银员线程给张三线程商品。使用 Exchanger 完全可以模拟这个过程，具体代码如下。

```
// Exchanger 模拟张三到超市买商品
public class ExchangerTest {
    public static void main(String[] args){
        Exchanger<String> exchanger = new Exchanger<>();
        new Thread(() -> {
            try {
                String exchangeResult = exchanger.exchange("付款信息");
                Thread.sleep(500);
                System.out.println("张三收到收银员递过来的--" + exchangeResult);
            } catch (InterruptedException e) {
                e.printStackTrace();
            }
        }).start();
        new Thread(() -> {
            try {
                String exchangeResult = exchanger.exchange("商品");
                System.out.println("收银员收到张三的--" + exchangeResult);
            } catch (InterruptedException e) {
                e.printStackTrace();
            }
        }).start();
    }
}
```

在 ExchangerTest 类的 main()方法中，首先创建了一个泛型为 String 类型的 Exchanger 对象 exchanger。随后分别创建并启动了两个线程，每个线程都会使用 Exchanger 的 exchange()方法进行数据交换。

这里的一个线程模拟张三向收银员付款，并等待收银员递过来商品。另一个线程模拟收银员等待张三付款，然后将商品交给张三。

为了模拟张三先付款、收银员后给商品这个流程，在模拟张三的线程中，会在输出日志之前让当前线程休眠 500 毫秒。

至此，整个案例实现完毕。

3. 案例测试

运行 ExchangerTest 类的 main()方法，输出结果如下。

```
收银员收到张三的--付款信息
张三收到收银员递过来的--商品
```

从输出结果可以看出，张三用付款信息向收银员换来了商品，而收银员用商品向张三换来了付款信息，整个过程符合预期。

第 10 章

锁工具类

在 Java 1.5 版本之前，只能通过 synchronized 关键字同步共享资源的临界区。从 Java 1.5 版本开始，JDK 提供了显式锁功能，使用显式锁能够实现与 synchronized 关键字类似的同步功能，但是在使用上需要显式地获取锁和释放锁。显式锁比 synchronized 关键字更加灵活，也提供了可中断和超时等多种 synchronized 关键字不具备的特性。

10.1 Lock 接口

Lock 接口是 Java 从 JDK1.5 版本开始提供的显式锁接口，在使用 Lock 接口时，需要显式地获取锁和释放锁，Lock 比 synchronized 具有更高的灵活性和可控制性。

10.1.1 概述及核心方法

在 Java1.5 版本之前，只能使用 synchronized 关键字实现多个线程对同一共享资源的同步操作。随着 JDK 版本的不断迭代和升级，synchronized 关键字进行了一系列的优化，包括自旋锁、偏向锁、轻量级锁、重量级锁、锁升级、锁粗化、锁消除等，但它仍然存在不少弊端，体现在如下几方面。

（1）当同步的临界区资源支持多个线程同时进行读操作时，使用 synchronized 关键字对临界区加锁，会造成无论对于读线程还是写线程，同一时刻都只能有一个线程进入临界区对数据进行读写操作的情况，影响临界区数据的读性能。

（2）由于 synchronized 是 JVM 底层提供的锁机制，所以在使用 synchronized 关键字时，无法确定当前线程是否真正获取到锁。

（3）使用 synchronized 对临界区加锁，如果线程在操作临界区的资源时被阻塞，只要线程不释放锁，则其他所有要操作临界区资源的线程都将被阻塞。

上述 synchronized 关键字的弊端可以使用 JDK 提供的 Lock 锁消除。Lock 接口位于

java.util.concurrent.locks 包下，除了提供 synchronized 锁的功能，还具备一些 synchronized 锁不具备的特性。

（1）非阻塞地获取锁资源。某一线程在尝试获取 Lock 锁时，如果锁没有被其他线程占有，则直接获取到锁并返回，如果锁已经被其他线程占有，则直接返回，不会被阻塞。

（2）支持响应中断机制。在使用 Lock 接口获取锁时，获取到的锁支持响应中断机制。如果获取到锁的线程被中断，则会抛出被中断的异常，同时释放锁资源。

（3）支持超时机制。在使用 Lock 接口获取锁时，可以指定一个超时时间，如果线程在超时时间内获取到锁，则立即返回。如果超过超时时间仍未获取到锁，那么线程也会返回。

注意：Lock 锁的核心原理，可参见《深入理解高并发编程：核心原理与案例实战》一书。

Lock 锁接口位于 java.util.concurrent.locks 包下，代码如下。

```
public interface Lock {
    void lock();
    void lockInterruptibly() throws InterruptedException;
    boolean tryLock();
    boolean tryLock(long time, TimeUnit unit) throws InterruptedException;
    void unlock();
    Condition newCondition();
}
```

其中，每种方法的具体含义如下。

- void lock()：获取锁，如果获取锁成功则立即返回，如果获取锁失败则阻塞等待。
- void lockInterruptibly()：可中断式获取锁，如果获取锁成功，则立即返回，如果获取锁失败，则阻塞等待。在阻塞等待的过程中，如果有其他线程中断当前线程，则当前线程会响应中断信号。
- boolean tryLock()：非阻塞式尝试获取锁，如果获取锁成功，则立即返回，如果获取锁失败，则返回 false。
- boolean tryLock(long time, TimeUnit unit)：非阻塞式尝试获取锁，如果获取锁成功，则立即返回，如果获取锁失败或者超时，则返回 false。
- void unlock()：释放锁。
- Condition newCondition()：获取与 Lock 锁绑定的 Condition 对象，一般用于线程间的通信操作。

在使用 Lock 锁时，通常需要将 unlock()方法放在 finally 代码块中，保证锁被释放，防止死锁的发生。通常可以按照如下方式使用 Lock 锁。

```
//创建Lock锁对象
Lock lock = new XXXLock();
//进行加锁操作
```

```
lock.lock();
try{
    //处理业务操作
    doBusiness();
}catch(Exception e){
    //处理异常
}finally{
    //释放锁
    lock.unlock();
}
```

通常在使用 Lock 锁时，首先创建 Lock 锁对象，调用 lock()方法进行加锁。然后在 try 代码块中处理业务，在 catch 代码块中处理异常。最后在 finally 代码块中释放锁资源。

注意：上述代码片段使用了 Lock 锁的标准模板，在实际工作过程中，可参照上述代码片段使用 Lock 锁。

10.1.2 使用案例

1. 案例需求

使用 Lock 锁实现在同一时刻只能有一个线程访问临界区资源。

2. 案例实现

Lock 锁可以实现多个线程之间的互斥操作，能够保证在同一时刻只有一个线程访问临界区资源，所以，使用 Lock 锁能够轻松实现案例需求。案例程序的实现步骤如下。

（1）在 mykit-concurrent-chapter10 工程的 io.binghe.concurrent.chapter10 包下创建 LockTask 类，实现 Runnable 接口，表示使用 Lock 锁执行的任务类，代码如下。

```
public class LockTask implements Runnable{
    private Lock lock;
    public LockTask(Lock lock) {
        this.lock = lock;
    }
    @Override
    public void run() {
        lock.lock();
        try{
            System.out.println(Thread.currentThread().getName() + "--线程获取到锁");
            System.out.println(Thread.currentThread().getName() + "--线程执行任务");
        }finally {
            System.out.println(Thread.currentThread().getName() + "--线程释放锁");
            lock.unlock();
        }
    }
}
```

LockTask 类实现了 Runnable 接口，在 LockTask 类中，首先定义了一个 Lock 类型的成员变量 lock，表示 Lock 锁；然后在构造方法中为 Lock 对象赋值；随后在 run()方法中使用 lock 对象加锁；接下来在 try 代码块中输出获取到锁和执行任务的日志；最后在 finally 代码块中输出释放锁的日志，并使用 lock 对象释放锁。

（2）在 io.binghe.concurrent.chapter10 包下创建 LockTest，并在 main()方法中创建 5 个线程进行测试，代码如下。

```
public class LockTest {
    public static void main(String[] args){
        Lock lock = new ReentrantLock();
        IntStream.range(0, 5).forEach((i) -> {
            new Thread(new LockTask(lock)).start();
        });
    }
}
```

3. 案例测试

运行 LockTest 类的 main()方法，输出结果如下。

```
Thread-0--线程获取到锁
Thread-0--线程执行任务
Thread-0--线程释放锁
Thread-1--线程获取到锁
Thread-1--线程执行任务
Thread-1--线程释放锁
Thread-2--线程获取到锁
Thread-2--线程执行任务
Thread-2--线程释放锁
Thread-3--线程获取到锁
Thread-3--线程执行任务
Thread-3--线程释放锁
Thread-4--线程获取到锁
Thread-4--线程执行任务
Thread-4--线程释放锁
```

每个线程都是在前一个线程完整地输出获取到锁、执行任务和释放锁的日志后，才会完整地输出获取到锁、执行任务和释放锁的日志。说明使用 Lock 锁能够实现在同一时刻只能有一个线程访问临界区资源，符合预期。

10.2 Condition 接口

Condition 接口提供了类似 Object 的监视器方法，与 Lock 锁配合使用可以实现线程间的通信，也就是可以实现线程间的等待与通知模式。

10.2.1 概述及核心方法

Java 的 Object 方法中统一提供了 wait()、notify() 与 notifyAll() 方法实现对象的等待与通知机制，Object 的这些方法必须与 synchronized 配合使用，如下所示。

```
private final Object obj = new Object();
public void testWait(){
   synchronized(obj){
      //#######省略其他代码#######
      obj.wait();
   }
}
public void testNotify(){
   synchronized(obj){
      //#######省略其他代码#######
      obj.notify();
   }
}
public void testNotifyAll(){
   synchronized(obj){
      //#######省略其他代码#######
      obj.notifyAll();
   }
}
```

Condition 接口也提供了类似的方法，但是 Condition 接口需要和 Lock 锁配合使用。通常，在使用 Condition 接口时，不会直接创建 Condition 接口的对象，而是通过 Lock 接口的如下方法创建并返回 Condition 接口的对象。

```
//获取与Lock锁绑定的Condition对象
Condition newCondition();
```

调用 Lock 锁对象的 newCondition() 方法，就能够生成一个与当前 Lock 锁绑定的 Condition 对象，利用 Condition 对象就可以实现线程的等待与通知机制，如下所示。

```
private Lock lock = new XXXLock();
private Condition condition = lock.newCondition();
public void testAwait(){
   lock.lock();
   try{
      condition.await();
   }finally{
      lock.unlock();
   }
}
public void testSignal(){
   lock.lock();
   try{
      condition.signal();
   }finally{
```

```
      lock.unlock();
   }
}
public void testSignalAll(){
   lock.lock();
   try{
      condition.signalAll();
   }finally{
      lock.unlock();
   }
}
```

某一线程调用 Condition 的 await()方法，当前线程会释放锁并阻塞等待。如果其他线程调用了 Condition 的 signal()方法或者 signalAll()方法，通知当前线程，当前线程就会被唤醒从 await()方法返回，继续向下执行，并且当前线程在返回前会再次获取锁。

注意：线程调用 Condition 的 await()方法被阻塞时，会释放 Lock 锁。其他线程调用 Condition 的 signal()方法和 signalAll()方法唤醒被阻塞的线程时，被阻塞的线程在被唤醒前会再次获取 Lock 锁。

Condition 位于 JDK 的 java.util.concurrent.locks 包下，代码如下。

```
public interface Condition {
   void await() throws InterruptedException;
   void awaitUninterruptibly();
   long awaitNanos(long nanosTimeout) throws InterruptedException;
   boolean await(long time, TimeUnit unit) throws InterruptedException;
   boolean awaitUntil(Date deadline) throws InterruptedException;
   void signal();
   void signalAll();
}
```

Condition 中各方法的含义如下。

- void await()：阻塞当前线程，直到被其他线程唤醒或中断。

- void awaitUninterruptibly()：阻塞当前线程，直到被其他线程唤醒，阻塞期间不可被其他线程中断。

- long awaitNanos(long nanosTimeout)：阻塞当前线程，直到被其他线程唤醒、中断或超时。

- boolean await(long time, TimeUnit unit)：阻塞当前线程，直到被其他线程唤醒、中断或超时。

- boolean awaitUntil(Date deadline)：阻塞当前线程，直到被其他线程唤醒、中断或达到某个时间点。

- void signal()：唤醒等待中的某个线程。

- void signalAll()：唤醒所有等待的线程。

10.2.2 使用案例

1. 案例需求

使用 Condition 接口结合 Lock 实现线程的阻塞与唤醒。

2. 案例实现

使用 Condition 的 await()方法可以阻塞线程，使用 Condition 的 signal()方法和 signalAll()方法可以唤醒线程，而 Condition 需要结合 Lock 使用。所以，通过 Condition 接口结合 Lock 可以实现线程的阻塞与唤醒。案例程序的具体实现步骤如下。

（1）在 io.binghe.concurrent.chapter10 包下创建 ConditionTask 类，实现 Runnable 接口，在 ConditionTask 类中阻塞线程并等待其他线程唤醒当前线程，代码如下。

```java
public class ConditionTask implements Runnable {
    private Lock lock;
    private Condition condition;
    public ConditionTask(Lock lock, Condition condition) {
        this.lock = lock;
        this.condition = condition;
    }
    @Override
    public void run() {
        lock.lock();
        try {
            System.out.println(Thread.currentThread().getName() + "--线程被阻塞: " + new Date());
            //线程阻塞时，会释放锁
            condition.await();
            System.out.println(Thread.currentThread().getName() + "--线程被唤醒: " + new Date());
        } catch (InterruptedException e) {
            e.printStackTrace();
        }finally {
            lock.unlock();
        }
    }
}
```

首先在 ConditionTask 类中定义一个 Lock 类型的成员变量和一个 Condition 类型的成员变量，在 ConditionTask 类的构造方法中为成员变量赋值；然后在 run()方法中获取 Lock 锁，在 try 代码块中输出线程被阻塞的日志；接下来调用 Condition 的 await()方法阻塞当前线程并释放锁，当线程被唤醒时会再次获取 Lock 锁并输出线程被唤醒的日志；最后在 finally 代码块中释放 Lock 锁。

（2）在 io.binghe.concurrent.chapter10 包下创建 ConditionTest 类，在 ConditionTest 类的 main()

方法中创建并启动子线程，子线程被阻塞后，使用 main 线程唤醒子线程，代码如下。

```java
public class ConditionTest {
    public static void main(String[] args) throws InterruptedException {
        Lock lock = new ReentrantLock();
        Condition condition = lock.newCondition();
        //创建并启动子线程
        new Thread(new ConditionTask(lock, condition)).start();
        //主线程休眠 1s
        Thread.sleep(1000);
        //主线程获取锁
        lock.lock();
        try{
            System.out.println(Thread.currentThread().getName() + "--线程唤醒被阻塞的线程开始: " + new Date());
            //唤醒被阻塞的线程
            condition.signal();
            System.out.println(Thread.currentThread().getName() + "--线程唤醒被阻塞的线程结束: " + new Date());
        }finally {
            lock.unlock();
        }
    }
}
```

在 ConditionTest 的 main()方法中，首先创建一个 Lock 锁对象，并使用 Lock 锁对象创建一个 Condition 对象。然后传入 Lock 锁对象和 Condition 对象创建一个子线程并启动。子线程启动后，由于在 run()方法中调用了 Condition 的 await()方法，所以子线程会被阻塞并释放 Lock 锁。

接下来，main()方法的主线程休眠 1s，随后主线程获取 Lock 锁，在 try 代码块中输出唤醒被阻塞的线程开始的日志，调用 Condition 的 signal()方法唤醒阻塞的线程，随后输出唤醒被阻塞的线程结束的日志。最后在 finally 代码块中释放 Lock 锁。

3. 案例测试

运行 ConditionTest 的 main()方法，输出结果如下。

```
Thread-0--线程被阻塞: Sat Sep 17 15:09:32 CST 2022
main--线程唤醒被阻塞的线程开始: Sat Sep 17 15:09:33 CST 2022
main--线程唤醒被阻塞的线程结束: Sat Sep 17 15:09:33 CST 2022
Thread-0--线程被唤醒: Sat Sep 17 15:09:33 CST 2022
```

从输出结果可以看出，子线程调用 Condition 的 await()方法被阻塞后，成功被主线程调用 condition 的 signal()方法唤醒，符合预期。

10.3 ReentrantLock 可重入锁

ReentrantLock 是 Java 从 JDK 1.5 版本开始提供的一种可重入锁。

10.3.1 概述及核心方法

ReentrantLock 实现了 Lock 接口，内部基于 AQS 实现。同一个线程能够多次获取到同一个 ReentrantLock 锁，如下所示。

```
Lock lock = new ReentrantLock();
public void mutilLock(){
   lock.lock();
   lock.lock();
   try{
      //处理业务方法
      doBusiness();
   }finally{
      lock.unlock();
      lock.unlock();
   }
}
```

一个线程可以多次通过 ReentrantLock 的 lock()方法获取锁，但是要想完全释放锁，就必须调用相同次数的 unlock()方法。ReentrantLock 对同一个线程的可重入性是通过内部的计数器实现的，同一个线程调用 lock()方法加锁，计数器就会加 1，同一个线程调用 unlock()方法释放锁，计数器就会减 1。只有内部的计数器减为 0，才算真正释放了锁。

ReentrantLock 对同一个线程是可重入的，对不同的线程是独占的。一个线程获取到 ReentrantLock 锁，其他线程如果再次抢占 ReentrantLock 锁，就会被阻塞，直到获取到 ReentrantLock 锁的线程释放锁，其他线程才能抢占到锁。

在 ReentrantLock 的内部，有两个抽象类，分别为 NonFairSync 和 FairSync，二者都继承了 Sync 类，而 Sync 类继承了 AQS。NonFairSync 类表示的是非公平模式，FairSync 类表示的是公平模式。

ReentrantLock 提供了可以传入一个 boolean 类型参数的构造方法，通过这个 boolean 类型的参数来指定 ReentrantLock 锁是否是公平锁，默认是非公平锁。

也就说，ReentrantLock 锁是一种基于 AQS 实现的、对于单线程可重入、对于多线程排他、支持公平与非公平两种模式的显式锁。

注意：有关锁的分类与 ReentrantLock 锁的核心原理，读者可参见《深入理解高并发编程：核心原理与案例实战》一书。

1. 初始化

在 ReentrantLock 中，提供了如下构造方法进行初始化。

```
//构造方法①
public ReentrantLock() {
   sync = new NonfairSync();
```

```
}
//构造方法②
public ReentrantLock(boolean fair) {
    sync = fair ? new FairSync() : new NonfairSync();
}
```

（1）构造方法①是默认的无参构造方法，默认实现非公平锁模式。

（2）构造方法②传入是否为公平锁的标识，根据传入的标识确定是公平锁模式还是非公平锁模式。

注意：ReentrantLock 类中比较重要的内部类有 Sync、FairSync 和 NonFairSync，FairSync 类和 NonFairSync 类继承了 Sync 类，而 Sync 类又继承了 AQS 类。事实上，FairSync 类和 NonFairSync 类分别实现了 ReentrantLock 的公平锁与非公平锁模式，并且底层是基于 AQS 实现的。有关公平锁与非公平的核心原理与 Sync、FairSync 和 NonFairSync 类的代码解析，读者可以参考《深入理解高并发编程：核心原理与案例实战》一书。

2. 加锁核心方法

ReentrantLock 类提供了如下比较重要的加锁方法。

```
public void lock() {
    sync.lock();
}
public void lockInterruptibly() throws InterruptedException {
    sync.acquireInterruptibly(1);
}
public boolean tryLock() {
    return sync.nonfairTryAcquire(1);
}
public boolean tryLock(long timeout, TimeUnit unit)
        throws InterruptedException {
    return sync.tryAcquireNanos(1, unit.toNanos(timeout));
}
```

每种方法的说明如下。

- void lock()：获取锁，如果锁未被其他线程获取，则当前线程获取锁并立即返回，同时将内部的锁计数器设置为 1。如果当前线程再次获取锁，则内部的计数器加 1 并立即返回。如果在当前线程获取锁时，锁已经被其他线程获取，则当前线程被阻塞挂起，直到获取到锁。

- void lockInterruptibly()：功能与 void lock() 方法类似，只是在锁被其他线程获取、当前线程进入阻塞状态后，可以被中断，也就是当前线程可以获取到中断信号。

- boolean tryLock()：尝试获取锁，获取成功则立即返回 true；获取失败则立即返回 false。线程不会阻塞。

- boolean tryLock(long timeout, TimeUnit unit)：尝试获取锁，如果在 timeout 时间内获取成功，则立即返回 true。如果超过 timeout 时间仍未获取到锁，则返回 false。

注意：ReentrantLock 的加锁方法内部会调用 Sync 类的方法，具体根据公平锁或非公平锁调用 FairSync 类的方法或 NonFairSync 类的方法，Sync 类继承了 AQS，内部还会调用 AQS 的方法。在线程首次调用上述加锁方法获取锁成功后，内部的锁计数器会被设置为 1，如果后续相同的线程再次调用上述方法获取锁成功，内部的锁计数器就会加 1。当前线程需要调用相同次数的解锁操作，将内部的锁计数器减为 0，才能真正释放锁。

3. 释放锁核心方法

ReentrantLock 类提供了如下比较重要的释放锁方法。

```
//释放锁
public void unlock() {
    sync.release(1);
}
```

void unlock()方法是释放锁的方法，同一个线程调用 void unlock()方法释放锁的次数需要与加锁的次数相同，才能真正释放锁。

10.3.2 使用案例

1. 案例需求

验证 ReentrantLock 锁对同一个线程的可重入性，对多个线程的排他性。

2. 案例实现

ReentrantLock 本身对同一个线程具有可重入性，对多个线程具有排他性。体现在同一个线程可以多次获取 ReentrantLock 锁，而当某个线程获取到 ReentrantLock 锁时，如果其他线程获取 ReentrantLock 锁就会失败，要么阻塞，要么返回 false。案例程序的具体实现步骤如下。

（1）在 io.binghe.concurrent.chapter10 包下创建 ReentrantLockTask 类，实现 Runnable 接口，使用 ReentrantLock 实现同一个线程的可重入性，代码如下。

```
public class ReentrantLockTask implements Runnable{
    private Lock lock;
    public ReentrantLockTask(Lock lock) {
        this.lock = lock;
    }
    @Override
    public void run() {
        lock.lock();
        String threadName = Thread.currentThread().getName();
        System.out.println(threadName + "--线程第一次加锁");
        lock.lock();
```

```
            System.out.println(threadName + "--线程第二次加锁");
            try{
                System.out.println(threadName + "--线程执行业务逻辑方法");
            }finally {
                System.out.println(threadName + "--线程第一次释放锁");
                lock.unlock();
                System.out.println(threadName + "--线程第二次释放锁");
                lock.unlock();
            }
        }
    }
}
```

ReentrantLockTask 类实现了 Runnable 接口，在 ReentrantLockTask 类中，首先定义了 Lock 成员变量，并通过 ReentrantLockTask 类的构造方法为其赋值。然后在 run()方法中两次调用 lock() 方法加锁并输出加锁日志。接下来，在 try 代码块中输出执行业务逻辑方法的日志。最后在 finally 代码块中输出释放锁的日志并两次调用 unlock()释放锁。

（2）在 io.binghe.concurrent.chapter10 包下创建 ReentrantLockTest 类进行测试，代码如下。

```
public class ReentrantLockTest {
    public static void main(String[] args){
        Lock lock = new ReentrantLock();
        IntStream.range(0, 2).forEach((i) -> {
            new Thread(new ReentrantLockTask(lock)).start();
        });
    }
}
```

ReentrantLockTest 类的代码比较简单，这里不再赘述其实现逻辑。

3. 案例测试

运行 ReentrantLockTest 类的 main()方法，输出结果如下。

```
Thread-0--线程第一次加锁
Thread-0--线程第二次加锁
Thread-0--线程执行业务逻辑方法
Thread-0--线程第一次释放锁
Thread-0--线程第二次释放锁
Thread-1--线程第一次加锁
Thread-1--线程第二次加锁
Thread-1--线程执行业务逻辑方法
Thread-1--线程第一次释放锁
Thread-1--线程第二次释放锁
```

从输出结果可以看出，Thread-0 线程在连续两次获取锁后输出了执行业务逻辑的方法，随后在连续两次释放锁后，Thread-1 线程才重复 Thread-0 线程的操作。说明 ReentrantLock 对同一个线程具有可重入性，对多个线程具有排他性，符合预期。

10.4 ReadWriteLock 读写锁

ReadWriteLock 是 Java 从 JDK 1.5 版本开始提供的一种读写锁，使用 ReadWriteLock 可以分别创建一个读锁和一个写锁。

10.4.1 概述及核心方法

Java 的 ReadWriteLock 是一个接口，位于 JDK 的 java.util.concurrent.locks 包下，代码如下。

```
public interface ReadWriteLock {
    //获取读锁
    Lock readLock();
    //获取写锁
    Lock writeLock();
}
```

两种方法的含义如下。

- Lock readLock()：获取读锁，读锁在多个线程之间共享，多个线程可以同时获取读锁。
- Lock writeLock()：获取写锁，写锁在多个线程之间不共享，多个线程不能同时获取写锁。

ReadWriteLock 接口的实现类是 java.util.concurrent.locks 包下的 ReentrantReadWriteLock 类，ReentrantReadWriteLock 类是支持可重入的读写锁，内部的实现依赖了 ReentrantReadWriteLock 类的内部类 Sync，而 Sync 类继承了 AQS，所以 ReentrantReadWriteLock 的实现仍然依赖 AQS 的实现。

ReentrantReadWriteLock 与 ReentrantLock 一样，支持公平锁与非公平锁两种模式。

> 注意：有关 ReadWriteLock、锁降级与 AQS 的核心原理与核心代码解析，读者可参考《深入理解高并发编程：核心原理与案例实战》一书。

1. 初始化

ReentrantReadWriteLock 类是 ReadWriteLock 接口的实现类，ReentrantReadWriteLock 类提供了如下构造方法进行初始化。

```
//构造方法①
public ReentrantReadWriteLock() {
    this(false);
}
//构造方法②
public ReentrantReadWriteLock(boolean fair) {
    sync = fair ? new FairSync() : new NonfairSync();
    readerLock = new ReadLock(this);
    writerLock = new WriteLock(this);
}
```

其中，每种构造方法的含义如下。

- 构造方法①是默认的无参构造方法，内部会直接调用构造方法②传入 false，实现非公平锁模式。
- 构造方法②传入是否是公平锁的标识来实现公平模式或非公平模式，同时实例化读锁和写锁。

2. 获取读写锁

ReentrantReadWriteLock 实现了在 ReadWriteLock 接口中获取读锁和写锁的方法，代码如下。

```
//获取写锁
public ReentrantReadWriteLock.WriteLock writeLock() {
    return writerLock;
}
//获取读锁
public ReentrantReadWriteLock.ReadLock readLock() {
    return readerLock;
}
```

获取写锁的方法 writeLock() 返回的类型是 ReentrantReadWriteLock.WriteLock，获取读锁的方法 readLock() 返回的类型是 ReentrantReadWriteLock.ReadLock。

ReentrantReadWriteLock.WriteLock 类与 ReentrantReadWriteLock.ReadLock 类的定义如下。

```
public static class WriteLock implements Lock, java.io.Serializable {
    //#############省略其他代码###############
}
public static class ReadLock implements Lock, java.io.Serializable {
    //#############省略其他代码###############
}
```

无论是 ReentrantReadWriteLock.WriteLock 类还是 ReentrantReadWriteLock.ReadLock 类，都实现了 Lock 接口。也就是说，ReentrantReadWriteLock 的读写锁本质上还是 Lock 锁，适用 Lock 接口的方法。

10.4.2 使用案例

1. 案例需求

验证 ReadWriteLock 的读锁是共享的，而写锁是独占的。

2. 案例实现

ReadWriteLock 提供的读锁可以在多个线程间共享，多个线程可以同时获取读锁，而 ReadWriteLock 提供的写锁在同一时刻只能被一个线程获取。案例程序代码如下。

```
public class ReadWriteLockTest {
    private ReadWriteLock lock = new ReentrantReadWriteLock();
    private Lock readLock = lock.readLock();
    private Lock writeLock = lock.writeLock();
```

```java
    public void read(){
        readLock.lock();
        try{
            System.out.println(Thread.currentThread().getName() + "--线程获取到读锁");
            Thread.sleep(200);
        }catch (InterruptedException e){
            e.printStackTrace();
        }finally {
            System.out.println(Thread.currentThread().getName() + "--线程释放读锁");
            readLock.unlock();
        }
    }
    public void write() {
        writeLock.lock();
        try{
            System.out.println(Thread.currentThread().getName() + "--线程获取到写锁");
            Thread.sleep(200);
        }catch (InterruptedException e){
            e.printStackTrace();
        }finally {
            System.out.println(Thread.currentThread().getName() + "--线程释放写锁");
            writeLock.unlock();
        }
    }
    public static void main(String[] args){
        ReadWriteLockTest readWriteLockTest = new ReadWriteLockTest();
        IntStream.range(0, 5).forEach((i) -> {
            new Thread(() -> {
                readWriteLockTest.read();
            }).start();
        });
        IntStream.range(0, 5).forEach((i) -> {
            new Thread(() -> {
                readWriteLockTest.write();
            }).start();
        });
    }
}
```

在 ReadWriteLockTest 类中，首先创建了一个 ReadWriteLock 类型的成员变量 lock，并使用 lock 创建了读锁 readLock 和写锁 writeLock。然后在 read()方法中使用 readLock 加锁和释放锁，并输出相关的日志。接下来，在 write()方法中使用 writeLock 加锁和释放锁，并输出相关的日志。为了使效果更加明显，在 read()方法和 write()方法中获取读锁和写锁成功后，让线程休眠 200 毫秒后再释放锁。

最后，在 main()方法中创建 ReadWriteLockTest 类的对象，分别在 5 次循环中创建线程调用 ReadWriteLockTest 的 read()方法和 write()方法。

3. 案例测试

运行 ReadWriteLockTest 类的代码，输出结果如下。

```
Thread-0--线程获取到读锁
Thread-2--线程获取到读锁
Thread-4--线程获取到读锁
Thread-1--线程获取到读锁
Thread-3--线程获取到读锁
Thread-2--线程释放读锁
Thread-0--线程释放读锁
Thread-1--线程释放读锁
Thread-4--线程释放读锁
Thread-3--线程释放读锁
Thread-5--线程获取到写锁
Thread-5--线程释放写锁
Thread-6--线程获取到写锁
Thread-6--线程释放写锁
Thread-7--线程获取到写锁
Thread-7--线程释放写锁
Thread-8--线程获取到写锁
Thread-8--线程释放写锁
Thread-9--线程获取到写锁
Thread-9--线程释放写锁
```

从输出结果可以看出，多个线程可以同时获取到读锁，同一时刻只能有一个线程获取到写锁，说明读锁是共享的，写锁是独占的。符合预期。

10.5　StampedLock 读写锁

StampedLock 是 Java 从 JDK 1.8 版本开始提供的一种读写锁，不允许重入。

10.5.1　概述及核心方法

StampedLock 比 ReadWriteLock 读写锁多了乐观读的功能，支持写锁、读锁和乐观读三种模式，其中写锁是独占锁，读锁是共享锁。StampedLock 只支持非公平锁的实现，支持多个线程同时获取读锁，同一时间只允许一个线程获取写锁。StampedLock 提供的乐观读在多个线程读取共享变量时，也会允许一个线程对共享变量进行写操作。

StampedLock 在获取读锁和写锁成功后都会返回一个 Long 型的返回值，在释放锁时需要传入这个返回值。在使用 StampedLock 时需要注意，StampedLock 不支持条件变量、不支持重入，使用不当会引起 CPU 占用 100%的问题。

值得一提的是，StampedLock 支持锁的升级与降级。

注意：这里只是对 StampedLock 进行了简单的介绍，有关 StampedLock 读写锁的核心原理、StampedLock 锁的升级与降级，读者可以参考《深入理解高并发编程：核心原理与案例实战》一书。

1. 获取写锁

在 StampedLock 中，对外提供的获取写锁的主要方法如下。

（1）long writeLock()方法。获取 StampedLock 的写锁，如果锁被另一个线程获取，则当前线程会被阻塞，直到获取写锁成功并返回票据。代码如下。

```
public long writeLock() {
    long s, next;
    return ((((s = state) & ABITS) == 0L &&
            U.compareAndSwapLong(this, STATE, s, next = s + WBIT)) ?
            next : acquireWrite(false, 0L));
}
```

（2）long tryWriteLock()方法。尝试以非公平的方式获取写锁，不会阻塞线程。如果获取锁成功则返回相应的票据；如果获取锁失败则返回 0。代码如下。

```
public long tryWriteLock() {
    long s, next;
    return ((((s = state) & ABITS) == 0L &&
            U.compareAndSwapLong(this, STATE, s, next = s + WBIT)) ?
            next : 0L);
}
```

（3）long tryWriteLock(long time, TimeUnit unit)方法。在指定的时间内尝试获取写锁，如果获取成功则立即返回相应的票据。如果获取写锁失败，或者超时，或者被其他线程中断则返回 0。代码如下。

```
public long tryWriteLock(long time, TimeUnit unit)
    throws InterruptedException {
    long nanos = unit.toNanos(time);
    if (!Thread.interrupted()) {
        long next, deadline;
        if ((next = tryWriteLock()) != 0L)
            return next;
        if (nanos <= 0L)
            return 0L;
        if ((deadline = System.nanoTime() + nanos) == 0L)
            deadline = 1L;
        if ((next = acquireWrite(true, deadline)) != INTERRUPTED)
            return next;
    }
    throw new InterruptedException();
}
```

（4）long writeLockInterruptibly()方法。获取 StampedLock 的写锁，如果锁被另一个线程获取，则当前线程被阻塞，直到获取写锁成功并返回票据，或者被其他线程中断返回。代码如下。

```
public long writeLockInterruptibly() throws InterruptedException {
    long next;
    if (!Thread.interrupted() &&
        (next = acquireWrite(true, 0L)) != INTERRUPTED)
        return next;
    throw new InterruptedException();
}
```

2. 释放写锁

在 StampedLock 中，对外提供的释放写锁的主要方法如下。

（1）void unlockWrite(long stamp)方法。如果传入的票据与锁的状态匹配则释放写锁。代码如下。

```
public void unlockWrite(long stamp) {
    WNode h;
    if (state != stamp || (stamp & WBIT) == 0L)
        throw new IllegalMonitorStateException();
    state = (stamp += WBIT) == 0L ? ORIGIN : stamp;
    if ((h = whead) != null && h.status != 0)
        release(h);
}
```

（2）boolean tryUnlockWrite()方法。如果当前线程占有写锁则释放写锁。与 void unlockWrite(long stamp)方法相比，boolean tryUnlockWrite()方法不需要传入票据信息，减少了检验票据是否正确的逻辑。代码如下。

```
public boolean tryUnlockWrite() {
    long s; WNode h;
    if (((s = state) & WBIT) != 0L) {
        state = (s += WBIT) == 0L ? ORIGIN : s;
        if ((h = whead) != null && h.status != 0)
            release(h);
        return true;
    }
    return false;
}
```

（3）void unlock(long stamp)方法。既可以释放写锁也可以释放读锁。代码如下。

```
public void unlock(long stamp) {
    long a = stamp & ABITS, m, s; WNode h;
    while (((s = state) & SBITS) == (stamp & SBITS)) {
        if ((m = s & ABITS) == 0L)
            break;
        else if (m == WBIT) {
            if (a != m)
```

```
            break;
        state = (s += WBIT) == 0L ? ORIGIN : s;
        if ((h = whead) != null && h.status != 0)
            release(h);
        return;
    }
    else if (a == 0L || a >= WBIT)
        break;
    else if (m < RFULL) {
        if (U.compareAndSwapLong(this, STATE, s, s - RUNIT)) {
            if (m == RUNIT && (h = whead) != null && h.status != 0)
                release(h);
            return;
        }
    }
    else if (tryDecReaderOverflow(s) != 0L)
        return;
}
throw new IllegalMonitorStateException();
}
```

3. 获取读锁

在 StampedLock 中，对外提供的获取读锁的主要方法如下。

（1）long readLock()方法。获取 StampedLock 读锁，如果锁被其他线程占有，则当前线程阻塞，直到获取到锁并返回相应的票据。代码如下。

```
public long readLock() {
    long s = state, next;
    return ((whead == wtail && (s & ABITS) < RFULL &&
            U.compareAndSwapLong(this, STATE, s, next = s + RUNIT)) ?
            next : acquireRead(false, 0L));
}
```

（2）long tryReadLock()方法。尝试以非公平的方式获取读锁，不会阻塞线程，如果获取锁成功则返回相应的票据；如果获取锁失败则返回 0。代码如下。

```
public long tryReadLock() {
    for (;;) {
        long s, m, next;
        if ((m = (s = state) & ABITS) == WBIT)
            return 0L;
        else if (m < RFULL) {
            if (U.compareAndSwapLong(this, STATE, s, next = s + RUNIT))
                return next;
        }
        else if ((next = tryIncReaderOverflow(s)) != 0L)
            return next;
    }
}
```

（3）long tryReadLock(long time, TimeUnit unit)方法。在给定的时间内获取读锁，如果获取锁成功，则立即返回相应的票据。如果获取锁失败，或者超时，或者被其他线程中断，则返回0。代码如下。

```
public long tryReadLock(long time, TimeUnit unit)
    throws InterruptedException {
    long s, m, next, deadline;
    long nanos = unit.toNanos(time);
    if (!Thread.interrupted()) {
        if ((m = (s = state) & ABITS) != WBIT) {
            if (m < RFULL) {
                if (U.compareAndSwapLong(this, STATE, s, next = s + RUNIT))
                    return next;
            }
            else if ((next = tryIncReaderOverflow(s)) != 0L)
                return next;
        }
        if (nanos <= 0L)
            return 0L;
        if ((deadline = System.nanoTime() + nanos) == 0L)
            deadline = 1L;
        if ((next = acquireRead(true, deadline)) != INTERRUPTED)
            return next;
    }
    throw new InterruptedException();
}
```

（4）long readLockInterruptibly()方法。获取 StampedLock 读锁，如果锁被其他线程占有，则当前线程阻塞，直到获取到锁并返回相应的票据，或者被其他线程中断才返回。代码如下。

```
public long readLockInterruptibly() throws InterruptedException {
    long next;
    if (!Thread.interrupted() &&
        (next = acquireRead(true, 0L)) != INTERRUPTED)
        return next;
    throw new InterruptedException();
}
```

4. 释放读锁

在 StampedLock 中，对外提供的释放读锁的主要方法如下。

（1）void unlockRead(long stamp)方法。如果传入的票据与锁的状态匹配则释放读锁。代码如下。

```
public void unlockRead(long stamp) {
    long s, m; WNode h;
    for (;;) {
        if (((s = state) & SBITS) != (stamp & SBITS) ||
            (stamp & ABITS) == 0L || (m = s & ABITS) == 0L || m == WBIT)
```

```
            throw new IllegalMonitorStateException();
        if (m < RFULL) {
            if (U.compareAndSwapLong(this, STATE, s, s - RUNIT)) {
                if (m == RUNIT && (h = whead) != null && h.status != 0)
                    release(h);
                break;
            }
        }
        else if (tryDecReaderOverflow(s) != 0L)
            break;
    }
}
```

（2）boolean tryUnlockRead()方法。如果当前线程占有读锁则尝试解锁。与 void unlockRead(long stamp)方法相对，boolean tryUnlockRead()方法不需要传入票据信息，减少了检验票据正确性的逻辑。代码如下。

```
public boolean tryUnlockRead() {
    long s, m; WNode h;
    while ((m = (s = state) & ABITS) != 0L && m < WBIT) {
        if (m < RFULL) {
            if (U.compareAndSwapLong(this, STATE, s, s - RUNIT)) {
                if (m == RUNIT && (h = whead) != null && h.status != 0)
                    release(h);
                return true;
            }
        }
        else if (tryDecReaderOverflow(s) != 0L)
            return true;
    }
    return false;
}
```

注意：void unlock(long stamp)方法也可以释放读锁。

5. 乐观读

在 StampedLock 中，对外提供的乐观读的主要方法如下。

（1）long tryOptimisticRead()方法。如果目前未持有写锁，则返回当前的锁版本作为票据，获取乐观锁成功。如果目前已经持有写锁，则返回 0，获取乐观锁失败。代码如下。

```
public long tryOptimisticRead() {
    long s;
    return (((s = state) & WBIT) == 0L) ? (s & SBITS) : 0L;
}
```

注意：在当前线程通过 long tryOptimisticRead()方法获取锁版本后，只要有其他线程获取或者释放了写锁，锁版本就会发生改变，导致当前线程通过 long tryOptimisticRead()方法获取的锁版本和真正的锁版本不匹配。

（2）boolean validate(long stamp)方法。如果在通过 long tryOptimisticRead()方法获取锁版本期间未获取或释放过写锁则返回 true，说明锁版本未发生改变；否则返回 false。代码如下。

```
public boolean validate(long stamp) {
   U.loadFence();
   return (stamp & SBITS) == (state & SBITS);
}
```

6. 锁升降级

在 StampedLock 中，对外提供的锁升降级的主要方法如下。

（1）long tryConvertToWriteLock(long stamp)方法。将指定票据的锁升级为写锁，具体的逻辑是：如果当前线程不持有锁，则获取写锁；如果当前线程持有写锁，则不进行任何操作；如果当前线程持有一个读锁，则释放读锁，获取写锁。代码如下。

```
public long tryConvertToWriteLock(long stamp) {
   long a = stamp & ABITS, m, s, next;
   while (((s = state) & SBITS) == (stamp & SBITS)) {
      if ((m = s & ABITS) == 0L) {
         if (a != 0L)
            break;
         if (U.compareAndSwapLong(this, STATE, s, next = s + WBIT))
            return next;
      }
      else if (m == WBIT) {
         if (a != m)
            break;
         return stamp;
      }
      else if (m == RUNIT && a != 0L) {
         if (U.compareAndSwapLong(this, STATE, s,
                          next = s - RUNIT + WBIT))
            return next;
      }
      else
         break;
   }
   return 0L;
}
```

（2）long tryConvertToReadLock(long stamp)方法。将指定票据的锁降级为读锁。具体的逻辑是：如果当前线程未持有锁，则获取读锁；如果当前线程持有读锁，则不进行任何操作；如果当前线程持有写锁，则释放写锁获取读锁。代码如下。

```java
public long tryConvertToReadLock(long stamp) {
    long a = stamp & ABITS, m, s, next; WNode h;
    while (((s = state) & SBITS) == (stamp & SBITS)) {
        if ((m = s & ABITS) == 0L) {
            if (a != 0L)
                break;
            else if (m < RFULL) {
                if (U.compareAndSwapLong(this, STATE, s, next = s + RUNIT))
                    return next;
            }
            else if ((next = tryIncReaderOverflow(s)) != 0L)
                return next;
        }
        else if (m == WBIT) {
            if (a != m)
                break;
            state = next = s + (WBIT + RUNIT);
            if ((h = whead) != null && h.status != 0)
                release(h);
            return next;
        }
        else if (a != 0L && a < WBIT)
            return stamp;
        else
            break;
    }
    return 0L;
}
```

（3）long tryConvertToOptimisticRead(long stamp)方法。将指定票据的锁降级为乐观读。主要的逻辑就是验证当前锁的版本、持有的锁状态与传入的票据是否匹配，如果匹配则释放一次锁并返回锁版本；如果不匹配或者指定票据的锁状态发生错误，则返回0。代码如下。

```java
public long tryConvertToOptimisticRead(long stamp) {
    long a = stamp & ABITS, m, s, next; WNode h;
    U.loadFence();
    for (;;) {
        if (((s = state) & SBITS) != (stamp & SBITS))
            break;
        if ((m = s & ABITS) == 0L) {
            if (a != 0L)
                break;
            return s;
        }
        else if (m == WBIT) {
            if (a != m)
                break;
            state = next = (s += WBIT) == 0L ? ORIGIN : s;
            if ((h = whead) != null && h.status != 0)
                release(h);
```

```
            return next;
        }
        else if (a == 0L || a >= WBIT)
            break;
        else if (m < RFULL) {
            if (U.compareAndSwapLong(this, STATE, s, next = s - RUNIT)) {
                if (m == RUNIT && (h = whead) != null && h.status != 0)
                    release(h);
                return next & SBITS;
            }
        }
        else if ((next = tryDecReaderOverflow(s)) != 0L)
            return next & SBITS;
    }
    return 0L;
}
```

10.5.2 StampedLock 使用案例

1. 案例需求

验证 StampedLock 在线程读数据的过程中也允许某个线程对数据进行写操作。

2. 案例实现

StampedLock 提供的乐观读在多个线程读取共享变量时，允许一个线程对共享变量进行写操作。案例程序的实现步骤如下。

（1）在 mykit-concurrent-chapter10 工程下的 io.binghe.concurrent.chapter10 包下新建 StampedLockTest 类，并在 StampedLockTest 类中定义两个成员变量，一个是 int 类型的成员变量 count，表示共享变量，一个是 StampedLock 类型的成员变量 stampedLock，表示 StampedLock 锁，代码如下。

```
public class StampedLockTest {
    //共享变量
    private int count = 0;
    //StampedLock 锁
    private StampedLock stampedLock = new StampedLock();
}
```

（2）在 StampedLockTest 类中，创建一个 write()方法，在 write()方法中获取 StampedLock 写锁，并对共享变量的值进行加 1 操作，代码如下。

```
//对共享变量进行写操作
public void write() {
    long stamp = stampedLock.writeLock();
    System.out.println(Thread.currentThread().getName()+"--写线程修改共享变量的值开始");
    try{
```

```
        count += 1;
    }finally {
        stampedLock.unlockWrite(stamp);
    }
    System.out.println(Thread.currentThread().getName()+"--写线程修改共享变量的值结束
");
}
```

在 write()方法中，首先获取 StampedLock 写锁，输出写线程修改共享变量的值开始的日志，然后对共享变量的值进行加 1 操作，最后释放 StampedLock 写锁，并输出写线程修改共享变量的值结束的日志。

（3）在 StampedLockTest 类中，创建一个 optimisticRead()方法，代码如下。

```
//StampedLock 提供的乐观读，在读的过程中也允许某个线程进行写操作
public void optimisticRead(){
    long stamp = stampedLock.tryOptimisticRead();
    int result = count;
    System.out.println(Thread.currentThread().getName()+ "--线程检测共享变量的值是否被
修改(true 无修改, false 有修改): " + stampedLock.validate(stamp));
    System.out.println(Thread.currentThread().getName()+ "--线程读取到的共享变量的值:
" + result);
    try {
        TimeUnit.SECONDS.sleep(2);
    } catch (InterruptedException e) {
        e.printStackTrace();
    }
    System.out.println(Thread.currentThread().getName()+ "--线程 2s 后检测共享变量的值
是否被修(true 无修改, false 有修改): " + stampedLock.validate(stamp));
    if(!stampedLock.validate(stamp)){
        System.out.println("其他线程已经修改了共享变量的值");
        //将乐观读升级为悲观读
        stamp = stampedLock.readLock();
        try{
            System.out.println(Thread.currentThread().getName() +
"--线程从乐观读升级为悲观读");
            result = count;
            System.out.println(Thread.currentThread().getName() +
"--线程从乐观读升级为悲观读后的共享变量的值: "+result);
        }finally {
            stampedLock.unlockRead(stamp);
        }
    }
    System.out.println(Thread.currentThread().getName()
+"--线程读取到的最终的共享变量的值: "+result);
}
```

optimisticRead()方法的主要逻辑是首先以乐观读的方式读取共享变量的值，然后输出共享变量是否被修改并共享变量的值。随后让线程休眠 2s，再次输出共享变量是否被修改的日志。

如果检测到共享变量已经被修改，则输出其他线程已经修改了共享变量的值，将乐观读升级为悲观读，输出日志并重新读取共享变量的值后，再次输出相关的日志。

（4）在 StampedLockTest 类中创建一个 main()方法，代码如下。

```
public static void main(String[] args) throws InterruptedException {
    StampedLockTest stampedLockTest = new StampedLockTest();
    //读线程
    new Thread(() -> {
        stampedLockTest.optimisticRead();
    },"Thread-Read").start();
    TimeUnit.SECONDS.sleep(1);
    System.out.println("1s 后启动写线程修改共享变量的值");
    //写线程
    new Thread(() -> {
        stampedLockTest.write();
    },"Thread-Write").start();
}
```

在 main()方法中，首先创建一个 StampedLockTest 对象，然后创建并启动读线程，在读线程的 run()方法中调用 optimisticRead()方法。接下来，让主线程休眠 1s 后创建并启动写线程。最后在写线程的 run()方法中调用 write()方法，对共享变量进行写操作。

3. 案例测试

运行 StampedLockTest 类的 main()方法，输出结果如下。

```
Thread-Read--线程检测共享变量的值是否被修改(true 无修改，false 有修改)：true
Thread-Read--线程读取到的共享变量的值：0
1s 后启动写线程修改共享变量的值
Thread-Write--写线程修改共享变量的值开始
Thread-Write--写线程修改共享变量的值结束
Thread-Read--线程 2s 后检测共享变量的值是否被修改(true 无修改，false 有修改)：false
其他线程已经修改了共享变量的值
Thread-Read--线程从乐观读升级为悲观读
Thread-Read--线程从乐观读升级为悲观读后的共享变量的值：1
Thread-Read--线程读取到的最终的共享变量的值：1
```

从输出结果可以看出，当以 StampedLock 乐观读的方式读取共享变量的值时，如果没有其他线程修改过共享变量的值，则调用 StampedLock 的 validate()方法会检测出没有其他线程修改过共享变量的值。否则，会检测出有其他线程修改过共享变量的值。如果在乐观读期间，有其他线程修改共享变量的值，则需要将乐观读升级为悲观读，再次读取共享变量的值。

整个案例程序验证了 StampedLock 在线程读数据的过程中也允许某个线程对数据进行写操作，符合预期。

10.6 锁性能对比案例

前面的章节主要介绍了 ReentrantLock 可重入锁、ReadWriteLock 读写锁和 StampedLock 读写锁相关的知识，本节将对这几种锁做性能基准测试。

10.6.1 案例需求

对比 ReentrantLock 可重入锁、ReadWriteLock 读写锁和 StampedLock 读写锁对数据进行读写操作的性能，并输出结果。

10.6.2 案例实现

这里使用 JMH 对 ReentrantLock 可重入锁、ReadWriteLock 读写锁和 StampedLock 读写锁进行基准性能测试，具体的实现步骤如下。

（1）在 mykit-concurrent-chapter10 工程下的 io.binghe.concurrent.chapter10 包下创建 LockPerformance 类，在 LockPerformance 类中分别创建 ReentrantLock 可重入锁对象、ReadWriteLock 读写锁对象和 StampedLock 锁对象，并创建相关锁对象对应的读写数据的方法。程序代码如下。

```java
@State(Scope.Group)
public class LockPerformance {
    private int count = 0;
    //ReentrantLock
    private final Lock lock = new ReentrantLock();
    //ReadWriteLock
    private final ReadWriteLock readWriteLock = new ReentrantReadWriteLock();
    //读锁
    private final Lock readLock = readWriteLock.readLock();
    //写锁
    private final Lock writeLock = readWriteLock.writeLock();
    //StampedLock
    private final StampedLock stampedLock = new StampedLock();
    //使用ReentrantLock写数据
    public void lockWrite(){
        lock.lock();
        try{
            count++;
        }finally {
            lock.unlock();
        }
    }
    //使用ReentrantLock读数据
    public int lockRead(){
        lock.lock();
        try {
```

```java
        return count;
    }finally {
        lock.unlock();
    }
}
//使用ReadWriteLock的写锁写数据
public void readWriteLockWrite(){
    writeLock.lock();
    try{
        count++;
    }finally {
        writeLock.unlock();
    }
}
//使用ReadWriteLock的读锁读数据
public int readWriteLockRead(){
    readLock.lock();
    try{
        return count;
    }finally {
        readLock.unlock();
    }
}
//使用StampedLock写数据
public void stampedLockWrite(){
    long stamped = stampedLock.writeLock();
    try {
        count++;
    }finally {
        stampedLock.unlockWrite(stamped);
    }
}
//使用StampedLock读数据
public int stampedLockRead(){
    long stamped = stampedLock.readLock();
    try {
        return count;
    }finally {
        stampedLock.unlockRead(stamped);
    }
}
//使用StampedLock乐观读数据
public int stampedLockOptimisticRead(){
    long stamped = stampedLock.tryOptimisticRead();
    if (!stampedLock.validate(stamped)){
        stamped = stampedLock.readLock();
        try {
            return count;
        }finally {
            stampedLock.unlockRead(stamped);
```

```
        }
    }
    return count;
  }
}
```

上述代码逻辑比较简单。

（2）在 io.binghe.concurrent.chapter10 包下创建 LockPerformanceTest 类，LockPerformanceTest 类的逻辑主要是调用 LockPerformance 类的方法并使用 JMH 进行基准性能测试，代码如下。

```
@Warmup(iterations = 20)
@Measurement(iterations = 20)
@Fork(1)
@BenchmarkMode(Mode.AverageTime)
@OutputTimeUnit(TimeUnit.MICROSECONDS)
@State(Scope.Group)
public class LockPerformanceTest {
    @Benchmark
    @Group("reentrantLock")
    @GroupThreads(value = 5)
    public void lockWrite(LockPerformance lockPerformance){
        lockPerformance.lockWrite();
    }
    @Benchmark
    @Group("reentrantLock")
    @GroupThreads(value = 5)
    public void lockRead(LockPerformance lockPerformance, Blackhole blackhole){
        blackhole.consume(lockPerformance.lockRead());
    }
    @Benchmark
    @Group("readWriteLock")
    @GroupThreads(value = 5)
    public void readWriteLockWrite(LockPerformance lockPerformance){
        lockPerformance.readWriteLockWrite();
    }
    @Benchmark
    @Group("readWriteLock")
    @GroupThreads(value = 5)
    public void readWriteLockRead(LockPerformance lockPerformance, Blackhole blackhole){
        blackhole.consume(lockPerformance.readWriteLockRead());
    }
    @Benchmark
    @Group("stampedLock")
    @GroupThreads(value = 5)
    public void stampedLockWrite(LockPerformance lockPerformance){
        lockPerformance.stampedLockWrite();
    }
    @Benchmark
```

```java
    @Group("stampedLock")
    @GroupThreads(value = 5)
    public void stampedLockRead(LockPerformance lockPerformance, Blackhole blackhole){
        blackhole.consume(lockPerformance.stampedLockRead());
    }
    @Benchmark
    @Group("stampedLockOptimistic")
    @GroupThreads(value = 5)
    public void stampedLockOptimisticWrite(LockPerformance lockPerformance){
        lockPerformance.stampedLockWrite();
    }
    @Benchmark
    @Group("stampedLockOptimistic")
    @GroupThreads(value = 5)
    public void stampedLockOptimisticRead(LockPerformance lockPerformance, Blackhole blackhole){
        blackhole.consume(lockPerformance.stampedLockOptimisticRead());
    }
    public static void main(String[] args) throws RunnerException {
        Options options = new OptionsBuilder().
                include(LockPerformanceTest.class.getSimpleName())
                .build();
        new Runner(options).run();
    }
}
```

LockPerformanceTest 的代码逻辑也比较简单。

10.6.3 案例测试

运行 LockPerformanceTest 类的 main()方法，输出结果如下。

```
Benchmark                                                    Mode  Cnt    Score     Error   Units
readWriteLock                                                avgt   20    0.870 ±   0.036   us/op
readWriteLock:readWriteLockRead                              avgt   20    1.382 ±   0.068   us/op
readWriteLock:readWriteLockWrite                             avgt   20    0.357 ±   0.007   us/op
reentrantLock                                                avgt   20    0.331 ±   0.002   us/op
reentrantLock:lockRead                                       avgt   20    0.359 ±   0.005   us/op
reentrantLock:lockWrite                                      avgt   20    0.303 ±   0.003   us/op
stampedLock                                                  avgt   20   53.250 ±   1.991   us/op
stampedLock:stampedLockRead                                  avgt   20  106.284 ±   3.974   us/op
stampedLock:stampedLockWrite                                 avgt   20    0.215 ±   0.008   us/op
stampedLockOptimistic                                        avgt   20    0.159 ±   0.007   us/op
stampedLockOptimistic:stampedLockOptimisticRead              avgt   20    0.097 ±   0.020   us/op
stampedLockOptimistic:stampedLockOptimisticWrite             avgt   20    0.220 ±   0.011   us/op
```

可以将上述结果按照读数据和写数据进行归类，如下所示。

（1）读数据的输出结果如下。

```
Benchmark                                              Mode  Cnt    Score    Error  Units
readWriteLock:readWriteLockRead                        avgt   20    1.382 ±  0.068  us/op
reentrantLock:lockRead                                 avgt   20    0.359 ±  0.005  us/op
stampedLock:stampedLockRead                            avgt   20  106.284 ±  3.974  us/op
stampedLockOptimistic:stampedLockOptimisticRead        avgt   20    0.097 ±  0.020  us/op
```

在读数据时，各锁的耗时从小到大依次为 StampedLock 的乐观读、ReentrantLock 锁、ReadWriteLock 读锁、StampedLock 悲观读锁，这也是各锁的性能从高到低的排序。

（2）写数据时输出结果如下。

```
Benchmark                                              Mode  Cnt    Score    Error  Units
readWriteLock:readWriteLockWrite                       avgt   20    0.357 ±  0.007  us/op
reentrantLock:lockWrite                                avgt   20    0.303 ±  0.003  us/op
stampedLock:stampedLockWrite                           avgt   20    0.215 ±  0.008  us/op
stampedLockOptimistic:stampedLockOptimisticWrite       avgt   20    0.220 ±  0.011  us/op
```

在写数据时，各锁的耗时从小到大依次为 StampedLock 写锁、StampedLock 乐观读下的写锁、ReentrantLock 写锁、ReadWriteLock 写锁，这也是各锁的性能从高到低的排序。

注意： 上述基准性能测试程序在不同的机器上运行时输出结果可能略有不同。在实际工作过程中使用 JMH 进行基准性能测试时需要不断调整和优化 JMH 基准性能测试参数，通过反复测试找到程序性能的临界点。

更多有关 JMH 基准性能测试的知识，读者可以关注"冰河技术"微信公众号阅读相关文章。

第 11 章

无锁原子类

Java 从 JDK 1.5 版本开始，基于 CAS 实现了一系列的无锁原子类，这些无锁原子类全部位于 JDK 下的 java.util.concurrent.atomic 包中。

11.1 无锁原子类概述及分类

Java 提供的无锁原子类比单纯使用锁来操作数据的性能更高，大部分无锁原子类底层使用 Unsafe 的 CAS 算法来操作数据，不会涉及线程的阻塞与唤醒操作。

在 Java 中，除了使用锁实现线程安全，还可以使用 CAS 算法。CAS 是一种非阻塞算法，使用计算机底层的原子机器指令确保数据在并发环境下的一致性。非阻塞算法在可伸缩性和活跃性上比阻塞算法更有优势，在竞争同一临界区的数据时可以使用多个线程，不会发生阻塞，也不会发生死锁。

本质上，锁是一种悲观的策略，它假设每次操作临界区的数据时都会发生线程安全问题。因此，在使用锁时，同一时刻只能有一个线程进入临界区操作数据。而 CAS 算法是一种乐观策略，线程每次获取数据时都认为其他线程不会修改数据，所以不会加锁。但是线程在更新数据时会判断当前数据是否被其他线程修改过，如果修改过则重新获取临界区的数据进行操作。

Java 从 JDK 1.5 版本开始提供了一系列线程安全的原子类，这些原子类底层基本上都基于 Unsafe 的 CAS 实现了线程安全，比单纯使用锁来保证线程安全的性能更高。从 JDK 1.8 版本开始，又新增了 Double 类型与 Long 类型的累加器原子类，性能得到了进一步提升。

注意：无锁原子类的方法本身是线程安全的，但是在高并发环境下存在竞态条件或多种方法之间的调用场景时，使用无锁原子类处理数据可能出现线程不安全的问题。有关竞态条件的问题读者可参考本书第 3 章的内容。

另外，大部分无锁原子类底层都是基于 Unsafe 类的 CAS 操作实现的，有关 CAS 的核心原理、ABA 问题与解决方案以及 Unsafe 类的核心方法详解，读者可以参考《深入理解高并发编程：核心原理与案例实战》一书。

根据无锁原子类操作的具体数据类型不同，可以将无锁原子类划分为操作基本类型的原子类、操作引用类型的原子类、操作字段类型的原子类、操作数组类型的原子类和累加器类型的原子类。

每种类型的原子类与其对应的 JDK 实现类如表 11-1 所示。

表 11-1 原子类与其对应的 JDK 实现类

具体类型	JDK 的实现类
操作基本类型的原子类	AtomicInteger、AtomicLong、AtomicBoolean
操作引用类型的原子类	AtomicReference、AtomicStampedReference、AtomicMarkableReference
操作字段类型的原子类	AtomicIntegerFieldUpdater、AtomicLongFieldUpdater、AtomicReferenceFieldUpdater
操作数组类型的原子类	AtomicIntegerArray、AtomicLongArray、AtomicReferenceArray
累加器类型的原子类	DoubleAccumulator、DoubleAdder、LongAccumulator、LongAdder

后续章节会对这些具体类型的原子类进行简单的介绍。

11.2 操作基本类型的原子类

Java 从 JDK1.5 版本开始，专门提供了操作基本数据类型的原子类，这些原子类全部位于 JDK 的 java.util.concurrent.atomic 包下。

11.2.1 概述

操作基本类型的原子类主要包括 AtomicInteger、AtomicLong 和 AtomicBoolean。AtomicInteger 是 Java 提供的专门以线程安全的方式更新 Integer 类型整数值的原子类，AtomicLong 是更新 Long 类型整数值的原子类，而 AtomicBoolean 是更新 Boolean 类型值的原子类。3 个类都是线程安全的，并且 AtomicInteger 类与 AtomicLong 类提供的方法基本一致，底层都是基于 Unsafe 类的 CAS 实现的。

AtomicBoolean 类提供的方法虽然比 AtomicInteger 类和 AtomicLong 类少，但是总体实现机制与它们一致，底层也是基于 Unsafe 类的 CAS 实现的。

注意：本节后续的内容以 AtomicInteger 类为例介绍操作基本类型原子类的核心方法，关于 AtomicLong 类与 AtomicBoolean 类的方法，读者可自行对比 AtomicInteger 类的方法进行分析。

11.2.2 AtomicInteger 类核心方法解析

AtomicInteger 是 JDK 提供的专门操作 Integer 类型整数值的原子类，继承自 Java 的 Number 类，其内部提供了非常丰富的以原子方式操作 int 类型数据的方法。

（1）AtomicInteger 类的成员变量。AtomicInteger 类提供了一个使用 volatile 修饰的 int 类型的成员变量 value，AtomicInteger 类提供的原子操作的方法本质上修改的都是这个 value 的值，代码如下。

```
private volatile int value;
```

在使用 volatile 关键字修改 value 时，一旦某个线程修改了 value 的值，其他线程立刻就能看到修改后的 value 值。

（2）Unsafe 类的引用。AtomicInteger 类底层都是基于 Unsafe 类的 CAS 实现对 value 值的原子更新操作的，AtomicInteger 类中获取了 Unsafe 类的对象，并通过 value 值在内存中的偏移量来直接操作内存中的 value 值。AtomicInteger 类获取 Unsafe 类的对象以及 value 值在内存中偏移量的代码如下。

```
private static final Unsafe unsafe = Unsafe.getUnsafe();
private static final long valueOffset;
static {
    try {
        valueOffset = unsafe.objectFieldOffset
            (AtomicInteger.class.getDeclaredField("value"));
    } catch (Exception ex) { throw new Error(ex); }
}
```

（3）构造方法。AtomicInteger 类提供了两种构造方法，一种是默认的无参构造方法，一种是传入一个 int 类型数据的构造方法，代码如下。

```
//构造方法①
public AtomicInteger() {
}
//构造方法②
public AtomicInteger(int initialValue) {
    value = initialValue;
}
```

其中，构造方法①是默认的无参构造方法，在调用此构造方法创建 AtomicInteger 类对象时，内部的 value 值被初始化为 0。构造方法②是传入一个 int 类型数据的构造方法，在调用此构造方法创建 AtomicInteger 类对象时，内部的 value 值被初始化为传入的 int 类型的值。

注意：在调用构造方法②创建 AtomicInteger 对象时，可以传入一个负整数值。

（4）int get()方法。直接返回 AtomicInteger 类中的 value 值，代码如下。

```
public final int get() {
    return value;
```

（5）void set(int newValue)方法。将 AtomicInteger 类中的 value 值直接设置为传入的 newValue 的值，代码如下。

```
public final void set(int newValue) {
    value = newValue;
}
```

（6）void lazySet(int newValue)方法。将 AtomicInteger 类中的 value 值直接设置为传入的 newValue 的值，代码如下。

```
public final void lazySet(int newValue) {
    unsafe.putOrderedInt(this, valueOffset, newValue);
}
```

（7）int getAndSet(int newValue)方法。将 AtomicInteger 类中的 value 值设置为传入的 newValue 的值，并返回原来的 value 值，代码如下。

```
public final int getAndSet(int newValue) {
    return unsafe.getAndSetInt(this, valueOffset, newValue);
}
```

（8）boolean compareAndSet(int expect, int update)方法。当 AtomicInteger 内存中的 value 值为 expect 时，将 value 值更新为 update，如果更新成功则返回 true，否则返回 false，代码如下。

```
public final boolean compareAndSet(int expect, int update) {
    return unsafe.compareAndSwapInt(this, valueOffset, expect, update);
}
```

（9）boolean weakCompareAndSet(int expect, int update)方法。本质上与 boolean compareAndSet(int expect, int update)方法的功能相同，代码如下。

```
public final boolean weakCompareAndSet(int expect, int update) {
    return unsafe.compareAndSwapInt(this, valueOffset, expect, update);
}
```

（10）int getAndIncrement()方法。将 value 的值原子性地加 1，并返回原来的 value 值，代码如下。

```
public final int getAndIncrement() {
    return unsafe.getAndAddInt(this, valueOffset, 1);
}
```

（11）int getAndDecrement()方法。将 value 的值原子性地减 1，并返回原来的 value 值，代码如下。

```
public final int getAndDecrement() {
    return unsafe.getAndAddInt(this, valueOffset, -1);
}
```

（12）int getAndAdd(int delta)方法。将 value 的值原子性地增加 delta，并返回原来的 value 值，代码如下。

```
public final int getAndAdd(int delta) {
    return unsafe.getAndAddInt(this, valueOffset, delta);
}
```

（13）int incrementAndGet()方法。将 value 的值自增 1，并返回自增后的值，代码如下。

```
public final int incrementAndGet() {
    return unsafe.getAndAddInt(this, valueOffset, 1) + 1;
}
```

（14）int decrementAndGet()方法。将 value 的值自减 1，并返回自减后的值，代码如下。

```
public final int decrementAndGet() {
    return unsafe.getAndAddInt(this, valueOffset, -1) - 1;
}
```

（15）int addAndGet(int delta)方法。将 value 的值增加 delta，并返回增加后的值，代码如下。

```
public final int addAndGet(int delta) {
    return unsafe.getAndAddInt(this, valueOffset, delta) + delta;
}
```

（16）int getAndUpdate(IntUnaryOperator updateFunction)方法。根据一定的规则更新 value 的值，并返回更新前的 value 值，代码如下。

```
public final int getAndUpdate(IntUnaryOperator updateFunction) {
    int prev, next;
    do {
        prev = get();
        next = updateFunction.applyAsInt(prev);
    } while (!compareAndSet(prev, next));
    return prev;
}
```

（17）int updateAndGet(IntUnaryOperator updateFunction)方法。根据一定的规则更新 value 的值，并返回更新后的 value 值，代码如下。

```
public final int updateAndGet(IntUnaryOperator updateFunction) {
    int prev, next;
    do {
        prev = get();
        next = updateFunction.applyAsInt(prev);
    } while (!compareAndSet(prev, next));
    return next;
}
```

（18）int getAndAccumulate(int x, IntBinaryOperator accumulatorFunction)方法。根据一定的规则更新 value 的值，并返回更新前的 value 值，代码如下。

```
public final int getAndAccumulate(int x, IntBinaryOperator accumulatorFunction) {
    int prev, next;
    do {
        prev = get();
        next = accumulatorFunction.applyAsInt(prev, x);
```

```
    } while (!compareAndSet(prev, next));
    return prev;
}
```

（19）int accumulateAndGet(int x, IntBinaryOperator accumulatorFunction)方法。根据一定的规则更新 value 的值，并返回更新后的 value 值，代码如下。

```
public final int accumulateAndGet(int x, IntBinaryOperator accumulatorFunction) {
    int prev, next;
    do {
        prev = get();
        next = accumulatorFunction.applyAsInt(prev, x);
    } while (!compareAndSet(prev, next));
    return next;
}
```

（20）String toString()方法。将当前 value 的值转换为 String 类型并返回，代码如下。

```
public String toString() {
    return Integer.toString(get());
}
```

（21）int intValue()方法。直接返回当前的 value 值，代码如下。

```
public int intValue() {
    return get();
}
```

（22）long longValue()方法。将 value 的值强转为 long 类型并返回，代码如下。

```
public long longValue() {
    return (long)get();
}
```

（23）float floatValue()方法。将 value 的值强转为 float 类型并返回，代码如下。

```
public float floatValue() {
    return (float)get();
}
```

（24）double doubleValue()方法。将 value 的值强转为 double 类型并返回，代码如下。

```
public double doubleValue() {
    return (double)get();
}
```

11.3 操作引用类型的原子类

基本类型的原子类只能操作 Java 的基本数据类型，并且只能更新一个基本数据类型的变量，如果要同时操作多个变量，或者更新一个对象的多个属性，就需要使用操作引用类型的原子类。

11.3.1 概述

引用类型的原子类能够原子性地更新某个对象的一个或多个属性，如果需要同时操作多个变量的原子性，那么可以将多个变量封装到一个对象中，使用引用类型的原子类原子性地更新整个对象。

操作引用类型的原子类是 Java 从 JDK 1.5 版本开始提供的原子类，主要包括 AtomicReference、AtomicStampedReference 和 AtomicMarkableReference 3 个类，这 3 个类都是带有泛型的原子类。其中 AtomicReference 是最基础的引用类型原子类，AtomicStampedReference 是带有印戳的引用原子类，而 AtomicMarkableReference 是带有修改标志的引用原子类。

在使用 AtomicReference 类更新对象引用时可能产生 ABA 问题，为了解决这个问题，JDK 提供了与 AtomicReference 功能基本一致的 AtomicStampedReference 类与 AtomicMarkableReference 类。

AtomicStampedReference 类比 AtomicReference 类多了 stamp 邮戳，在使用 AtomicStampedReference 类更新数据时，不只要对比内存中的数据与传入的期望数据是否相等，还要对比内存中的 stamp 邮戳与传入的期望 stamp 邮戳是否相等，只有相等时才能更新内存中的数据和 stamp 邮戳，从而解决 ABA 问题。

AtomicMarkableReference 类比 AtomicReference 类多了一个标志位，可以原子性地更新一个 boolean 类型的标志位与一个引用类型的对象，从而解决 ABA 问题。

注意：关于 ABA 问题产生的原因与解决方案，读者可以参考《深入理解高并发编程：核心原理与案例实战》一书。由于 AtomicStampedReference 类与 AtomicMarkableReference 类提供的方法与 AtomicReference 类基本一致，本节后续的内容将以 AtomicReference 类为例介绍引用类型原子类的核心方法，读者可自行根据 AtomicReference 类的方法对比分析 AtomicStampedReference 类与 AtomicMarkableReference 类的方法。

11.3.2 AtomicReference 类核心方法解析

AtomicReference 类是 JDK 提供的专门用于原子性更新引用类型对象的原子类，其内部提供了丰富的操作引用类型对象的方法。

（1）成员变量与 Unsafe 类的引用。AtomicReference 类的成员变量 value 和 Unsafe 类的引用与 AtomicInteger 类的成员变量 value 和 Unsafe 类的引用作用基本相同，只是 AtomicReference 类的成员变量是泛型类型，而 AtomicInteger 类的成员变量是 int 类型。

（2）构造方法。AtomicReference 类提供了两种构造方法，一种是默认的无参构造方法，一种是传入一个泛型类型参数的构造方法，代码如下。

```
//构造方法①，默认无参构造方法，将value的值初始化为null
public AtomicReference() {
}
//构造方法②，传入一个泛型类型参数的构造方法，value的初始值为传入的泛型参数
public AtomicReference(V initialValue) {
    value = initialValue;
}
```

在调用默认的无参构造方法创建AtomicReference对象时，默认的value值为null。在调用传入一个泛型类型参数的构造方法创建AtomicReference对象时，value的值为传入的泛型类型参数的值。

（3）V get()方法。直接返回value的值，代码如下。

```
public final V get() {
    return value;
}
```

（4）void set(V newValue)方法。将value的值直接更新为newValue的值，代码如下。

```
public final void set(V newValue) {
    value = newValue;
}
```

（5）void lazySet(V newValue)方法。将value的值更新为newValue的值，代码如下。

```
public final void lazySet(V newValue) {
    unsafe.putOrderedObject(this, valueOffset, newValue);
}
```

（6）boolean compareAndSet(V expect, V update)方法。如果AtomicReference内存中的value值是expect，则将其更新为update。更新成功则返回true，更新失败则返回false。代码如下。

```
public final boolean compareAndSet(V expect, V update) {
    return unsafe.compareAndSwapObject(this, valueOffset, expect, update);
}
```

（7）boolean weakCompareAndSet(V expect, V update)方法。与boolean compareAndSet(V expect, V update)方法的功能相同，代码如下。

```
public final boolean weakCompareAndSet(V expect, V update) {
    return unsafe.compareAndSwapObject(this, valueOffset, expect, update);
}
```

（8）V getAndSet(V newValue)。将value值设置为newValue的值，并返回原来的value值。代码如下。

```
@SuppressWarnings("unchecked")
public final V getAndSet(V newValue) {
    return (V)unsafe.getAndSetObject(this, valueOffset, newValue);
}
```

（9）V getAndUpdate(UnaryOperator<V> updateFunction)方法。根据一定的规则更新value

值,并返回更新前的 value 值。代码如下。

```
public final V getAndUpdate(UnaryOperator<V> updateFunction) {
    V prev, next;
    do {
        prev = get();
        next = updateFunction.apply(prev);
    } while (!compareAndSet(prev, next));
    return prev;
}
```

(10) V updateAndGet(UnaryOperator<V> updateFunction)方法。根据一定的规则更新 value 值,并返回更新后的 value 值。代码如下。

```
public final V updateAndGet(UnaryOperator<V> updateFunction) {
    V prev, next;
    do {
        prev = get();
        next = updateFunction.apply(prev);
    } while (!compareAndSet(prev, next));
    return next;
}
```

(11) V getAndAccumulate(V x, BinaryOperator<V> accumulatorFunction)方法。根据一定的规则更新 value 值,并返回更新前的 value 值。代码如下。

```
public final V getAndAccumulate(V x, BinaryOperator<V> accumulatorFunction) {
    V prev, next;
    do {
        prev = get();
        next = accumulatorFunction.apply(prev, x);
    } while (!compareAndSet(prev, next));
    return prev;
}
```

(12) V accumulateAndGet(V x, BinaryOperator<V> accumulatorFunction)方法。根据一定的规则更新 value 值,并返回更新后的 value 值。代码如下。

```
public final V accumulateAndGet(V x, BinaryOperator<V> accumulatorFunction) {
    V prev, next;
    do {
        prev = get();
        next = accumulatorFunction.apply(prev, x);
    } while (!compareAndSet(prev, next));
    return next;
}
```

(13) String toString()方法。将泛型类型的 value 转换成 String 类型,并返回。代码如下。

```
public String toString() {
    return String.valueOf(get());
}
```

11.4 操作字段类型的原子类

操作引用类型的原子类只能传入一个完整的引用对象进行更新，如果只是想更新某个类中的某个字段，则可以使用 JDK 提供的专门操作字段类型的原子类。

11.4.1 概述

基本类型的原子类能够以原子性的方式更新一个基本数据类型的变量值，引用类型的原子类能够以原子性的方式更新一个完整的引用类型对象。虽然使用引用类型的原子类也能够更新对象中的字段值，但是当类中的字段比较多，只需要更新其中的某一个字段时，使用引用类型的原子类会显得比较臃肿。

Java 从 JDK 1.5 版本开始提供了操作字段类型的原子类，专门用于以原子性的方式更新类中的某个字段的值。这些操作字段类型的原子类包括 AtomicIntegerFieldUpdater、AtomicLongFieldUpdater、AtomicReferenceFieldUpdater，这 3 个类都能实现更新操作的原子性，并且 3 个类都在 JDK 中被定义成抽象类，在使用它们对某个类中的字段进行原子更新时，需要先调用各自类的 newUpdater() 方法来指定要更新的类和字段名称，如果使用的是 AtomicReferenceFieldUpdater 类，还要指定字段的类型。

AtomicIntegerFieldUpdater 类专门用于以原子性的方式更新类中整型变量的值。AtomicLongFieldUpdater 类专门用于以原子性的方式更新类中长整型变量的值。AtomicReferenceFieldUpdater 类专门用于以原子性的方式更新类中引用类型变量的值。

在使用操作字段类型的原子类更新某个类中的字段时，需要注意以下几点。

（1）类的静态变量不能被原子性地更新。

（2）父类的成员变量不能被原子性地更新。

（3）如果类的成员变量不能被直接访问，则不能被原子性地更新。

（4）如果类的成员变量被 final 关键字修饰，则不能被原子性地更新。

（5）如果类的成员变量没有使用 volatile 关键字修饰，则不能被原子性地更新。

注意：操作字段类型的原子类中的 3 个类提供的方法基本相同，本节后续的内容将以 AtomicReferenceFieldUpdater 类为例介绍操作字段类型原子类的核心方法，读者可自行根据 AtomicReferenceFieldUpdater 类的方法对比分析 AtomicIntegerFieldUpdater 类与 AtomicLongFieldUpdater 类的方法。

11.4.2 AtomicReferenceFieldUpdater 类核心方法解析

AtomicReferenceFieldUpdater 类是 JDK 提供的专门用于以原子性的方式更新类中引用类型

变量值的原子类，其内部提供了丰富的操作某个类中引用类型字段的方法。除了实现了几个完整的操作类中引用类型的字段，AtomicReferenceFieldUpdater 类中还定义了一些抽象方法，这些抽象方法由 AtomicReferenceFieldUpdater 类的内部类 AtomicReferenceFieldUpdaterImpl 实现。

（1）AtomicReferenceFieldUpdater<U,W> newUpdater(Class<U> tclass, Class<W> vclass, String fieldName)方法。位于 AtomicReferenceFieldUpdater 类中，是一个静态方法，由于 AtomicReferenceFieldUpdater 类是一个抽象类，因此在使用 AtomicReferenceFieldUpdater 类更新某个类的引用类型字段时，需要先调用此方法创建 AtomicReferenceFieldUpdater 类的对象，实际上就是创建 AtomicReferenceFieldUpdater 类的子类 AtomicReferenceFieldUpdaterImpl 类的对象，代码如下。

```
public static <U,W> AtomicReferenceFieldUpdater<U,W> newUpdater(Class<U> tclass,
Class<W> vclass, String fieldName) {
    return new AtomicReferenceFieldUpdaterImpl<U,W>
        (tclass, vclass, fieldName, Reflection.getCallerClass());
}
```

其中，方法各参数的含义如下。

- tclass：要更新的字段所在类的 Class 对象。
- vclass：要更新的字段的 Class 对象。
- fieldName：要更新的字段的名称。

（2）构造方法。位于 AtomicReferenceFieldUpdater 类中，被 protected 修饰，只能在子类中被调用，代码如下。

```
protected AtomicReferenceFieldUpdater() {
}
```

（3）V getAndSet(T obj, V newValue)方法。位于 AtomicReferenceFieldUpdater 类中，将 obj 对象中名称为 fieldName 值的字段值更新为 newValue 的值，并返回更新前的字段值。代码如下。

```
public V getAndSet(T obj, V newValue) {
    V prev;
    do {
        prev = get(obj);
    } while (!compareAndSet(obj, prev, newValue));
    return prev;
}
```

（4）V getAndUpdate(T obj, UnaryOperator<V> updateFunction)方法。位于 AtomicReferenceFieldUpdater 类中，按照某种规则更新 obj 对象中名称为 fieldName 值的字段值的值，并返回更新前的字段值。代码如下。

```
public final V getAndUpdate(T obj, UnaryOperator<V> updateFunction) {
    V prev, next;
    do {
```

```
      prev = get(obj);
      next = updateFunction.apply(prev);
   } while (!compareAndSet(obj, prev, next));
   return prev;
}
```

（5）V updateAndGet(T obj, UnaryOperator<V> updateFunction)方法。位于 AtomicReferenceFieldUpdater 类中，按照某种规则更新 obj 对象中名称为 fieldName 值的字段值，并返回更新后的字段值。代码如下。

```
public final V updateAndGet(T obj, UnaryOperator<V> updateFunction) {
   V prev, next;
   do {
      prev = get(obj);
      next = updateFunction.apply(prev);
   } while (!compareAndSet(obj, prev, next));
   return next;
}
```

（6）V getAndAccumulate(T obj, V x, BinaryOperator<V> accumulatorFunction)方法。位于 AtomicReferenceFieldUpdater 类中，按照某种规则更新 obj 对象中名称为 fieldName 值的字段值，并返回更新前的字段值。代码如下。

```
public final V getAndAccumulate(T obj, V x, BinaryOperator<V> accumulatorFunction)
{
   V prev, next;
   do {
      prev = get(obj);
      next = accumulatorFunction.apply(prev, x);
   } while (!compareAndSet(obj, prev, next));
   return prev;
}
```

（7）V accumulateAndGet(T obj, V x, BinaryOperator<V> accumulatorFunction)方法。位于 AtomicReferenceFieldUpdater 类中，按照某种规则更新 obj 对象中名称为 fieldName 值的字段值，并返回更新后的字段值。代码如下。

```
public final V accumulateAndGet(T obj, V x, BinaryOperator<V> accumulatorFunction)
{
   V prev, next;
   do {
      prev = get(obj);
      next = accumulatorFunction.apply(prev, x);
   } while (!compareAndSet(obj, prev, next));
   return next;
}
```

（8）boolean compareAndSet(T obj, V expect, V update)方法。位于 AtomicReferenceFieldUpdaterImpl 类中，如果 obj 对象中名称为 fieldName 值的字段值与 expect

相等，则将字段值更新为update。更新成功则返回true，更新失败则返回false。代码如下。

```
public final boolean compareAndSet(T obj, V expect, V update) {
    accessCheck(obj);
    valueCheck(update);
    return U.compareAndSwapObject(obj, offset, expect, update);
}
```

（9）boolean weakCompareAndSet(T obj, V expect, V update)方法。位于AtomicReferenceFieldUpdaterImpl类中，功能与boolean compareAndSet(T obj, V expect, V update)方法相同，代码如下。

```
public final boolean weakCompareAndSet(T obj, V expect, V update) {
    accessCheck(obj);
    valueCheck(update);
    return U.compareAndSwapObject(obj, offset, expect, update);
}
```

（10）void set(T obj, V newValue)方法。位于AtomicReferenceFieldUpdaterImpl类中，将obj对象中名称为fieldName值的字段值更新为newValue。代码如下。

```
public final void set(T obj, V newValue) {
    accessCheck(obj);
    valueCheck(newValue);
    U.putObjectVolatile(obj, offset, newValue);
}
```

（11）void lazySet(T obj, V newValue)方法。位于AtomicReferenceFieldUpdaterImpl类中，将obj对象中名称为fieldName值的字段值更新为newValue。代码如下。

```
public final void lazySet(T obj, V newValue) {
    accessCheck(obj);
    valueCheck(newValue);
    U.putOrderedObject(obj, offset, newValue);
}
```

（12）V get(T obj)方法。位于AtomicReferenceFieldUpdaterImpl类中，获取obj对象中名称为fieldName值的字段值。代码如下。

```
public final V get(T obj) {
    accessCheck(obj);
    return (V)U.getObjectVolatile(obj, offset);
}
```

（13）V getAndSet(T obj, V newValue)。位于AtomicReferenceFieldUpdaterImpl类中，将obj对象中名称为fieldName值的字段值更新为newValue，并返回更新前的字段值。代码如下。

```
public final V getAndSet(T obj, V newValue) {
    accessCheck(obj);
    valueCheck(newValue);
    return (V)U.getAndSetObject(obj, offset, newValue);
}
```

> **注意**：在 AtomicReferenceFieldUpdater 类中，如下方法会被定义成抽象方法，最终会在子类 AtomicReferenceFieldUpdaterImpl 中实现。
> - boolean compareAndSet(T obj, V expect, V update)。
> - boolean weakCompareAndSet(T obj, V expect, V update)。
> - void set(T obj, V newValue)。
> - void lazySet(T obj, V newValue)。
> - V get(T obj)。

11.5 操作数组类型的原子类

除了操作基本类型的原子类、操作引用类型的原子类和操作字段类型的原子类，Java 从 JDK 1.5 版本开始还提供了操作数组类型的原子类。

11.5.1 概述

操作基本类型的原子类、操作引用类型的原子类和操作字段类型的原子类，本质上更新的都是一个基本类型的变量、引用类型的变量和某个类中的某个字段。如果需要原子性地更新一个数组中的元素，可以将数组中的元素封装成操作基本类型的原子类、操作引用类型的原子类，这样在并发程序中就能够以原子性的方式更新数组中的每个元素。

其实，在实际开发过程中，没必要将操作基本类型的原子类和操作引用类型的原子类当作数组中的每个元素的类型。Java 从 JDK 1.5 版本开始，专门提供了操作数组的原子类。

操作数组类型的原子类主要包括 AtomicIntegerArray、AtomicLongArray 和 AtomicReferenceArray，这 3 个类内部都是基于 Unsafe 的 CAS 实现的。其中，AtomicIntegerArray 类专门以原子方式更新整型数组中的元素，AtomicLongArray 类专门以原子方式更新长整型数组中的元素，而 AtomicReferenceArray 类专门以原子方式更新引用类型数组中的元素。

> **注意**：操作数组类型的原子类中的 3 个类提供的方法基本相同，本节后续的内容将以 AtomicLongArray 类为例介绍操作数组类型原子类的核心方法，读者可自行根据 AtomicLongArray 类的方法对比分析 AtomicIntegerArray 类与 AtomicReferenceArray 类的方法。

11.5.2 AtomicLongArray 类核心方法解析

AtomicLongArray 类是 JDK 提供的专门用于原子性更新长整型数组中元素的原子类，其内部提供了丰富的操作长整型数组中元素的方法。

（1）构造方法。AtomicLongArray 类中包含两种构造方法，一种是传入数组长度的构造方法，一种是传入长整型数组的构造方法，代码如下。

```
//构造方法①，传入数组的长度，初始化一个长度为length的array数组
//并且数组中的每个元素都是0
public AtomicLongArray(int length) {
    array = new long[length];
}
//构造方法②，传入一个数组，内部array数组的长度与传入的array数组相等
//并且内部array数组的元素都从传入的array数组中复制而来
public AtomicLongArray(long[] array) {
    // Visibility guaranteed by final field guarantees
    this.array = array.clone();
}
```

（2）int length()方法。直接返回内部array数组的长度，代码如下。

```
public final int length() {
    return array.length;
}
```

（3）long get(int i)方法。获取array数组索引下标为i处的值，代码如下。

```
public final long get(int i) {
    return getRaw(checkedByteOffset(i));
}
```

（4）void set(int i, long newValue)方法。将array数组索引下标为i处的值设置为newValue，代码如下。

```
public final void set(int i, long newValue) {
    unsafe.putLongVolatile(array, checkedByteOffset(i), newValue);
}
```

（5）void lazySet(int i, long newValue)方法。功能与void set(int i, long newValue)方法基本相同，代码如下。

```
public final void lazySet(int i, long newValue) {
    unsafe.putOrderedLong(array, checkedByteOffset(i), newValue);
}
```

（6）long getAndSet(int i, long newValue)方法。将array数组索引下标为i处的值设置为newValue，并且返回索引下标为i处更新前的值。代码如下。

```
public final long getAndSet(int i, long newValue) {
    return unsafe.getAndSetLong(array, checkedByteOffset(i), newValue);
}
```

（7）boolean compareAndSet(int i, long expect, long update)方法。如果array数组索引下标为i处的值等于expect，则将其更新为update，更新成功则返回true，更新失败则返回false。代码如下。

```
public final boolean compareAndSet(int i, long expect, long update) {
    return compareAndSetRaw(checkedByteOffset(i), expect, update);
```

（8）boolean weakCompareAndSet(int i, long expect, long update)方法。功能与 boolean compareAndSet(int i, long expect, long update)方法基本相同，代码如下。

```
public final boolean weakCompareAndSet(int i, long expect, long update) {
    return compareAndSet(i, expect, update);
}
```

（9）long getAndIncrement(int i)方法。将 array 数组索引下标为 i 处的值自增 1，并返回自增前的值。代码如下。

```
public final long getAndIncrement(int i) {
    return getAndAdd(i, 1);
}
```

（10）long getAndDecrement(int i)方法。将 array 数组索引下标为 i 处的值自减 1，并返回自减前的值。代码如下。

```
public final long getAndDecrement(int i) {
    return getAndAdd(i, -1);
}
```

（11）long getAndAdd(int i, long delta)方法。将 array 数组索引下标为 i 处的值增加 delta，并返回增加前的值。代码如下。

```
public final long getAndAdd(int i, long delta) {
    return unsafe.getAndAddLong(array, checkedByteOffset(i), delta);
}
```

（12）long incrementAndGet(int i)方法。将 array 数组索引下标为 i 处的值自增 1，并返回自增后的值。代码如下。

```
public final long incrementAndGet(int i) {
    return getAndAdd(i, 1) + 1;
}
```

（13）long decrementAndGet(int i)方法。将 array 数组索引下标为 i 处的值自减 1，并返回自减后的值。代码如下。

```
public final long decrementAndGet(int i) {
    return getAndAdd(i, -1) - 1;
}
```

（14）long addAndGet(int i, long delta)方法。将 array 数组索引下标为 i 处的值增加 delta，并返回增加后的值。代码如下。

```
public long addAndGet(int i, long delta) {
    return getAndAdd(i, delta) + delta;
}
```

（15）long getAndUpdate(int i, LongUnaryOperator updateFunction)。根据某种规则更新 array 数组索引下标为 i 处的值，并返回更新前的值。代码如下。

```
public final long getAndUpdate(int i, LongUnaryOperator updateFunction) {
    long offset = checkedByteOffset(i);
    long prev, next;
    do {
        prev = getRaw(offset);
        next = updateFunction.applyAsLong(prev);
    } while (!compareAndSetRaw(offset, prev, next));
    return prev;
}
```

（16）long updateAndGet(int i, LongUnaryOperator updateFunction)方法。根据某种规则更新 array 数组索引下标为 i 处的值，并返回更新后的值。代码如下。

```
public final long updateAndGet(int i, LongUnaryOperator updateFunction) {
    long offset = checkedByteOffset(i);
    long prev, next;
    do {
        prev = getRaw(offset);
        next = updateFunction.applyAsLong(prev);
    } while (!compareAndSetRaw(offset, prev, next));
    return next;
}
```

（17）long getAndAccumulate(int i, long x, LongBinaryOperator accumulatorFunction)方法。根据某种规则更新 array 数组索引下标为 i 处的值，并返回更新前的值。代码如下。

```
public final long getAndAccumulate(int i, long x, LongBinaryOperator accumulatorFunction) {
    long offset = checkedByteOffset(i);
    long prev, next;
    do {
        prev = getRaw(offset);
        next = accumulatorFunction.applyAsLong(prev, x);
    } while (!compareAndSetRaw(offset, prev, next));
    return prev;
}
```

（18）long accumulateAndGet(int i, long x, LongBinaryOperator accumulatorFunction)方法。根据某种规则更新 array 数组索引下标为 i 处的值，并返回更新后的值。代码如下。

```
public final long accumulateAndGet(int i, long x, LongBinaryOperator accumulatorFunction) {
    long offset = checkedByteOffset(i);
    long prev, next;
    do {
        prev = getRaw(offset);
        next = accumulatorFunction.applyAsLong(prev, x);
    } while (!compareAndSetRaw(offset, prev, next));
    return next;
}
```

（19） String toString()方法。将内部数组 array 转换成 Spring 类型并返回，代码如下。

```
public String toString() {
   int iMax = array.length - 1;
   if (iMax == -1)
      return "[]";

   StringBuilder b = new StringBuilder();
   b.append('[');
   for (int i = 0; ; i++) {
      b.append(getRaw(byteOffset(i)));
      if (i == iMax)
         return b.append(']').toString();
      b.append(',').append(' ');
   }
}
```

11.6 累加器类型的原子类

Java 从 JDK 1.5 版本开始提供了操作基本数据类型的原子类，但是这些原子类在极端情况下还是会产生性能问题。为了进一步提升操作基本类型原子类的性能，Java 从 JDK 1.8 版本开始，提供了性能更高的累加器类型的原子类。

11.6.1 概述

Java 从 JDK 1.5 版本开始提供的 AtomicInteger 与 AtomicLong 类通过 Unsafe 的 CAS 实现了非阻塞的原子性操作。但是，在高并发环境下使用 AtomicInteger 或 AtomicLong 类更新某个变量时，会存在大量的线程同时竞争更新这个变量的现象，而同一时间只能有一个线程通过 CAS 更新变量成功，此时就会有大量的线程由于通过 CAS 更新变量失败而不断尝试 CAS 操作，极大地浪费了 CPU 的资源。整个过程如图 11-1 所示。

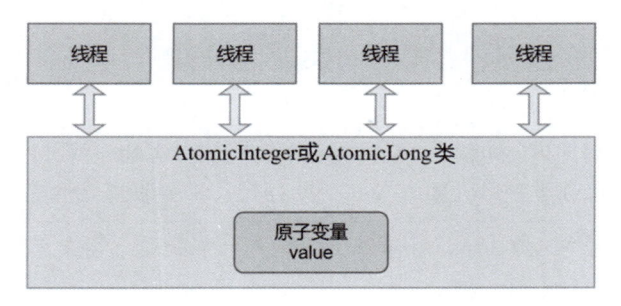

图 11-1　大量线程竞争同一个原子变量

为了进一步提升性能，Java 从 JDK 1.8 版本开始专门提供了累加器类型的原子类，包括 DoubleAccumulator、DoubleAdder、LongAccumulator 和 LongAdder。这些累加器类型的原子类可以将一个变量分解成多个变量，从而让多个线程竞争同一资源变成多个线程竞争多个资源，进一步提升性能，如图 11-2 所示。

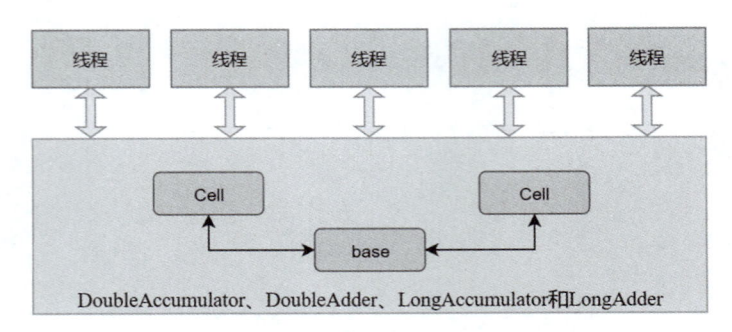

图 11-2　多个线程竞争累加器类型的原子类

使用累加器类型的原子类更新变量时，会维护一个 Cell 类型的数组，数组中维护着多个 Cell 类型的元素，每个 Cell 类型的元素内部会维护一个 double 类型或者 long 类型的变量，这些变量的初始值都为 0。线程在竞争资源时，会竞争数组中的多个 Cell 类型的元素。在获取累加器类型原子类中的值时，会将 Cell 数组中所有元素的 value 值进行累加，再加上 base 变量的值后返回结果。

如果一个线程竞争某个 Cell 类型的元素失败了，那么这个线程不会在原来的 Cell 变量上一直循环尝试 CAS 操作，而是在其他 Cell 变量上尝试 CAS 操作，这种设计增加了 CAS 操作成功的可能性。

相比 JDK 1.5 提供的操作基本类型的原子类，累加器类型的原子类具有更高的性能。

值得一提的是，累加器类型的原子类在实现 Cell 类时使用了非常巧妙的设计，具体体现在如下几个方面。

（1）如果 Cell 类型的数组为空且并发线程较少，则直接在 base 变量上进行累加操作。

（2）Cell 数组的长度为 2 的 N 次方，Cell 类在实现上是一个改进的 AtomicLong，这种设计基本解决了伪共享的问题，也就是缓存的争用问题。

（3）在 Cell 类上添加了 @sun.misc.Contended 注解，对 Cell 类进行字节填充，这样能够防止 Cell 数组中的多个 Cell 类型的元素共享一个缓存行，进一步提升性能。

注意：有关缓存争用、伪共享、字节填充、缓存行等知识，读者可参考《深入理解高并发编程：核心原理与案例实战》一书。

另外，在累加器类型的原子类提供的 4 个类中，XxxAccumulator 与 XxxAdder 提供的功能类似，但是 XxxAccumulator 比 XxxAdder 的功能更加强大，主要体现在如下两个方面。

（1）XxxAccumulator 累加的初始值可以自定义。

（2）XxxAccumulator 的运算规则可以自定义。

本章后续的内容将以 LongAdder 类为例，介绍累加器类型原子类的核心方法。

11.6.2 LongAdder 类核心方法解析

LongAdder 类是 JDK 提供的以原子方式累加 long 型变量的原子类，其内部提供了丰富的操作 long 型变量的方法。接下来，就对 LongAdder 类的方法进行简单的介绍。

（1）构造方法。在 LongAdder 类中，只提供了一个默认的无参构造方法，在创建 LongAdder 类的对象时，初始化的总和值为 0。代码如下。

```java
public LongAdder() {
}
```

（2）void add(long x)方法。原子性地增加 x 的值。代码如下。

```java
public void add(long x) {
    Cell[] as; long b, v; int m; Cell a;
    if ((as = cells) != null || !casBase(b = base, b + x)) {
        boolean uncontended = true;
        if (as == null || (m = as.length - 1) < 0 ||
            (a = as[getProbe() & m]) == null ||
            !(uncontended = a.cas(v = a.value, v + x)))
            longAccumulate(x, null, uncontended);
    }
}
```

（3）void increment()方法。调用 void add(long x)方法实现原子性地增加 1。代码如下。

```java
public void increment() {
    add(1L);
}
```

（4）void decrement()方法。调用 void add(long x)方法实现原子性地减少 1。代码如下。

```java
public void decrement() {
    add(-1L);
}
```

（5）long sum()方法。累加 base 变量与 Cell 数组中所有不为空的元素的 value 值，并返回累加结果。代码如下。

```java
public long sum() {
    Cell[] as = cells; Cell a;
    long sum = base;
    if (as != null) {
        for (int i = 0; i < as.length; ++i) {
            if ((a = as[i]) != null)
                sum += a.value;
        }
    }
}
```

```
        return sum;
}
```

（6）void reset()方法。将 base 变量的值与 Cell 数组中每个元素的 value 值都重置为 0。代码如下。

```
public void reset() {
    Cell[] as = cells; Cell a;
    base = 0L;
    if (as != null) {
        for (int i = 0; i < as.length; ++i) {
            if ((a = as[i]) != null)
                a.value = 0L;
        }
    }
}
```

（7）long sumThenReset()方法。累加 base 变量与 Cell 数组中所有不为空的元素的 value 值，累加后将 base 变量的值与 Cell 数组中每个元素的 value 值都重置为 0，并返回最终的累加结果。代码如下。

```
public long sumThenReset() {
    Cell[] as = cells; Cell a;
    long sum = base;
    base = 0L;
    if (as != null) {
        for (int i = 0; i < as.length; ++i) {
            if ((a = as[i]) != null) {
                sum += a.value;
                a.value = 0L;
            }
        }
    }
    return sum;
}
```

（8）String toString()方法。将 long 类型的累加结果值转换成 String 类型并返回。代码如下。

```
public String toString() {
    return Long.toString(sum());
}
```

（9）long longValue()方法。直接返回 long 类型的累加结果。代码如下。

```
public long longValue() {
    return sum();
}
```

（10）int intValue()方法。将 long 类型的累加结果值转换成 int 类型并返回。代码如下。

```
public int intValue() {
    return (int)sum();
}
```

（11）float floatValue()方法。将 long 类型的累加结果值转换成 float 类型并返回。代码如下。

```
public float floatValue() {
    return (float)sum();
}
```

（12）double doubleValue()方法。将 long 类型的累加结果值转换成 double 类型并返回。代码如下。

```
public double doubleValue() {
    return (double)sum();
}
```

11.7 性能对比案例

synchronized 锁、ReentrantLock 锁和原子类在大部分场景下能够确保操作的原子性，并保证线程的安全性。

11.7.1 锁与基本类型原子类性能对比案例

本节以 synchronized 锁、ReentrantLock 锁和 AtomicInteger 原子类为例，简单对比三者在更新数据时的性能。

1. 案例需求

综合对比 synchronized 锁、ReentrantLock 锁和 AtomicInteger 原子类更新 int 类型数据的性能，并输出结果。

2. 案例实现

synchronized 锁、ReentrantLock 锁和 AtomicInteger 原子类在更新 int 类型数据的时候，都能够保证原子性与线程安全，synchronized 锁和 ReentrantLock 锁是 JDK 提供的可重入锁，而 AtomicInteger 是 JDK 提供的专门用于更新 int 类型数据的原子类。这里，还是使用 JMH 对 synchronized 锁、ReentrantLock 锁和 AtomicInteger 原子类更新 int 类型数据进行基准性能测试。案例程序实现的具体步骤如下。

（1）在 mykit-concurrent-chapter11 的 io.binghe.concurrent.chapter1 包下新建 IntPerformance 类，在 IntPerformance 类中分别实现 synchronized 锁、ReentrantLock 锁和 AtomicInteger 原子类在更新 int 类型数据时的原子性，代码如下。

```
@State(Scope.Group)
public class IntPerformance {
    //参与测试的 int 类型的值
    private int count;
    //ReentrantLock 锁
    private final Lock lock = new ReentrantLock();
```

```
    //AtomicInteger 原子类
    private AtomicInteger atomicInteger = new AtomicInteger();
    //使用 ReentrantLock 加锁
    public void lockCount(){
        lock.lock();
        try{
            count ++;
        }finally {
            lock.unlock();
        }
    }
    //使用 synchronized 加锁
    public void syncCount(){
        synchronized (this){
            count ++;
        }
    }
    //使用 AtomicInteger 原子类
    public void atomicCount(){
        atomicInteger.incrementAndGet();
    }
}
```

IntPerformance 类实现的逻辑比较简单。

（2）在 io.binghe.concurrent.chapter11 包下新建 IntPerformanceTest 类，主要调用 IntPerformance 类中的方法交由 JMH 进行基准测试，代码如下。

```
@Warmup(iterations = 20)
@Measurement(iterations = 20)
@Fork(1)
@BenchmarkMode(Mode.AverageTime)
@OutputTimeUnit(TimeUnit.MICROSECONDS)
@State(Scope.Group)
public class IntPerformanceTest {
    @Benchmark
    @Group("reentrantLock")
    @GroupThreads(10)
    public void lockCount(IntPerformance performance){
        performance.lockCount();
    }
    @Benchmark
    @Group("synchronized")
    @GroupThreads(10)
    public void syncCount(IntPerformance performance){
        performance.syncCount();
    }
    @Benchmark
    @Group("atomicInteger")
    @GroupThreads(10)
```

```
    public void atomicCount(IntPerformance performance){
        performance.atomicCount();
    }
    public static void main(String[] args) throws RunnerException {
        Options options = new OptionsBuilder()
                        .include(IntPerformanceTest.class.getSimpleName())
                        .build();
        new Runner(options).run();
    }
}
```

3. 案例测试

运行 IntPerformanceTest 类的 main()方法，输出结果如下。

```
Benchmark                              Mode  Cnt  Score   Error    Units
IntPerformanceTest.atomicInteger       avgt   20  0.142 ± 0.004   us/op
IntPerformanceTest.reentrantLock       avgt   20  0.238 ± 0.009   us/op
IntPerformanceTest.synchronized        avgt   20  0.843 ± 0.035   us/op
```

从输出结果可以看出，性能最好的是 AtomicInteger 原子类，平均耗时 0.142μs，误差 ±0.004μs。其次为 ReentrantLock 锁，平均耗时 0.238μs，误差 ±0.009μs。最后是 synchronized 锁，平均耗时 0.843μs，误差 ±0.035μs。

所以，在 synchronized 锁、ReentrantLock 锁和 AtomicInteger 原子类三者的性能对比中，AtomicInteger 原子类性能最好，其次是 ReentrantLock 锁，而 synchronized 锁是三者中性能最差的。

注意：上述基准测试程序在不同的设备上运行，可能得出的最终数据略有差异。

11.7.2 锁与引用类型原子类性能对比案例

本节以 synchronized 锁、ReentrantLock 锁和 AtomicReference 原子类为例，对比三者在更新数据时的性能。

1. 案例需求

综合对比 synchronized 锁、ReentrantLock 锁和 AtomicReference 原子类更新对象类型数据的性能，并输出结果。

2. 案例实现

使用 synchronized 锁、ReentrantLock 锁和 AtomicReference 原子类分别更新 User 类中整数类型的 age 字段的值，整个案例程序的实现步骤如下。

（1）在 io.binghe.concurrent.chapter11 包下新建 User 类，作为要更新的对象类，User 类中定义了一个 String 类型的成员变量 name 和 int 类型的成员变量 age，代码如下。

```java
public class User implements Serializable {
    private static final long serialVersionUID = 6641419594499446434L;
    private String name;
    private int age;
    public User() {
    }
    public User(String name, int age) {
        this.name = name;
        this.age = age;
    }
    //############省略 getter 和 setter 方法###############
}
```

（2）在 io.binghe.concurrent.chapter11 包下新建 UserPerformance 类，使用 synchronized 锁、ReentrantLock 锁和 AtomicReference 原子类更新 User 对象中成员变量的值。为了使 synchronized 锁和 ReentrantLock 锁之间互不影响，在 UserPerformance 类中，针对 synchronized 锁和 ReentrantLock 锁分别创建了不同的 User 对象。代码如下。

```java
@State(Scope.Group)
public class UserPerformance {
    //User 对象实例，用于 synchronized 锁
    private User syncUser = new User("冰河", 18);
    //User 对象实例，用于 ReentrantLock 锁
    private User lockUser = new User("冰河", 18);
    //ReentrantLock 锁
    private Lock lock = new ReentrantLock();
    //AtomicReference 原子类
    private AtomicReference<User> atomicReference = new AtomicReference<>(new User("冰河", 18));
    //在 synchronized 锁下更新 User 对象
    public void syncUser(){
        synchronized (this){
            final User user = this.syncUser;
            final User newUser = new User(user.getName(), user.getAge() + 10);
            this.syncUser = newUser;
        }
    }
    //在 ReentrantLock 锁下更新 User 对象
    public void lockUser(){
        lock.lock();
        try{
            final User user = this.lockUser;
            final User newUser = new User(user.getName(), user.getAge() + 10);
            this.lockUser = newUser;
        }finally {
            lock.unlock();
        }
    }
    //AtomicReference 原子类更新 User 对象
```

```java
public void atomicUser(){
    final User user = atomicReference.get();
    final User newUser = new User(user.getName(), user.getAge() + 10);
    atomicReference.compareAndSet(user, newUser);
}
```

（3）在 io.binghe.concurrent.chapter11 包下新建 UserPerformanceTest 类，主要是调用 UserPerformance 类中的方法，将其交由 JMH 进行基准性能测试，代码如下。

```java
@Warmup(iterations = 20)
@Measurement(iterations = 20)
@Fork(1)
@BenchmarkMode(Mode.AverageTime)
@OutputTimeUnit(TimeUnit.MICROSECONDS)
@State(Scope.Group)
public class UserPerformanceTest {
    @Benchmark
    @Group("reentrantLock")
    @GroupThreads(10)
    public void lockCount(UserPerformance performance){
        performance.lockUser();
    }
    @Benchmark
    @Group("synchronized")
    @GroupThreads(10)
    public void syncCount(UserPerformance performance){
        performance.syncUser();
    }
    @Benchmark
    @Group("atomicReference")
    @GroupThreads(10)
    public void atomicCount(UserPerformance performance){
        performance.atomicUser();
    }
    public static void main(String[] args) throws RunnerException {
        Options options = new OptionsBuilder()
                    .include(UserPerformanceTest.class.getSimpleName())
                    .build();
        new Runner(options).run();
    }
}
```

3. 案例测试

运行 UserPerformanceTest 类的 main()方法，输出结果如下。

```
Benchmark                                Mode  Cnt  Score   Error  Units
UserPerformanceTest.atomicReference      avgt   20  0.272 ± 0.005  us/op
UserPerformanceTest.reentrantLock        avgt   20  0.246 ± 0.011  us/op
UserPerformanceTest.synchronized         avgt   20  0.833 ± 0.058  us/op
```

从输出结果可以看出，ReentrantLock 锁耗时最短，平均耗时 0.246μs，误差 ± 0.011μs。其次是 AtomicReference 原子类，平均耗时 0.272μs，误差 ± 0.005μs。耗时最长的是 synchronized 锁，平均耗时 0.833μs，误差 ± 0.058μs。

所以，在 synchronized 锁、ReentrantLock 锁和 AtomicReference 原子类三者的性能对比中，ReentrantLock 锁性能最好，其次是 AtomicReference 原子类，而 synchronized 锁是三者中性能最差的。

注意：上述基准测试程序在不同的设备上运行，可能得出的最终数据略有差异。

11.7.3 锁与字段类型原子类性能对比案例

本节，以 synchronized 锁、ReentrantLock 锁和 AtomicReferenceFieldUpdater 原子类为例，简单对比下三者在更新数据时的性能。

1. 案例需求

综合对比 synchronized 锁、ReentrantLock 锁和 AtomicReferenceFieldUpdater 原子类更新类中某个引用类型的字段值的性能，并输出结果。

2. 案例实现

分别使用 synchronized 锁、ReentrantLock 锁和 AtomicReferenceFieldUpdater 原子类更新 User 类中 String 类型的 name 字段的值，整个案例的实现步骤如下。

（1）本案例程序继续使用 11.8.2 节案例中的 User 类，将 User 类中的 name 字段使用 public volatile 修饰，代码如下。

```
public volatile String name;
```

（2）在 io.binghe.concurrent.chapter11 包下新建 UserReferencePerformance 类，主要提供保证使用 synchronized 锁、ReentrantLock 锁和 AtomicReferenceFieldUpdater 原子类的原子性的方法，代码如下。

```
@State(Scope.Group)
public class UserReferencePerformance {
    //User 对象实例，用于 synchronized 锁
    private User syncUser = new User("冰河", 18);
    //User 对象实例，用于 ReentrantLock 锁
    private User lockUser = new User("冰河", 18);
    //User 对象实例，用于 AtomicReferenceFieldUpdater 原子类
    private User atomicUser = new User("冰河", 18);
    //ReentrantLock 锁
    private Lock lock = new ReentrantLock();
    //AtomicReferenceFieldUpdater 原子类
private AtomicReferenceFieldUpdater<User, String> updater =
 AtomicReferenceFieldUpdater.newUpdater(User.class, String.class, "name");
```

```
//在 synchronized 锁下更新 User 对象
public void syncUser(){
    synchronized (this){
        final User user = this.syncUser;
        final User newUser = new User("冰河 001", user.getAge());
        this.syncUser = newUser;
    }
}
//在 ReentrantLock 锁下更新 User 对象
public void lockUser(){
    lock.lock();
    try{
        final User user = this.lockUser;
        final User newUser = new User("冰河 001", user.getAge());
        this.lockUser = newUser;
    }finally {
        lock.unlock();
    }
}
//AtomicReference 原子类更新 User 对象
public void atomicUser(){
    updater.compareAndSet(atomicUser, "冰河", "冰河 001");
}
}
```

（3）在 io.binghe.concurrent.chapter11 包下新建 UserReferencePerformanceTest 类，在 UserReferencePerformanceTest 类中调用 UserReferencePerformance 类的方法，并交由 JMH 进行基准性能测试。代码如下。

```
@Warmup(iterations = 20)
@Measurement(iterations = 20)
@Fork(1)
@BenchmarkMode(Mode.AverageTime)
@OutputTimeUnit(TimeUnit.MICROSECONDS)
@State(Scope.Group)
public class UserReferencePerformanceTest {
    @Benchmark
    @Group("reentrantLock")
    @GroupThreads(10)
    public void lockCount(UserReferencePerformance performance){
        performance.lockUser();
    }

    @Benchmark
    @Group("synchronized")
    @GroupThreads(10)
    public void syncCount(UserReferencePerformance performance){
        performance.syncUser();
    }
```

```java
@Benchmark
@Group("atomicReference")
@GroupThreads(10)
public void atomicCount(UserReferencePerformance performance){
    performance.atomicUser();
}

public static void main(String[] args) throws RunnerException {
    Options options = new OptionsBuilder()
                .include(UserReferencePerformanceTest.class.getSimpleName())
                .build();
    new Runner(options).run();
}
}
```

3. 案例测试

运行 UserReferencePerformanceTest 类的 main()方法，输出结果如下。

Benchmark	Mode	Cnt	Score		Error	Units
UserReferencePerformanceTest.atomicReference	avgt	20	0.164	±	0.002	us/op
UserReferencePerformanceTest.reentrantLock	avgt	20	0.317	±	0.014	us/op
UserReferencePerformanceTest.synchronized	avgt	20	0.904	±	0.192	us/op

从输出结果可以看出，AtomicReferenceFieldUpdater 原子类耗时最短，平均耗时 0.164μs，误差 ± 0.002μs。其次是 ReentrantLock 锁，平均耗时 0.317μs，误差 ± 0.014μs。耗时最长的是 synchronized 锁，平均耗时 0.904μs，误差 ± 0.192μs。

所以，在 synchronized 锁、ReentrantLock 锁和 AtomicReferenceFieldUpdater 原子类三者的性能对比中，AtomicReferenceFieldUpdater 原子类性能最好，其次是 ReentrantLock 锁，而 synchronized 锁是三者中性能最差的。

注意：上述基准测试程序在不同的设备上运行，得出的数据可能略有差异。

11.7.4 AtomicLong 与 LongAdder 性能对比案例

本节以 AtomicLong 与 LongAdder 为例，对比操作基本类型的原子类与累加器类型的原子类的性能。

1. 案例需求

对比 AtomicLong 与 LongAdder 对 long 型变量进行自增 1 操作时的性能，并输出性能对比结果。

2. 案例实现

AtomicLong 与 LongAdder 都能够原子性地累加 long 型变量的值，二者在高并发环境下的性能存在差异，案例的具体实现步骤如下。

(1)在 mykit-concurrent-chapter11 工程下的 io.binghe.concurrent.chapter11 包下新建 LongPerformance 类,在 LongPerformance 类中分别定义 AtomicLong 类型的常量与 LongAdder 类型的常量,并分别提供通过 AtomicLong 与 LongAdder 自增 long 型数据的方法,代码如下。

```
@State(Scope.Group)
public class LongPerformance {
    //AtomicLong
    private final AtomicLong atomicLong = new AtomicLong();
    //LongAdder
    private final LongAdder longAdder = new LongAdder();
    //使用AtomicInteger原子类
    public void atomicCount(){
        atomicLong.incrementAndGet();
    }
    //使用LongAdder
    public void adderCount(){
        longAdder.increment();
    }
}
```

(2)在 io.binghe.concurrent.chapter11 包下新建 LongPerformanceTest 类,在 LongPerformanceTest 类中主要是调用 LongPerformance 类的方法,并交由 JMH 进行基准性能测试,代码如下。

```
@Warmup(iterations = 20)
@Measurement(iterations = 20)
@Fork(1)
@BenchmarkMode(Mode.AverageTime)
@OutputTimeUnit(TimeUnit.MICROSECONDS)
@State(Scope.Group)
public class LongPerformanceTest {
    @Benchmark
    @Group("atomicLong")
    @GroupThreads(50)
    public void atomicCount(LongPerformance performance){
        performance.atomicCount();
    }
    @Benchmark
    @Group("longAdder")
    @GroupThreads(50)
    public void adderCount(LongPerformance performance){
        performance.adderCount();
    }
    public static void main(String[] args) throws RunnerException {
        Options options = new OptionsBuilder()
                        .include(LongPerformanceTest.class.getSimpleName())
                        .build();
        new Runner(options).run();
```

```
        }
}
```

3. 案例测试

运行 LongPerformanceTest 类的 main() 方法,输出结果如下。

```
Benchmark                              Mode  Cnt  Score    Error   Units
LongPerformanceTest.atomicLong         avgt   20  0.739 ±  0.018   us/op
LongPerformanceTest.longAdder          avgt   20  0.093 ±  0.002   us/op
```

由输出结果可以看出,LongAdder 耗时比较短,平均耗时 0.093μs,误差 ± 0.002μs。AtomicLong 耗时较长,平均耗时 0.739μs,误差 ± 0.018μs。所以,在高并发环境下,使用 LongAdder 自增 long 型的数据会比 AtomicLong 性能高。

第 12 章

线程工具类

Java 从 JDK 1.0 版本开始陆续提供了一些线程工具类,从广义上讲,本书介绍的所有 JDK 的工具类都是线程工具类,但笔者更愿意将这些工具类进行细分为创建并启动线程的 Thread 类、保证线程数据隔离与安全的 ThreadLocal 类和实现分组与合并的 Fork/Join 框架。

12.1 Thread 类

Thread 类是 Java 从 JDK 1.0 版本开始提供的创建并启动线程的工具类,使用 Thread 类能够非常方便地创建并启动线程。Thread 类创建的线程与操作系统内部的线程是一一对应的。

12.1.1 继承关系

Thread 类位于 JDK 的 java.lang 包下,实现了 Runnable 接口,如图 12-1 所示。

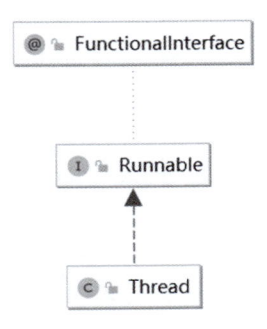

图 12-1 Thread 类的继承关系

Thread 类实现了 Runnable 接口,而 Runnable 接口从 JDK 1.8 版本开始被标注了 @FunctionalInterface 注解,被标注了@FunctionalInterface 注解的接口是一个函数式接口。Runnable 接口在 JDK 1.8 版本中的代码如下。

```
@FunctionalInterface
public interface Runnable {
    public abstract void run();
}
```

Runnable 接口的代码比较简单，只提供了一个 run()抽象方法供实现类实现。

接下来，再来看下标注到 Runnable 接口上的@FunctionalInterface 注解的代码，如下所示。

```
@Documented
@Retention(RetentionPolicy.RUNTIME)
@Target(ElementType.TYPE)
public @interface FunctionalInterface {
}
```

通过@FunctionalInterface 注解的代码可以看出，@FunctionalInterface 注解主要声明并标记在类上，并在程序运行时生效。

12.1.2 定义

Thread 类在定义上直接实现了 Runnable 接口，并实现了 run()方法，整体定义结构如下。

```
public class Thread implements Runnable{
@Override
    public void run() {

    }
}
```

在 Thread 类的整体结构定义上，实现了 Runnable 接口，并覆写了 Runnable 接口的 run()方法。

12.1.3 核心代码解析

1. 注册本地资源

在 Thread 类的开始部分，定义了一个静态本地方法 registerNatives()，这种方法主要用于注册本地系统的资源，并且会在 Thread 类的静态代码块中调用 registerNatives()方法，代码如下。

```
private static native void registerNatives();
static {
    registerNatives();
}
```

2. 核心成员变量

在 Thread 类中，定义了比较多的控制线程运行状态的成员变量，这些成员变量在线程运行的过程中起到了至关重要的作用，Thread 类的核心成员变量代码如下。

```
//当前线程的名称
private volatile String name;
```

```java
//当前线程的优先级
private int              priority;
private Thread           threadQ;
private long             eetop;
//标记当前线程是否是单步线程
private boolean    single_step;
//标记当前线程是否在后台运行
private boolean    daemon = false;
//Java 虚拟机的状态
private boolean    stillborn = false;
//在线程中执行的 Runnable 任务
private Runnable target;
//当前线程所在的线程组
private ThreadGroup group;
//当前线程的类加载器
private ClassLoader contextClassLoader;
//当前线程的访问控制上下文
private AccessControlContext inheritedAccessControlContext;
//如果当前线程是匿名线程，则此字段是为匿名线程生成线程名称的编号
private static int threadInitNumber;
//与当前线程关联的 ThreadLocal 信息
ThreadLocal.ThreadLocalMap threadLocals = null;
//与当前线程关联的 InheritableThreadLocal 信息
ThreadLocal.ThreadLocalMap inheritableThreadLocals = null;
//当前线程初始化时向 JVM 申请的堆栈大小
private long stackSize;
//线程终止后 JVM 中的私有状态
private long nativeParkEventPointer;
//当前线程的 id
private long tid;
//用于生成线程 id
private static long threadSeqNumber;
//当前线程的状态，初始化为 0，代表当前线程还未启动
private volatile int threadStatus = 0;
//由（私有）java.util.concurrent.locks.LockSupport.setBlocker 设置
//使用 java.util.concurrent.locks.LockSupport.getBlocker 获取
volatile Object parkBlocker;
//线程被标记为中断状态后，会调用 Interruptible 接口的 interrupt()方法
private volatile Interruptible blocker;
//当前线程的内部锁
private final Object blockerLock = new Object();
//线程拥有的最小优先级
public final static int MIN_PRIORITY = 1;
//线程拥有的默认优先级
public final static int NORM_PRIORITY = 5;
//线程拥有的最大优先级
public final static int MAX_PRIORITY = 10;
```

上述成员变量代码标注了详细的注释信息，读者可根据注释信息理解 Thread 类中每个成员

变量的具体含义。

从 Thread 类的成员变量上可以看出，Thread 本质上是一个线程对象，而在这个线程对象中定义了一个 Runnable 接口类型的成员变量 target，这个 target 成员变量就是要在 Thread 线程中执行的具体任务。

3. 线程的状态定义

在 Thread 类的内部定义了一个枚举类型类 State，State 类中定义了线程的执行状态，代码如下。

```
public enum State {
    //初始化状态
    NEW,
    //可运行状态，此时的可运行包括运行中的状态和就绪状态
    RUNNABLE,
    //线程阻塞状态
    BLOCKED,
    //等待状态
    WAITING,
    //超时等待状态
    TIMED_WAITING,
    //线程终止状态
    TERMINATED;
}
```

这个枚举类的各个状态本质上代表了整个线程的生命周期。

注意：关于线程的生命周期，读者可参考本书第 1 章的内容。

4. Thread 类的构造方法

Thread 类提供了比较多的重载的构造方法，代码如下。

```
public Thread() {
    init(null, null, "Thread-" + nextThreadNum(), 0);
}
public Thread(Runnable target) {
    init(null, target, "Thread-" + nextThreadNum(), 0);
}
Thread(Runnable target, AccessControlContext acc) {
    init(null, target, "Thread-" + nextThreadNum(), 0, acc, false);
}
public Thread(ThreadGroup group, Runnable target) {
    init(group, target, "Thread-" + nextThreadNum(), 0);
}
public Thread(String name) {
    init(null, null, name, 0);
}
public Thread(ThreadGroup group, String name) {
```

```
    init(group, null, name, 0);
}
public Thread(Runnable target, String name) {
    init(null, target, name, 0);
}
public Thread(ThreadGroup group, Runnable target, String name) {
    init(group, target, name, 0);
}
public Thread(ThreadGroup group, Runnable target, String name,
              long stackSize) {
    init(group, target, name, stackSize);
}
```

尽管 Thread 类的构造方法较多，但是有些构造方法并不是很常用，比较常用的构造方法的代码如下。

```
public Thread() {
    init(null, null, "Thread-" + nextThreadNum(), 0);
}
public Thread(Runnable target) {
    init(null, target, "Thread-" + nextThreadNum(), 0);
}
public Thread(String name) {
    init(null, null, name, 0);
}
public Thread(ThreadGroup group, String name) {
    init(group, null, name, 0);
}
public Thread(Runnable target, String name) {
    init(null, target, name, 0);
}
public Thread(ThreadGroup group, Runnable target, String name) {
    init(group, target, name, 0);
}
```

Thread 类的构造方法调用了 init() 方法进行初始化。

5. init() 方法解析

Thread 类中 init() 方法的主要作用是在调用 Thread 类的构造方法创建对象时，对 Thread 类对象进行一系列的初始化，init() 方法的代码如下。

```
//第①个 init()方法，直接调用了第②个 init()方法
private void init(ThreadGroup g, Runnable target, String name, long stackSize) {
    init(g, target, name, stackSize, null, true);
}
//第②个 init()方法，方法中对 Thread 对象进行了一系列的初始化
private void init(ThreadGroup g, Runnable target, String name,
                  long stackSize, AccessControlContext acc,
                  boolean inheritThreadLocals) {
    //如果传入的线程名称为空，则直接抛出空指针异常
```

```java
if (name == null) {
    throw new NullPointerException("name cannot be null");
}

this.name = name;
Thread parent = currentThread();
//获取系统安全管理器
SecurityManager security = System.getSecurityManager();
//当传入的线程组为空时
if (g == null) {
    //如果获取的系统安全管理器不为空
    if (security != null) {
        //则从系统安全管理器中获取一个线程分组
        g = security.getThreadGroup();
    }
    //如果线程分组为空，则从父线程获取一个线程分组
    if (g == null) {
        g = parent.getThreadGroup();
    }
}
//检查线程组的访问权限
g.checkAccess();
//如果获取的系统安全管理器不为空
if (security != null) {
    //则检查权限
    if (isCCLOverridden(getClass())) {
        security.checkPermission(SUBCLASS_IMPLEMENTATION_PERMISSION);
    }
}
g.addUnstarted();
this.group = g;
//当前线程继承父线程的相关属性
this.daemon = parent.isDaemon();
this.priority = parent.getPriority();
if (security == null || isCCLOverridden(parent.getClass()))
    this.contextClassLoader = parent.getContextClassLoader();
else
    this.contextClassLoader = parent.contextClassLoader;
this.inheritedAccessControlContext =
    acc != null ? acc : AccessController.getContext();
this.target = target;
setPriority(priority);
if (inheritThreadLocals && parent.inheritableThreadLocals != null)
    this.inheritableThreadLocals =
    ThreadLocal.createInheritedMap(parent.inheritableThreadLocals);

this.stackSize = stackSize;

//设置线程id
```

```
        tid = nextThreadID();
}
```

通过 init()方法的代码可以看出，在第①个 init()方法中直接调用了第②个 init()方法进行初始化。第②个 init()方法的方法体看起来比较长，但是核心逻辑比较简单。整体上就是获取系统安全管理器、验证各种权限、由当前创建出来的线程继承父线程相关的属性。

> **注意**：在 init()方法中有如下一行代码。
> `Thread parent = currentThread();`
> 上述代码表示将调用 init()方法的线程作为新建线程的父线程，也就是说，在 init()方法中，新建的线程会继承调用 init()方法的线程的部分属性。

6. run()方法解析

Thread 类实现了 Runnable 接口，自然就需要实现 Runnable 接口的 run()方法，run()方法的代码如下。

```
@Override
public void run() {
    if (target != null) {
        target.run();
    }
}
```

run()方法的实现非常简单，在 run()方法中，如果 target 对象不为空，就会直接调用 target 对象的 run()方法。所以，在 Thread 类中，会通过 run()方法执行任务。

另外，通过 run()方法的代码也能看出，直接调用 Thread 类的 run()方法并不会创建一个新的线程来执行任务，如果需要创建一个新的线程执行任务，就需要调用 Thread 类的 start()方法。

7. start()方法解析

Thread 类的 start()方法会启动一个线程。只有调用了 Thread 类的 start()方法后，才会在操作系统中新建一个对应的线程来执行任务。start()方法的代码如下。

```
public synchronized void start() {
  //如果线程不是初始化状态，则直接抛出异常
  if (threadStatus != 0)
     throw new IllegalThreadStateException();
  //添加当前启动的线程到线程组
  group.add(this);
  //标记线程是否已经启动
  boolean started = false;
  try {
     //调用本地方法启动线程
     start0();
     //将线程是否启动标记为 true
     started = true;
```

```
        } finally {
            try {
                //线程未启动成功
                if (!started) {
                    //将线程在线程组里标记为启动失败
                    group.threadStartFailed(this);
                }
            } catch (Throwable ignore) {
                /* do nothing. If start0 threw a Throwable then
                   it will be passed up the call stack */
            }
        }
    }
```

通过 start() 方法的代码可以看出，start() 方法使用了 synchronized 关键字修饰，说明 start() 方法是一个线程安全的同步方法，同一时刻，只能有一个线程执行 start() 方法的逻辑。start() 方法会在线程启动之前检查线程的状态，如果线程未处于初始化状态，就会直接抛出 IllegalThreadStateException 异常。所以，每个线程只能调用 start() 方法启动一次，多次调用 start() 方法启动会抛出 IllegalThreadStateException 异常。

细心的读者会发现，在 start() 方法中调用了 start0() 方法，start0() 方法的代码如下。

```
private native void start0();
```

start0() 方法是一个本地方法，会调用操作系统中新建线程的指令来新建线程。

在调用 start() 方法后，新建的线程如果没有分配到 CPU 资源，就会处于就绪状态，当有空闲的 CPU 时，这个线程就会获得 CPU 资源执行任务，此时线程就会转换为运行状态。

8. sleep()方法解析

sleep() 方法可以使当前线程休眠一定的时间（单位：毫秒），代码如下。

```
//方法①，调用方法②实现线程休眠
public static void sleep(long millis, int nanos)
throws InterruptedException {
    if (millis < 0) {
        throw new IllegalArgumentException("timeout value is negative");
    }

    if (nanos < 0 || nanos > 999999) {
        throw new IllegalArgumentException(
                            "nanosecond timeout value out of range");
    }

    if (nanos >= 500000 || (nanos != 0 && millis == 0)) {
        millis++;
    }

    sleep(millis);
```

```
}
```

//方法②,真正使线程休眠的方法
```
public static native void sleep(long millis) throws InterruptedException;
```
使线程真正休眠的是本地 sleep()方法。

注意：调用 Thread 类的 sleep()方法使线程休眠后，如果休眠的线程持有锁，则线程不会释放锁。

9. join()方法解析

如果线程 a 调用了线程 b 的 join()方法，则线程 a 会一直等待线程 b 执行完毕，或者超时。join()方法的代码如下。

```
//方法①,调用方法③实现线程等待
public final void join() throws InterruptedException {
    join(0);
}

//方法②,调用方法③实现线程等待
public final synchronized void join(long millis, int nanos)
throws InterruptedException {

    if (millis < 0) {
        throw new IllegalArgumentException("timeout value is negative");
    }

    if (nanos < 0 || nanos > 999999) {
        throw new IllegalArgumentException(
                    "nanosecond timeout value out of range");
    }

    if (nanos >= 500000 || (nanos != 0 && millis == 0)) {
        millis++;
    }

    join(millis);
}

//方法③,真正实现线程等待的方法
public final synchronized void join(long millis)
throws InterruptedException {
    long base = System.currentTimeMillis();
    long now = 0;

    if (millis < 0) {
        throw new IllegalArgumentException("timeout value is negative");
    }
```

```
if (millis == 0) {
    while (isAlive()) {
        wait(0);
    }
} else {
    while (isAlive()) {
        long delay = millis - now;
        if (delay <= 0) {
            break;
        }
        wait(delay);
        now = System.currentTimeMillis() - base;
    }
}
```

join()方法通常用于一个线程需要等待另一个线程执行完毕后，继续向下执行的场景。通常是创建并启动新线程的线程、调用新线程的 join()方法、等待新线程执行完毕，或者超时返回的场景。

10. interrupt()方法解析

interrupt()方法是一个通过设置线程中断标志位来中断线程的方法，代码如下。

```
public void interrupt() {
    if (this != Thread.currentThread())
        checkAccess();

    synchronized (blockerLock) {
        Interruptible b = blocker;
        if (b != null) {
            interrupt0();
            b.interrupt(this);
            return;
        }
    }
    interrupt0();
}
```

interrupt()方法内部调用 interrupt0()方法来设置线程中断标志位，interrupt0()方法的代码如下。

```
private native void interrupt0();
```

interrupt0()方法是一个本地方法，会调用操作系统的指令来设置线程中断标志位。

interrupt()方法是通过设置线程的中断标志位来中断线程的。在设置线程标志位时，可能会抛出 InteruptedExeption 异常，并清除当前线程的中断状态。这种中断线程的方式比较安全，能够使正在执行的任务执行完毕。Thread 类中推荐使用 interrupt()方法中断线程。

11. stop()方法与interrupt()方法的区别

stop()方法与interrupt()方法在某种程度上都能让线程停止运行，但是二者有着本质的区别。

（1）stop()方法。stop()方法会强制终止正在运行的线程，这种方式有一个致命的问题：当线程持有ReentrantLock锁、正在执行业务逻辑时，如果调用stop()方法强行终止线程，那么哪怕在程序中的finally代码块中调用了ReentrantLock的unlock()释放锁，实际上线程也不会执行到finally代码块，而是提前终止了。后续线程没有机会获取到ReentrantLock锁，也就没有机会执行获取锁后的业务逻辑。

这种问题是致命的，所以，在JDK中不推荐使用stop()方法强行终止线程。与stop()方法类似的还有suspend()方法和resume()方法，suspend()方法表示强行挂起线程，resume()方法表示恢复线程，在JDK中也不推荐使用这两种方法。

（2）interrupt()方法。interrupt()方法也能让线程停止运行，不过interrupt()方法不会强行终止线程，而是为线程设置一个中断标志位，通知线程后续可能中断，此时线程还能执行后续的一些操作，线程也可以无视interrupt()方法发出的通知。

如果调用一个线程的interrupt()方法中断这个线程，那么这个线程有两种方式接收中断通知，一种是触发异常，另一种是主动检测中断。

- 触发异常

假设有线程a处于等待或超时等待状态，如果此时其他线程调用线程a的interrupt()方法中断线程，那么会使线程a转换成可运行状态，同时线程a的代码会触发InterruptedException异常，并清除中断标志位。如果想中断线程a的执行，则可以在捕获到InterruptedException异常的catch代码块中再次调用线程a的interrupt()方法。

另外，在使用Java进行NIO编程、线程a处于可运行状态且阻塞在java.nio.channels.InterruptibleChannel上时，如果其他线程调用线程a的interrupt()方法，则线程a会触发java.nio.channels.ClosedByInterruptException异常；当线程a阻塞在java.nio.channels.Selector上时，如果其他线程调用线程a的interrupt()方法，则线程a的java.nio.channels.Selector会立即返回。

- 主动检测中断

当线程a处于可运行状态，并且未阻塞在某个I/O操作上时，如果其他线程调用了线程a的interrupt()方法来中断线程a，则需要线程a通过Thread类提供的isInterrupted()方法主动检测自己是否被中断。

12.2 ThreadLocal 类

ThreadLocal 是 Java 从 JDK 1.2 版本开始提供的用于线程数据隔离的工具类，ThreadLocal 支持本地变量，访问 ThreadLocal 的每个线程都会有变量的一个副本。当线程对 ThreadLocal 中的变量进行写操作时，实际上操作的是线程本地内存中的变量，这样就避免了线程安全问题。

注意：关于 ThreadLocal 类的核心原理、代码解析，以及使用案例，读者可以参考《深入理解高并发编程：核心原理与案例实战》一书。

12.3 Fork/Join 框架

Fork/Join 框架是 Java 从 JDK 1.7 版本开始提供的用于执行并行任务的框架，其基本思想与 Hadoop 的 MapReduce 类似，可以将一个比较大的任务拆分成多个处理逻辑相同的小任务，在汇总每个小任务的执行结果后得到最终的结果。

12.3.1 概述

Fork/Join 与线程池的 ThreadPoolExecutor 一样，实现了 Executor 和 ExecutorService 接口，内部使用一个无界队列来保存需要执行的任务。执行任务的线程可以通过构造函数传递进来，如果没有执行执行任务的线程数量，则默认的线程数量为 CPU 核数。

Fork/Join 框架内部涉及工作窃取算法，简单来说，工作窃取算法就是某个线程执行完自己队列中的任务后，去其他线程对应的队列中获取任务来执行，如图 12-2 所示。

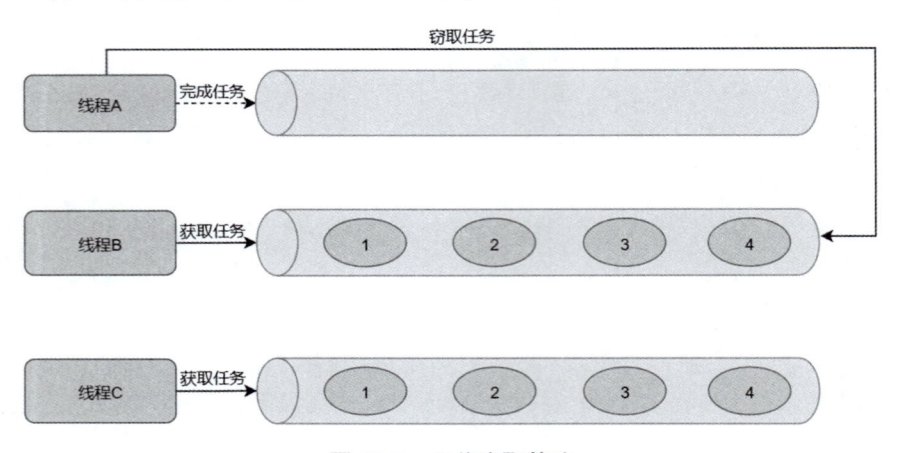

图 12-2 工作窃取算法

由图 12-2 可以看出，线程 A 执行完自己对应的队列中的任务后，会从线程 B 对应的队列的尾部获取任务执行。

例如，需要完成一个比较大的任务，可以将这个任务分割成几个互不依赖并且核心逻辑相同的子任务。为了减少线程之前的竞争，可以将每个子任务放到不同的队列中，并且为每个队列创建一个单独的线程来执行队列中的任务，线程与队列之间是一一对应的关系。

在执行任务的过程中，有的线程会先执行完自己队列中的任务，此时其他线程对应的队列中可能还存在未执行的任务。执行完自己队列中任务的线程就会去其他线程对应的队列中窃取一个任务来执行，此时，窃取任务的线程和被窃取任务的线程会访问同一个队列。为了减少窃取任务的线程与被窃取任务的线程之间的竞争，存放子任务的队列会采用双端队列，被窃取任务的线程会从双端队列的头部获取任务执行，而窃取任务的线程会从双端队列的尾部获取任务执行。

工作窃取算法会消耗更多的系统资源，例如会创建多个线程和多个双端队列。但是，工作窃取算法在一定程度上能充分利用线程进行并行计算，并且能减少线程之间的竞争。

12.3.2 核心类

JDK 提供的 Fork/Join 框架涉及的核心类包括 ForkJoinPool 类、ForkJoinWorkerThread 类、ForkJoinTask 类、RecursiveTask 类、RecursiveAction 类和 CountedCompleter 类。

- ForkJoinPool 类：实现了 Fork/Join 框架的线程池，可以使用 Executors.newWorkStealPool() 方法创建 ForkJoinPool。
- ForkJoinWorkerThread 类：Fork/Join 框架中运行的每一个线程。
- ForkJoinTask 类：内部封装了数据和对应的计算逻辑，能够支持细粒度的数据并行操作，主要包括 fork() 和 join() 两种方法，分别实现了任务的拆分与合并。
- RecursiveTask 类：ForkJoinTask 类的子类，实现了 Callable 接口，并提供返回结果。
- RecursiveAction 类：ForkJoinTask 类的子类，实现了 Runnable 接口，无返回结果。
- CountedCompleter 类：在任务执行完成后会触发执行一个自定义的任务。

注意：有关 Fork/Join 框架更深层次的内容，读者可以关注"冰河技术"微信公众号阅读相关文章。

另外，在 Java 8 中引入的并行流计算，内部就是采用 ForkJoinPool 实现的。有关 Java 8 并行流计算的内容，读者可以关注"冰河技术"微信公众号，回复"java8"或者数字"10"获取相关的技术资料。

12.4 线程工具类案例

Thread 类、ThreadLocal 类和 Fork/Join 框架都是在并发场景下使用得比较广泛的线程工具类，本节针对这 3 个线程工具类给出典型的实战案例。

12.4.1 Thread 类线程中断案例

1. 案例需求

验证在使用 Thread 类的 interrupt()方法中断一个处于休眠状态的线程时，如果没有捕获 InterruptedException 异常，并再次调用 interrupt()方法中断线程，则线程并不会真正退出。

2. 案例实现

在 12.1.3 节已经详细介绍了 Thread 类的 interrupt()方法中断线程的机制，这里，我们直接实现案例程序。具体步骤如下。

（1）在 mykit-concurrent-chapter12 工程的 io.binghe.concurrent.chapter12 包下创建 ThreadInterruptTask 类，实现 Runnable 接口，代码如下。

```java
public class ThreadInterruptTask implements Runnable {
    @Override
    public void run() {
        Thread currentThread = Thread.currentThread();
        while (true){
            if (currentThread.isInterrupted()){
                break;
            }
            try {
                Thread.sleep(100);
            } catch (InterruptedException e) {
                System.out.println("触发 InterruptedException 异常");
            }
        }
        System.out.println("线程被成功中断");
    }
}
```

在 ThreadInterruptTask 类中实现的 run()方法中，首先获取当前线程对象 currentThread，然后在 while(true)死循环中，调用 currentThread 的 isInterrupted()方法检测线程是否被中断，如果线程被中断，则直接退出死循环，输出线程被成功中断的日志。如果线程没有被中断，则休眠 100 毫秒后，继续执行循环操作。

（2）在 io.binghe.concurrent.chapter12 包下新建 ThreadInterruptTest 类，主要在 main 线程中中断新建的线程，代码如下。

```java
public class ThreadInterruptTest {
    public static void main(String[] args){
        Thread thread = new Thread(new ThreadInterruptTask());
        thread.start();
        try {
            Thread.sleep(200);
        } catch (InterruptedException e) {
```

```
            e.printStackTrace();
        }
        thread.interrupt();
    }
}
```

ThreadInterruptTest 的代码比较简单，在 main()方法中新建一个 Thread 线程，其中传入的 Runnable 接口的对象就是 ThreadInterruptTask 类的对象。在启动新建的线程后，主线程休眠 200 毫秒，通过调用新建的线程的 interrupt()方法来中断新建的线程。

3．案例测试

运行 ThreadInterruptTest 类的 main()方法，输出结果如下。

```
触发 InterruptedException 异常
```

程序只输出了触发 InterruptedException 异常的日志，说明线程并没有真正退出。实际上，在 IDEA、Eclipse 等开发环境，或者服务器上都能发现新建的线程并没有退出，Java 进程还在，符合预期。

那如何真正退出线程呢？这就需要在 ThreadInterruptTask 类的 run()方法中捕获 InterruptedException 的 catch 代码块中添加如下代码，再次设置线程的中断标志位。

```
currentThread.interrupt();
```

此时，在 ThreadInterruptTask 类的 run()方法中捕获 InterruptedException 的 catch 代码块的代码片段如下。

```
} catch (InterruptedException e) {
    System.out.println("触发 InterruptedException 异常");
    currentThread.interrupt();
}
```

再次运行 ThreadInterruptTest 类的 main()方法，输出结果如下。

```
触发 InterruptedException 异常
线程被成功中断
```

此时，同时输出了触发 InterruptedException 异常的日志和线程被成功中断的日志，说明新建的线程已经退出 while(true)死循环。实际上，在 IDEA、Eclipse 等开发环境，或者服务器上都能够发现线程已经退出，Java 进程也已经退出。

12.4.2　Fork/Join 框架分组合并案例

1．案例需求

使用 Fork/Join 框架实现 1~10000 的累加和。

2．案例实现

案例的需求比较简单，直接按照如下步骤实现案例程序。

（1）在 mykit-concurrent-chapter12 工程的 io.binghe.concurrent.chapter12 包下新建 ForkJoinTaskComputer 类，用于实现 Fork/Join 框架的分组与合并逻辑，代码如下。

```java
public class ForkJoinTaskComputer extends RecursiveTask<Integer> {
    //任务拆分的最小粒度
    private static final int MIN_COUNT = 2;
    //开始数字
    private int startNum;
    //结束数字
    private int endNum;
    public ForkJoinTaskComputer(int startNum, int endNum) {
        this.startNum = startNum;
        this.endNum = endNum;
    }
    @Override
    protected Integer compute() {
        //计算结果
        int sum = 0;
        int count = endNum - startNum;
        //达到可计算的范围
        if (count <= MIN_COUNT){
            for (int i = startNum; i <= endNum; i++){
                sum += i;
            }
        }else{
            //找到中间值
            int middleCount = (startNum + endNum) / 2;
            //生成子任务
            ForkJoinTaskComputer leftComputer =
new ForkJoinTaskComputer(startNum, middleCount);
            //生成子任务
            ForkJoinTaskComputer rightComputer =
new ForkJoinTaskComputer(middleCount + 1, endNum);

            //执行子任务
            leftComputer.fork();
            rightComputer.fork();

            //合并结果
            Integer leftResult = leftComputer.join();
            Integer rightResult = rightComputer.join();

            sum += (leftResult + rightResult);
        }
        return sum;
    }
}
```

ForkJoinTaskComputer 类在实现分组与合并时，主要使用了递归的思想，实现逻辑比较简单。

（2）在 io.binghe.concurrent.chapter12 包下新建 ForkJoinTest 类，用于测试使用 Fork/Join 框架实现 1~10000 的累加和，代码如下。

```
public class ForkJoinTest {
    private static final int MIN_COUNT = 1;
    private static final int MAX_COUNT = 10000;
    public static void main(String[] args) throws Exception {
        ForkJoinTaskComputer computer =
        new ForkJoinTaskComputer(MIN_COUNT, MAX_COUNT);
        ForkJoinPool forkJoinPool = new ForkJoinPool();
        ForkJoinTask<Integer> result = forkJoinPool.submit(computer);
        System.out.println("最终的计算结果为===>>> " + result.get());
    }
}
```

在 ForkJoinTest 类中主要将 ForkJoinTaskComputer 对象作为一个大任务提交到 ForkJoinPool 中计算，在创建 ForkJoinTaskComputer 对象时，传入的最小值为 1，最大值为 10000。在提交任务后，返回一个 ForkJoinTask 类型的 result 结果，通过 result 的 get()方法获取最终的计算结果。

3．案例测试

运行 ForkJoinTest 类的 main()方法，输出结果如下。

最终的计算结果为===>>> 50005000

输出了正确的结果，符合预期。

第 13 章

异步编程工具类

异步编程能够充分利用 CPU 的资源，使得主线程不用长时间阻塞在某个耗时较长的任务上，例如 I/O 操作、调用远程服务，以及其他复杂的计算等。为了更加方便地进行异步编程，JDK 提供了相应的异步编程工具类。

13.1 Callable 接口

Java 可以通过多线程的方式编写异步程序，单独通过 Thread 类的方式或通过 Thread 类结合 Runnable 接口的方式都能够创建子线程执行任务，但是无法返回结果。为了解决这个问题，Java 从 JDK 1.5 版本开始提供了 Callable 接口，使用 Callable 接口结合 Future 接口，就能够轻松实现从异步线程中获取返回结果。

13.1.1 概述

Callable 接口是 Java 从 JDK 1.5 版本开始提供的泛型接口类，从 JDK 1.8 版本开始，被声明为函数式接口，代码如下。

```
@FunctionalInterface
public interface Callable<V> {
    V call() throws Exception;
}
```

从 JDK 1.8 版本开始，如果接口中只声明了一种方法（除默认方法外），这个接口就可以被称为函数式接口。函数式接口上可以标注 @FunctionalInterface 注解，也可以不标注 @FunctionalInterface 注解。

> 注意：函数式接口是 Java 从 JDK 1.8 版本开始提供的新特性。

在 JDK 中，Callable 接口的实现类比较多，其中比较重要的包括 Executors 类的静态内部类 PrivilegedCallable、PrivilegedCallableUsingCurrentClassLoader、RunnableAdapter 和 Task 类下的

TaskCallable。如图 13-1 所示。

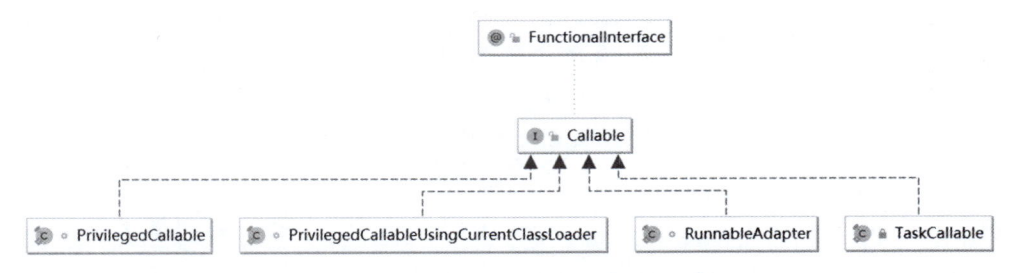

图 13-1　JDK 中 Callable 接口的重要实现类

13.1.2　PrivilegedCallable 实现类

PrivilegedCallable 类是 Callable 接口的一个特殊实现类，它表明 Callable 对象有某种特权可以访问系统的某种资源，PrivilegedCallable 类的代码如下。

```
static final class PrivilegedCallable<T> implements Callable<T> {
    private final Callable<T> task;
    private final AccessControlContext acc;
    PrivilegedCallable(Callable<T> task) {
        this.task = task;
        this.acc = AccessController.getContext();
    }
    public T call() throws Exception {
        try {
            return AccessController.doPrivileged(
                new PrivilegedExceptionAction<T>() {
                    public T run() throws Exception {
                        return task.call();
                    }
                }, acc);
        } catch (PrivilegedActionException e) {
            throw e.getException();
        }
    }
}
```

通过 PrivilegedCallable 类的代码可以看出，在 PrivilegedCallable 类中定义了一个 Callable 类型的成员变量 task。所以，可以将 PrivilegedCallable 类看成是对 Callable 接口的封装，并且 PrivilegedCallable 类也实现了 Callable 接口。

在 PrivilegedCallable 类中，除了 Callable 类型的成员变量，还有一个 AccessControlContext 类型的成员变量，代码如下。

```
private final Callable<T> task;
private final AccessControlContext acc;
```

其中，AccessControlContext 类是一个能够访问系统资源的上下文类，通过这个类可以访问系统的特定资源。通过 PrivilegedCallable 类的构造方法可以看出，在实例化 PrivilegedCallable 类的对象时，只需要传递 Callable 接口子类的对象，代码如下。

```
PrivilegedCallable(Callable<T> task) {
   this.task = task;
   this.acc = AccessController.getContext();
}
```

AccessControlContext 类的对象是通过 AccessController 类的 getContext()方法获取的，查看 AccessController 类的 getContext()方法，代码如下。

```
public static AccessControlContext getContext(){
   AccessControlContext acc = getStackAccessControlContext();
   if (acc == null) {
      return new AccessControlContext(null, true);
   } else {
      return acc.optimize();
   }
}
```

这里通过 getStackAccessControlContext()方法获取 AccessControlContext 对象实例，如果获取的 AccessControlContext 对象实例为空，则通过调用 AccessControlContext 类的构造方法实例化，否则，调用 AccessControlContext 对象实例的 optimize()方法返回 AccessControlContext 对象实例。

其中，getStackAccessControlContext()方法的代码如下。

```
private static native AccessControlContext getStackAccessControlContext();
```

getStackAccessControlContext()方法是一个本地方法，最终调用操作系统的方法来获取访问系统资源的上下文对象。

接下来，继续回到 PrivilegedCallable 类，查看 PrivilegedCallable 类的 call()方法，代码如下。

```
public T call() throws Exception {
   try {
      return AccessController.doPrivileged(
         new PrivilegedExceptionAction<T>() {
            public T run() throws Exception {
               return task.call();
            }
         }, acc);
   } catch (PrivilegedActionException e) {
      throw e.getException();
   }
}
```

在 PrivilegedCallable 类的 call()方法中，通过调用 AccessController.doPrivileged()方法，传递 PrivilegedExceptionAction 接口对象和 AccessControlContext 对象，最终返回泛型的实例对象。

AccessController.doPrivileged()方法的代码如下。

```
@CallerSensitive
public static native <T> T
   doPrivileged(PrivilegedExceptionAction<T> action,
               AccessControlContext context)
   throws PrivilegedActionException;
```

AccessController.doPrivileged()方法也是一个本地方法，会将 PrivilegedExceptionAction 接口对象和 AccessControlContext 对象实例传递给 AccessController.doPrivileged()方法执行。并且最终会在 PrivilegedExceptionAction 接口对象的 run()方法中调用 Callable 接口的 call()方法来执行业务逻辑，返回最终结果。

13.1.3　PrivilegedCallableUsingCurrentClassLoader 实现类

PrivilegedCallableUsingCurrentClassLoader 类表示在已经建立的特定访问控制和当前的类加载器下运行的 Callable 接口类型的类，代码如下。

```
static final class PrivilegedCallableUsingCurrentClassLoader<T> implements
Callable<T> {
    private final Callable<T> task;
    private final AccessControlContext acc;
    private final ClassLoader ccl;
    PrivilegedCallableUsingCurrentClassLoader(Callable<T> task) {
        SecurityManager sm = System.getSecurityManager();
        if (sm != null) {
            sm.checkPermission(SecurityConstants.GET_CLASSLOADER_PERMISSION);
            sm.checkPermission(new RuntimePermission("setContextClassLoader"));
        }
        this.task = task;
        this.acc = AccessController.getContext();
        this.ccl = Thread.currentThread().getContextClassLoader();
    }
    public T call() throws Exception {
        try {
            return AccessController.doPrivileged(
                new PrivilegedExceptionAction<T>() {
                    public T run() throws Exception {
                        Thread t = Thread.currentThread();
                        ClassLoader cl = t.getContextClassLoader();
                        if (ccl == cl) {
                            return task.call();
                        } else {
                            t.setContextClassLoader(ccl);
                            try {
                                return task.call();
                            } finally {
                                t.setContextClassLoader(cl);
                            }
                        }
```

```
                }
            }
        }, acc);
    } catch (PrivilegedActionException e) {
        throw e.getException();
    }
}
```

总体来看，PrivilegedCallableUsingCurrentClassLoader 类的功能还是比较简单的，首先定义了 3 个成员变量，代码如下。

```
private final Callable<T> task;
private final AccessControlContext acc;
private final ClassLoader ccl;
```

然后，通过构造方法注入 Callable 对象，在构造方法中获取系统安全管理器对象实例，通过系统安全管理器对象实例检查是否具有获取 ClassLoader 和设置 ContextClassLoader 的权限，并在构造方法中为 3 个成员变量赋值，代码如下。

```
PrivilegedCallableUsingCurrentClassLoader(Callable<T> task) {
    SecurityManager sm = System.getSecurityManager();
    if (sm != null) {
        sm.checkPermission(SecurityConstants.GET_CLASSLOADER_PERMISSION);
        sm.checkPermission(new RuntimePermission("setContextClassLoader"));
    }
    this.task = task;
    this.acc = AccessController.getContext();
    this.ccl = Thread.currentThread().getContextClassLoader();
}
```

接下来，通过调用 call()方法来执行具体的业务逻辑，代码如下。

```
public T call() throws Exception {
    try {
        return AccessController.doPrivileged(
            new PrivilegedExceptionAction<T>() {
                public T run() throws Exception {
                    Thread t = Thread.currentThread();
                    ClassLoader cl = t.getContextClassLoader();
                    if (ccl == cl) {
                        return task.call();
                    } else {
                        t.setContextClassLoader(ccl);
                        try {
                            return task.call();
                        } finally {
                            t.setContextClassLoader(cl);
                        }
                    }
                }
```

```
      }, acc);
    } catch (PrivilegedActionException e) {
      throw e.getException();
    }
}
```

在 call()方法中同样通过调用 AccessController 类的本地方法 doPrivileged()，传递 PrivilegedExceptionAction 接口的实例对象和 AccessControlContext 类的对象实例。

call()方法的具体执行逻辑是在 PrivilegedExceptionAction 对象的 run()方法中获取当前线程的 ContextClassLoader 对象，如果在构造方法中获取的 ClassLoader 对象与此处的 ContextClassLoader 对象是同一个对象，则直接调用 Callable 对象的 call()方法返回结果。否则，将当前线程的 ContextClassLoader 设置为在构造方法中获取的类加载器对象。接下来，调用 Callable 对象的 call()方法返回结果。最后将当前线程的 ContextClassLoader 重置为线程自己的类加载器对象。

13.1.4　RunnableAdapter 实现类

RunnableAdapter 类的逻辑比较简单，传递运行的任务和返回的结果对象，执行任务并返回结果，代码如下。

```
static final class RunnableAdapter<T> implements Callable<T> {
    final Runnable task;
    final T result;
    RunnableAdapter(Runnable task, T result) {
       this.task = task;
       this.result = result;
    }
    public T call() {
       task.run();
       return result;
    }
}
```

RunnableAdapter 类的代码比较简单，笔者不再赘述其实现逻辑。

13.1.5　TaskCallable 实现类

TaskCallable 类是 javafx.concurrent.Task 类的静态内部类，TaskCallable 类主要实现了 Callable 接口，并被定义为 FutureTask 的类，在这个类中允许通过拦截 call()方法来更新 task 任务的状态。代码如下。

```
private static final class TaskCallable<V> implements Callable<V> {
    private Task<V> task;
    private TaskCallable() { }
    @Override
    public V call() throws Exception {
```

```
            task.started = true;
            task.runLater(() -> {
                task.setState(State.SCHEDULED);
                task.setState(State.RUNNING);
            });
            try {
                final V result = task.call();
                if (!task.isCancelled()) {
                    task.runLater(() -> {
                        task.updateValue(result);
                        task.setState(State.SUCCEEDED);
                    });
                    return result;
                } else {
                    return null;
                }
            } catch (final Throwable th) {
                task.runLater(() -> {
                    task._setException(th);
                    task.setState(State.FAILED);
                });
                if (th instanceof Exception) {
                    throw (Exception) th;
                } else {
                    throw new Exception(th);
                }
            }
        }
    }
}
```

在 call()方法中，首先将 task 对象的 started 属性设置为 true，表示任务已经开始，并将任务的状态依次设置为 State.SCHEDULED 和 State.RUNNING，依次触发任务的调度事件和运行事件。代码如下。

```
task.started = true;
task.runLater(() -> {
    task.setState(State.SCHEDULED);
    task.setState(State.RUNNING);
});
```

然后，在 try 代码块中执行 Task 对象的 call()方法，返回泛型对象。如果任务没有被取消，则更新任务的缓存，将调用 call()方法返回的泛型对象绑定到 Task 对象的 ObjectProperty<V>对象中，其中，ObjectProperty<V>在 Task 类的定义的代码如下。

```
private final ObjectProperty<V> value = new SimpleObjectProperty<>(this, "value");
```

接下来，将任务的状态设置为成功。如下所示。

```
try {
    final V result = task.call();
    if (!task.isCancelled()) {
```

```
        task.runLater(() -> {
            task.updateValue(result);
            task.setState(State.SUCCEEDED);
        });
        return result;
    } else {
        return null;
    }
}
```

如果程序抛出了异常或者错误，则进入 catch()代码块，设置 Task 对象的 Exception 信息并将状态设置为 State.FAILED，也就是将任务标记为失败。最后，判断异常或错误的类型，如果是 Exception 类型的异常，则直接强转为 Exception 类型的异常并抛出。否则，将异常或者错误封装为 Exception 对象并抛出，如下所示。

```
catch (final Throwable th) {
    task.runLater(() -> {
        task._setException(th);
        task.setState(State.FAILED);
    });
    if (th instanceof Exception) {
        throw (Exception) th;
    } else {
        throw new Exception(th);
    }
}
```

13.2 异步编程接口

通过 Future 接口的 get()方法能够非常方便地获取异步线程的返回结果。

13.2.1 两种异步模型

Java 并发编程中有两种异步编程模型，一种直接以异步的形式并行执行任务，不需要返回任务的结果。一种是以异步的形式执行任务，需要返回结果。

1. 无返回结果的异步模型

无返回结果的异步任务可以直接放进线程或线程池中运行，无法直接获得任务的执行结果，可以使用回调方法获取任务的执行结果。

具体的方法是：定义一个回调接口，并在接口中定义接收任务执行结果的方法，具体逻辑在回调接口的实现类中实现。将回调接口与任务参数一同提交到线程或线程池中运行，运行结束后调用回调接口的方法，并将任务执行的结果传入方法中，执行回调接口实现类的逻辑来处理结果。

2. 有返回结果的异步模型

回调接口能够获取异步任务的结果，但使用起来略显复杂。JDK 提供了可以直接返回异步结果的方案。最常用的就是使用 Future 接口或者其实现类 FutureTask 来接收任务的返回结果。

注意：随书源码的 mykit-concurrent-chapter13 工程下的 io.binghe.concurrent.chapter13.call 包和 io.binghe.concurrent.chapter13.future 包下分别给出了无返回结果异步模型和有返回结果异步模型的完整案例程序，限于全书篇幅，这里不再粘贴案例代码，读者可自行下载随书源码后查阅。

13.2.2 Future 接口

Future 接口是 Java 从 JDK 1.5 版本开始提供的异步编程接口，代码如下。

```
public interface Future<V> {
    boolean cancel(boolean mayInterruptIfRunning);
    boolean isCancelled();
    boolean isDone();
    V get() throws InterruptedException, ExecutionException;
    V get(long timeout, TimeUnit unit)
        throws InterruptedException, ExecutionException, TimeoutException;
}
```

通过代码可以看出，Future 接口共定义了 5 种抽象方法，接下来，就对这 5 种方法进行简单的介绍。

（1）cancel(boolean)方法。取消任务的执行，接收一个 boolean 类型的参数，如果成功取消任务则返回 true，否则返回 false。如果任务已经完成、已经结束或者因其他原因不能取消，则返回 false，表示任务取消失败。如果任务未启动，调用了此方法并且结果返回 true（取消成功），则当前任务不再运行。如果任务已经启动，则根据当前传递的 boolean 类型的参数决定是否通过中断当前运行的线程来取消当前运行的任务。

（2）isCancelled()方法。判断任务在完成之前是否被取消，如果在任务完成之前被取消则返回 true，否则返回 false。这里需要注意一个细节：只有在任务未启动或者完成之前被取消，才会返回 true，表示任务已经被成功取消，其他情况都会返回 false。

（3）isDone()方法。判断任务是否已经完成，如果任务正常结束、抛出异常退出、被取消，那么返回 true，表示任务已经完成。

（4）get()方法。当任务完成时，直接返回任务的结果。当任务未完成时，等待任务完成并返回任务的结果。

（5）get(long, TimeUnit)方法。当任务完成时，直接返回任务的结果。当任务未完成时，等待任务完成，并设置超时等待时间。如果在超时时间内任务完成，则返回结果，否则抛出

TimeoutException 异常。

13.2.3　RunnableFuture 接口

Future 接口有一个重要的子接口——RunnableFuture 接口，RunnableFuture 接口不但继承了 Future 接口，也继承了 java.lang.Runnable 接口，代码如下。

```
public interface RunnableFuture<V> extends Runnable, Future<V> {
   void run();
}
```

RunnableFuture 接口比较简单，run()方法就是运行任务时调用的方法。

13.2.4　FutureTask 类

FutureTask 类是 RunnableFuture 接口的一个非常重要的实现类，它实现了 RunnableFuture 接口、Future 接口和 Runnable 接口的所有方法。FutureTask 类的代码比较多，这里不再粘贴它的完整代码，读者可自行到 JDK 的 java.util.concurrent 包下查看。这里直接将 FutureTask 类的代码拆解后进行详细分析。

（1）FutureTask 类中的变量与常量。在 FutureTask 类中，首先定义了一个状态变量 state，这个变量使用了 volatile 关键字修饰，然后定义了几个任务运行时的状态常量，代码如下。

```
private volatile int state;
private static final int NEW          = 0;
private static final int COMPLETING   = 1;
private static final int NORMAL       = 2;
private static final int EXCEPTIONAL  = 3;
private static final int CANCELLED    = 4;
private static final int INTERRUPTING = 5;
private static final int INTERRUPTED  = 6;
```

代码注释中给出了如下几个可能的状态变更流程。

```
NEW -> COMPLETING -> NORMAL
NEW -> COMPLETING -> EXCEPTIONAL
NEW -> CANCELLED
NEW -> INTERRUPTING -> INTERRUPTED
```

接下来，定义了如下几个成员变量。

```
private Callable<V> callable;
private Object outcome;
private volatile Thread runner;
private volatile WaitNode waiters;
```

每个变量的含义如下。

- callable：调用 call()方法执行具体任务。
- outcome：Object 类型，表示通过 get()方法获取的结果或者异常信息。

- runner：运行 Callable 的线程，运行期间会使用 CAS 保证线程安全。
- waiters：WaitNode 类型的变量，表示等待线程的堆栈，在 FutureTask 的实现中，会通过 CAS 结合此堆栈交换任务的运行状态。

最后，看一下 WaitNode 类的定义，代码如下。

```
static final class WaitNode {
    volatile Thread thread;
    volatile WaitNode next;
    WaitNode() { thread = Thread.currentThread(); }
}
```

WaitNode 类是 FutureTask 类的静态内部类，类中定义了一个 Thread 成员变量和指向下一个 WaitNode 节点的引用，通过构造方法将 thread 变量设置为当前线程。

（2）构造方法。FutureTask 类提供了两种构造方法，代码如下。

```
public FutureTask(Callable<V> callable) {
    if (callable == null)
        throw new NullPointerException();
    this.callable = callable;
    this.state = NEW;
}

public FutureTask(Runnable runnable, V result) {
    this.callable = Executors.callable(runnable, result);
    this.state = NEW;
}
```

FutureTask 类的构造方法的代码比较简单。

（3）是否取消与完成的方法。FutureTask 类中是否取消与完成的方法的代码如下。

```
public boolean isCancelled() {
    return state >= CANCELLED;
}

public boolean isDone() {
    return state != NEW;
}
```

这两种方法都是通过任务的状态来判定任务是已被取消还是已完成的。FutureTask 类中对状态常量的定义是有规律的，大于或等于 CANCELLED 的常量为 CANCELLED、INTERRUPTING 和 INTERRUPTED，这 3 个状态均可以表示线程已经被取消。当状态不等于 NEW 时，可以表示任务已经完成。

（4）取消方法。FutureTask 类的取消方法的代码如下。

```
public boolean cancel(boolean mayInterruptIfRunning) {
    if (!(state == NEW &&
          UNSAFE.compareAndSwapInt(this, stateOffset, NEW,
```

```
            mayInterruptIfRunning ? INTERRUPTING : CANCELLED)))
        return false;
    try {
        if (mayInterruptIfRunning) {
            try {
                Thread t = runner;
                if (t != null)
                    t.interrupt();
            } finally {
                UNSAFE.putOrderedInt(this, stateOffset, INTERRUPTED);
            }
        }
    } finally {
        finishCompletion();
    }
    return true;
}
```

在 cancel(boolean)方法中，首先判断任务的状态和 CAS 的操作结果，如果任务的状态不等于 NEW 或 CAS 的操作返回 false，则直接返回 false，表示任务取消失败。如下所示。

```
if (!(state == NEW &&
      UNSAFE.compareAndSwapInt(this, stateOffset, NEW,
          mayInterruptIfRunning ? INTERRUPTING : CANCELLED)))
    return false;
```

然后，在 try 代码块中判断是否可以通过中断当前任务所在的线程来取消任务，如果可以，则以一个 Thread 临时变量来指向运行任务的线程，当指向的变量不为空时，调用线程对象的 interrupt()方法来中断线程的运行。最后将线程标记为被中断的状态。如下所示。

```
try {
    if (mayInterruptIfRunning) {
        try {
            Thread t = runner;
            if (t != null)
                t.interrupt();
        } finally {
            UNSAFE.putOrderedInt(this, stateOffset, INTERRUPTED);
        }
    }
}
```

这里，发现变更任务状态使用的是 UNSAFE.putOrderedInt()方法，UNSAFE.putOrderedInt()方法是一个本地方法，代码如下。

```
public native void putOrderedInt(Object var1, long var2, int var4);
```

接下来，cancel(boolean)方法会进入 finally 代码块，如下所示。

```
finally {
    finishCompletion();
}
```

在 finally 代码块中调用了 finishCompletion()方法，表示结束运行的任务。finishCompletion()方法的代码如下所示。

```java
private void finishCompletion() {
    for (WaitNode q; (q = waiters) != null;) {
        if (UNSAFE.compareAndSwapObject(this, waitersOffset, q, null)) {
            for (;;) {
                Thread t = q.thread;
                if (t != null) {
                    q.thread = null;
                    LockSupport.unpark(t);
                }
                WaitNode next = q.next;
                if (next == null)
                    break;
                q.next = null; // unlink to help gc
                q = next;
            }
            break;
        }
    }
    done();
    callable = null;
}
```

在 finishCompletion()方法中，首先定义一个 for 循环，循环终止的条件是 waiters 为 null，在循环中判断 CAS 操作是否成功，如果成功则执行 if 条件中的逻辑。定义一个 for 自旋循环，在自旋循环体中唤醒 WaitNode 堆栈中的线程，使其运行完成。然后，当 WaitNode 堆栈中的线程执行完毕后，通过 break 退出外层 for 循环。最后调用 done()方法，done()方法的代码如下。

```java
protected void done() { }
```

done()方法是一个空的方法体，由子类实现具体的业务逻辑。当需要在具体的业务开发过程中取消任务时，执行一些额外的业务逻辑，就可以通过在子类中覆写 done()方法来实现。

（5）get()方法。FutureTask 类中实现了两个 get()方法，代码如下。

```java
public V get() throws InterruptedException, ExecutionException {
    int s = state;
    if (s <= COMPLETING)
        s = awaitDone(false, 0L);
    return report(s);
}

public V get(long timeout, TimeUnit unit)
    throws InterruptedException, ExecutionException, TimeoutException {
    if (unit == null)
        throw new NullPointerException();
    int s = state;
    if (s <= COMPLETING &&
```

```
      (s = awaitDone(true, unit.toNanos(timeout))) <= COMPLETING)
    throw new TimeoutException();
  return report(s);
}
```

无参数的 get()方法表示在任务未运行完成时会阻塞，直到返回任务结果。有参数的 get()方法表示在任务未运行完成，并且等待时间超出了超时时间时，会抛出 TimeoutException 异常。

两个 get()方法实现的主体逻辑基本相同，一个没有超时设置，一个有超时设置，主要的逻辑就是判断任务的当前状态是否小于或等于 COMPLETING，也就是说，任务是 NEW 状态还是 COMPLETING 状态，调用 awaitDone()方法，awaitDone()方法的代码如下。

```
private int awaitDone(boolean timed, long nanos)
    throws InterruptedException {
  final long deadline = timed ? System.nanoTime() + nanos : 0L;
  WaitNode q = null;
  boolean queued = false;
  for (;;) {
    if (Thread.interrupted()) {
      removeWaiter(q);
      throw new InterruptedException();
    }

    int s = state;
    if (s > COMPLETING) {
      if (q != null)
        q.thread = null;
      return s;
    }
    else if (s == COMPLETING) // cannot time out yet
      Thread.yield();
    else if (q == null)
      q = new WaitNode();
    else if (!queued)
      queued = UNSAFE.compareAndSwapObject(this, waitersOffset,
                                           q.next = waiters, q);
    else if (timed) {
      nanos = deadline - System.nanoTime();
      if (nanos <= 0L) {
        removeWaiter(q);
        return state;
      }
      LockSupport.parkNanos(this, nanos);
    }
    else
      LockSupport.park(this);
  }
}
```

在 awaitDone()方法中，最重要的就是 for 自旋循环，在循环中首先判断当前线程是否被中断，如果已经被中断，则调用 removeWaiter()将当前线程从堆栈中移除，并且抛出 InterruptedException 异常，如下所示。

```
if (Thread.interrupted()) {
    removeWaiter(q);
    throw new InterruptedException();
}
```

接下来，判断任务的当前状态是否完成，如果完成，并且堆栈句柄不为空，则将堆栈中的当前线程设置为空，返回当前任务的状态，如下所示。

```
int s = state;
if (s > COMPLETING) {
    if (q != null)
        q.thread = null;
    return s;
}
```

当任务的状态为 COMPLETING 时，当前线程让出 CPU 资源，如下所示。

```
else if (s == COMPLETING)
    Thread.yield();
```

如果堆栈为空，则创建堆栈对象，如下所示。

```
else if (q == null)
    q = new WaitNode();
```

如果 queued 变量为 false，则通过 CAS 操作为 queued 赋值。如果 awaitDone()方法传递的 timed 参数为 true，则计算超时时间。如果已超时，则在堆栈中移除当前线程并返回任务状态，如果未超时，则重置超时时间，如下所示。

```
else if (!queued)
    queued = UNSAFE.compareAndSwapObject(this, waitersOffset, q.next = waiters, q);
else if (timed) {
    nanos = deadline - System.nanoTime();
    if (nanos <= 0L) {
        removeWaiter(q);
        return state;
    }
    LockSupport.parkNanos(this, nanos);
}
```

如果不满足上述的所有条件，则将当前线程设置为等待状态，如下所示。

```
else
    LockSupport.park(this);
```

接下来，回到 get()方法中，当 awaitDone()方法返回结果，或者任务的状态不满足条件时，就调用 report()方法，将当前任务的状态传递到 report()方法中，并返回结果，如下所示。

```
return report(s);
```

report()方法的代码如下。

```
private V report(int s) throws ExecutionException {
    Object x = outcome;
    if (s == NORMAL)
        return (V)x;
    if (s >= CANCELLED)
        throw new CancellationException();
    throw new ExecutionException((Throwable)x);
}
```

report()方法的实现比较简单，先将 outcome 数据赋值给 x 变量，接下来判断接收到的任务状态，如果为 NORMAL，则将 x 强转为泛型类型返回；如果任务的状态大于或等于 CANCELLED，也就是任务已经取消，则抛出 CancellationException 异常；如果是其他情况则抛出 ExecutionException 异常。

（6）set()方法与 setException()方法。FutureTask 类的 set()方法与 setException()方法的代码如下。

```
protected void set(V v) {
    if (UNSAFE.compareAndSwapInt(this, stateOffset, NEW, COMPLETING)) {
        outcome = v;
        UNSAFE.putOrderedInt(this, stateOffset, NORMAL); // final state
        finishCompletion();
    }
}
protected void setException(Throwable t) {
    if (UNSAFE.compareAndSwapInt(this, stateOffset, NEW, COMPLETING)) {
        outcome = t;
        UNSAFE.putOrderedInt(this, stateOffset, EXCEPTIONAL); // final state
        finishCompletion();
    }
}
```

通过代码可以看出，set()方法与 setException()方法的整体逻辑基本相同，只是在设置任务状态时一个将状态设置为 NORMAL，一个将状态设置为 EXCEPTIONAL。

（7）run()方法与 runAndReset()方法。FutureTask 类的 run()方法的代码如下。

```
public void run() {
    if (state != NEW ||
        !UNSAFE.compareAndSwapObject(this, runnerOffset, null,
                                     Thread.currentThread()))
        return;
    try {
        Callable<V> c = callable;
        if (c != null && state == NEW) {
            V result;
            boolean ran;
            try {
                result = c.call();
                ran = true;
```

```
            } catch (Throwable ex) {
                result = null;
                ran = false;
                setException(ex);
            }
            if (ran)
                set(result);
        }
    } finally {
        runner = null;
        int s = state;
        if (s >= INTERRUPTING)
            handlePossibleCancellationInterrupt(s);
    }
}
```

只要使用了 Future 和 FutureTask，就必然会调用 run()方法来执行任务，掌握 run()方法的流程是非常有必要的。在 run()方法中，如果当前状态不是 NEW，或者 CAS 操作返回的结果为 false，则直接返回，不再执行后续逻辑，如下所示。

```
if (state != NEW ||
    !UNSAFE.compareAndSwapObject(this, runnerOffset, null, Thread.currentThread()))
    return;
```

接下来，在 try 代码块中，将成员变量 callable 赋值给一个临时变量 c，如果临时变量不等于 null，并且任务状态为 NEW，则调用 Callable 接口的 call()方法并接收结果，同时将 ran 变量设置为 true。当程序抛出异常时，将接收结果的变量设置为 null、ran 变量设置为 false，并调用 setException()方法将任务的状态设置为 EXCEPTIONA。如果 ran 变量为 true，则调用 set()方法，如下所示。

```
try {
    Callable<V> c = callable;
    if (c != null && state == NEW) {
        V result;
        boolean ran;
        try {
            result = c.call();
            ran = true;
        } catch (Throwable ex) {
            result = null;
            ran = false;
            setException(ex);
        }
        if (ran)
            set(result);
    }
}
```

接下来，程序会进入 finally 代码块中，如下所示。

```
finally {
    runner = null;
    int s = state;
    if (s >= INTERRUPTING)
        handlePossibleCancellationInterrupt(s);
}
```

这里，将 runner 设置为 null，如果任务的当前状态大于或等于 INTERRUPTING，也就是线程被中断了，则调用 handlePossibleCancellationInterrupt() 方法。handlePossibleCancellationInterrupt() 方法的代码如下。

```
private void handlePossibleCancellationInterrupt(int s) {
    if (s == INTERRUPTING)
        while (state == INTERRUPTING)
            Thread.yield();
}
```

handlePossibleCancellationInterrupt() 方法的实现比较简单，当任务的状态为 INTERRUPTING 时，使用 while() 循环，将当前线程占用的 CPU 资源释放。

runAndReset() 方法的逻辑与 run() 基本相同，只是 runAndReset() 方法会在 finally 代码块中将任务状态重置为 NEW。

（8）removeWaiter() 方法。removeWaiter() 方法主要使用自旋循环的方式移除 WaitNode 中的线程，代码如下。

```
private void removeWaiter(WaitNode node) {
    if (node != null) {
        node.thread = null;
        retry:
        for (;;) {
            for (WaitNode pred = null, q = waiters, s; q != null; q = s) {
                s = q.next;
                if (q.thread != null)
                    pred = q;
                else if (pred != null) {
                    pred.next = s;
                    if (pred.thread == null) // check for race
                        continue retry;
                }
                else if (!UNSAFE.compareAndSwapObject(this, waitersOffset, q, s))
                    continue retry;
            }
            break;
        }
    }
}
```

在 FutureTask 类的最后，有 UNSAFE 相关的代码。

13.3 CompletableFuture 类

CompletableFuture 是 Java 从 JDK 1.8 版本开始提供的非常强大的异步编程工具类，使用 CompletableFuture 类能够完成复杂的异步编程任务。

13.3.1 概述

在使用 Future 接口进行异步编程时需要注意，在使用无参数的 get()方法获取结果时可能长时间阻塞，所以，尽量使用 Future 带有超时的 get()方法获取异步任务的执行结果。另外，在使用 Future 接口进行异步编程时，如果任务已经完成，任务的状态就无法变更了。

Future 接口是 JDK 提供的一种非常优秀的用于获取异步任务执行结果的接口，但 Future 接口也存在如下限制。

（1）Future 接口只提供了 get()方法获取异步任务，并且 get()方法是阻塞的，无法并发执行多个任务。

（2）Future 接口无法做到运行多个任务，并在所有任务结束后执行特定的业务逻辑。也就是说，使用 Future 接口无法合并多个任务的执行结果。

（3）Future 接口无法做到执行完任务后执行其他逻辑，也就是说，使用 Future 接口无法完成多个任务的串行化处理。

（4）Future 接口无法很好地处理异常。

Future 接口的这些局限性在 CompletableFuture 类和 CompletionService 中得到了很好的解决。CompletableFuture 类除了解决了 Future 接口存在的不足，还实现了 CompletionStage 接口，以此实现了各种复杂的任务编排功能。

13.3.2 初始化

CompletableFuture 类提供了一个默认的无参构造方法，代码如下。

```
public CompletableFuture() {
}
```

但是这种构造方法并不常用，在实际开发过程中，更多的是使用 runAsync()方法和 supplyAsync()方法进行初始化，代码如下。

```
public static <U> CompletableFuture<U> supplyAsync(Supplier<U> supplier) {
    return asyncSupplyStage(asyncPool, supplier);
}
public static <U> CompletableFuture<U> supplyAsync(Supplier<U> supplier, Executor executor) {
    return asyncSupplyStage(screenExecutor(executor), supplier);
}
```

```
public static CompletableFuture<Void> runAsync(Runnable runnable) {
    return asyncRunStage(asyncPool, runnable);
}
public static CompletableFuture<Void> runAsync(Runnable runnable, Executor executor)
{
    return asyncRunStage(screenExecutor(executor), runnable);
}
```

runAsync()方法和 supplyAsync()方法都是静态方法，都会返回一个新的 CompletableFuture 对象，由于 Runnable 接口的 run()方法没有返回值，所以在传入 Runnable 接口的对象时，无法获取到异步任务的返回结果。由于 Supplier 接口的 get()方法有返回值，所以在传入 Supplier 接口的对象时，能够获取到异步任务的返回结果。

runAsync()方法和 supplyAsync()方法的区别是 runAsync()方法没有返回数据，而 supplyAsync()方法有返回数据。

另外，无论是 runAsync()方法还是 supplyAsync()方法，都提供了一个可以传入 Executor 线程池的方法，如果在调用 runAsync()方法或者 supplyAsync()方法时传入了自定义的线程池，就会使用自定义的线程池执行异步任务，否则，就使用默认的 ForkJoinPool 线程池执行任务。

ForkJoinPool 线程池默认创建的线程数等于 CPU 的核心数，也可以通过 JVM 参数进行修改，如下所示。

```
-Djava.util.concurrent.ForkJoinPool.common.parallelism=8
```

上述参数表示将 ForkJoinPool 线程池创建的线程数修改为 8。如果直接使用默认的 ForkJoinPool 线程池执行异步任务，就会导致所有任务共享一个线程池，如果有些任务执行很慢，就会影响其他任务的执行，进而影响整个系统的性能。

所以，在使用 CompletableFuture 类进行异步编程时，最好使用带有 Executor 参数的 runAsync()方法和 supplyAsync()方法为每个不同业务类型的异步任务指定不同的线程池，以避免业务之间相互干扰。

初始化示例代码如下。

```
//无返回结果的初始化
public void runInit(){
    CompletableFuture.runAsync(() -> {
        System.out.println("无返回结果的异步任务");
    });
}
//有返回结果的初始化
public String supplyInit() throws Exception {
    CompletableFuture<String> future = CompletableFuture.supplyAsync(() -> {
        return "binghe";
    });
    return future.get();
}
```

注意：由于CompletableFuture类实现了CompletionStage接口，后续在介绍串行执行任务、并行执行任务、AND聚合任务、OR聚合任务和处理结果的方法时，除并行执行任务外，其他的方法都会按照CompletionStage接口的代码进行介绍。

13.3.3 串行执行任务

CompletableFuture 类实现了 CompletionStage 接口，CompletionStage 接口中定义thenApply()、thenAccept()、thenRun()和thenCompose()这4个系列串行化执行任务的方法，代码如下。

```
//有返回处理结果的串行化执行任务
public <U> CompletionStage<U> thenApply(Function<? super T,? extends U> fn);
public <U> CompletionStage<U> thenApplyAsync(Function<? super T,? extends U> fn);
public <U> CompletionStage<U> thenApplyAsync (Function<? super T,? extends U> fn,
Executor executor);

//无返回处理结果的串行化执行任务
public CompletionStage<Void> thenAccept(Consumer<? super T> action);
public CompletionStage<Void> thenAcceptAsync(Consumer<? super T> action);
public CompletionStage<Void> thenAcceptAsync(Consumer<? super T> action, Executor
executor);

//无返回处理结果的串行化执行任务
public CompletionStage<Void> thenRun(Runnable action);
public CompletionStage<Void> thenRunAsync(Runnable action);
public CompletionStage<Void> thenRunAsync(Runnable action, Executor executor);

//有返回处理结果的串行化执行任务
public <U> CompletionStage<U> thenCompose (Function<? super T, ?
extends CompletionStage<U>> fn);
public <U> CompletionStage<U> thenComposeAsync (Function<? super T, ?
extends CompletionStage<U>> fn);
public <U> CompletionStage<U> thenComposeAsync (Function<? super T, ?
extends CompletionStage<U>> fn,
Executor executor);
```

每个系列的方法说明如下。

（1）thenApply()系列方法。接收Function类型以及Executor类型的参数，返回一个具有处理结果的CompletionStage对象，表示上一个任务正常执行完毕后再开始执行下一个任务，将上一个任务的执行结果作为下一个任务的执行参数，有返回值。

（2）thenAccept()系列方法。接收Consumer类型以及Executor类型的参数，返回一个没有处理结果的CompletionStage对象。与thenApply()系列的方法功能相同，只是无返回值。

（3）thenRun()系列方法。接收Runnable类型以及Executor类型的参数，返回一个没有处理结果的 CompletionStage 对象。表示无须关注上一个任务的执行结果，直接执行下一个任务。

（4）thenCompose()系列方法。接收 Function 类型以及 Executor 类型的参数，返回一个具有处理结果的 CompletionStage 对象。表示在某个任务执行完毕后，将该任务的执行结果作为方法的入参并执行指定的方法，最终生成一个新的 CompletableFuture 对象。

串行化执行任务的代码示例如下。

```java
public void serialization() throws Exception {
    CompletableFuture<String> future = CompletableFuture.supplyAsync(() -> {
        return "Hello";
    }).thenApply((s) -> s + " binghe")
            .thenApply((s) -> s.replace("binghe", "binghe001"));
    System.out.println(future.get());
}
```

13.3.4　并行执行任务

CompletionStage 接口并没有单独提供并行执行任务的方法，但是在 CompletableFuture 类提供了并行执行任务的方法，代码如下。

```java
public static CompletableFuture<Void> allOf(CompletableFuture<?>... cfs) {
    return andTree(cfs, 0, cfs.length - 1);
}

public static CompletableFuture<Object> anyOf(CompletableFuture<?>... cfs) {
    return orTree(cfs, 0, cfs.length - 1);
}
```

每种方法的说明如下。

（1）allOf()方法。当所有给定的 CompletableFuture 执行完成时，返回一个新的 CompletableFuture。

（2）anyOf()方法。只要有一个 CompletableFuture 执行完成，就返回一个新的 CompletableFuture。

并行化执行任务的代码示例如下。

```java
public void parallelization() throws Exception{
    CompletableFuture<String> future1 = CompletableFuture.supplyAsync(() -> {
        return "Hello binghe001";
    });
    CompletableFuture<String> future2 = CompletableFuture.supplyAsync(() -> {
        return "Hello binghe002";
    });
    CompletableFuture<Object> future = CompletableFuture.anyOf(future1, future2);
    System.out.println(future.get());
}
```

上述代码使用了 anyOf()，表示 future1 和 future2 只要有一个执行完毕就返回，如果希望 future1 和 future2 都执行完毕后再返回，那么可以将 anyOf() 方法修改成 allOf() 方法，不过 allOf() 方法不会返回处理结果。

13.3.5　AND 聚合任务

CompletionStage 接口提供的 AND 聚合任务方法主要是 thenCombine()、thenAcceptBoth() 和 runAfterBoth() 系列的方法，代码如下。

```
public <U,V> CompletionStage<V> thenCombine(CompletionStage<? extends U> other,
BiFunction<? super T,? super U,? extends V> fn);
public <U,V> CompletionStage<V> thenCombineAsync(CompletionStage<? extends U> other,
          BiFunction<? super T,? super U,? extends V> fn);
public <U,V> CompletionStage<V> thenCombineAsync(CompletionStage<? extends U> other,
 BiFunction<? super T,? super U,? extends V> fn, Executor executor);

public <U> CompletionStage<Void> thenAcceptBoth(CompletionStage<? extends U> other,
                BiConsumer<? super T, ? super U> action);
public <U> CompletionStage<Void> thenAcceptBothAsync(CompletionStage<? extends U> other,
                BiConsumer<? super T, ? super U> action);
public <U> CompletionStage<Void> thenAcceptBothAsync(CompletionStage<? extends U> other,
                BiConsumer<? super T, ? super U> action, Executor executor);

public CompletionStage<Void> runAfterBoth(CompletionStage<?> other, Runnable action);
public CompletionStage<Void> runAfterBothAsync(CompletionStage<?> other, Runnable action);
public CompletionStage<Void> runAfterBothAsync(CompletionStage<?> other, Runnable action, Executor executor);
```

每个系列的方法说明如下。

（1）thenCombine() 系列方法。传入 CompletionStage 类型和 BiFunction 类型以及 Executor 类型的参数，返回一个带有处理结果的 CompletionStage 对象。表示合并前两个任务正常完成的结果，有返回数据。

（2）thenAcceptBoth() 系列方法。传入 CompletionStage 类型和 BiConsumer 类型以及 Executor 类型的参数，返回一个没有处理结果的 CompletionStage 对象。表示合并前两个任务正常完成的结果，无返回数据。

（3）runAfterBoth() 系列方法。传入 CompletionStage 类型和 Runnable 类型以及 Executor 类型的参数，返回一个没有处理结果的 CompletionStage 对象。表示合并前两个任务正常完成的结果，无返回数据。

AND 聚合任务的代码示例如下。

```java
public void andTask() throws Exception {
    CompletableFuture<Integer> future1 = CompletableFuture.supplyAsync(() -> {
        int count = new Random().nextInt(10);
        System.out.println("任务 1 结果: " + count);
        return count;
    });
    CompletableFuture<Integer> future2 = CompletableFuture.supplyAsync(() -> {
        int count = new Random().nextInt(10);
        System.out.println("任务 2 结果: " + count);
        return count;
    });
    CompletableFuture<Integer> future = future1
        .thenCombine(future2, (x ,y) -> {
            return x + y;
        });
    System.out.println("组合后结果: " + future.get());
}
```

13.3.6　OR 聚合任务

CompletionStage 接口提供的 OR 聚合任务方法主要是 applyToEither()、acceptEither()和 runAfterEither()系列的方法，代码如下。

```java
public <U> CompletionStage<U> applyToEither(CompletionStage<? extends T> other,
                    Function<? super T, U> fn);
public <U> CompletionStage<U> applyToEitherAsync(CompletionStage<? extends T> other,
                    Function<? super T, U> fn);
public <U> CompletionStage<U> applyToEitherAsync(CompletionStage<? extends T> other,
                    Function<? super T, U> fn, Executor executor);
public CompletionStage<Void> acceptEither(CompletionStage<? extends T> other,
                    Consumer<? super T> action);
public CompletionStage<Void> acceptEitherAsync(CompletionStage<? extends T> other,
                    Consumer<? super T> action);
public CompletionStage<Void> acceptEitherAsync(CompletionStage<? extends T> other,
                    Consumer<? super T> action, Executor executor);
public CompletionStage<Void> runAfterEither(CompletionStage<?> other,
                    Runnable action);
public CompletionStage<Void> runAfterEitherAsync(CompletionStage<?> other,
                    Runnable action);
public CompletionStage<Void> runAfterEitherAsync(CompletionStage<?> other,
                    Runnable action, Executor executor);
```

每个系列的方法说明如下。

（1）applyToEither()系列方法。表示 CompletionStage 类型、Function 类型及 Executor 类型的参数，返回一个带有处理结果的 CompletionStage 对象。表示获得最先完成任务的结果，有返回值。

（2）acceptEither()系列方法。传入 CompletionStage 类型、Consumer 类型及 Executor 类型的参数，返回一个没有处理结果的 CompletionStage 对象。表示获得最先完成任务的结果，无返回值。

（3）runAfterEither()系列方法。传入 CompletionStage 类型、Runnable 类型及 Executor 类型的参数，返回一个没有处理结果的 CompletionStage 对象。表示获得最先完成任务的结果，无返回值。

OR 聚合任务的代码示例如下。

```java
public void orTask() throws Exception {
    CompletableFuture<Integer> future1 = CompletableFuture.supplyAsync(()-> {
        int time = new Random().nextInt(10);
        try {
            TimeUnit.SECONDS.sleep(time);
        } catch (InterruptedException e) {
            e.printStackTrace();
        }
        System.out.println("任务 1 结果:" + time);
        return time;
    });
    CompletableFuture<Integer> future2 = CompletableFuture.supplyAsync(() -> {
        int time = new Random().nextInt(10);
        try {
            TimeUnit.SECONDS.sleep(time);
        } catch (InterruptedException e) {
            e.printStackTrace();
        }
        System.out.println("任务 2 结果:" + time);
        return time;
    });
    CompletableFuture<Integer> future = future1.applyToEither(future2, (t) -> {
        System.out.println("最快返回的结果为结果: " + t);
        return t * 2;
    });
    System.out.println("最终的结果为===>>> " + future.get());
}
```

13.3.7 处理结果

CompletionStage 接口提供的处理结果的方法主要是 exceptionally()、whenComplete()和 handle()系列的方法，代码如下。

```java
public CompletionStage<T> exceptionally(Function<Throwable, ? extends T> fn);

public CompletionStage<T> whenComplete(BiConsumer<? super T, ? super Throwable> action);
public CompletionStage<T> whenCompleteAsync(BiConsumer<? super T, ? super Throwable>
```

```
action);
public CompletionStage<T> whenCompleteAsync(BiConsumer<? super T, ? super Throwable>
action, Executor executor);

public <U> CompletionStage<U> handle(BiFunction<? super T, Throwable, ? extends U>
fn);
public <U> CompletionStage<U> handleAsync(BiFunction<? super T, Throwable, ? extends
U> fn);
public <U> CompletionStage<U> handleAsync(BiFunction<? super T, Throwable, ? extends
U> fn,
                         Executor executor);
```

每个系列的方法说明如下。

（1）exceptionally()系列方法。传入 Function 类型的参数，返回一个带有处理结果的 CompletionStage 对象。表示处理上一个阶段的异常，并且有返回值。

（2）whenComplete()系列方法。传入 BiConsumer 类型及 Executor 类型的参数，返回一个没有处理结果的 CompletionStage 对象。表示获取上一个阶段的执行结果或异常信息，没有返回值。

（3）handle 系列方法。传入 BiFunction 类型及 Executor 类型的参数，返回一个带有处理结果的 CompletionStage 对象。获取上一个阶段的执行结果或异常信息，有返回值。

注意：exceptionally()方法相当于 try{}catch{}代码块中的 catch{}代码块，whenComplete()系列方法和 handle()系列方法相当于 try{} finally{}代码块中的 finally{}代码块，无论程序是否发生异常，都会执行 whenComplete()系列方法的 BiConsumer 回调函数和 handle()系列方法的 BiFunction 回调函数。whenComplete()系列方法与 handle()系列方法的区别是 whenComplete()系列方法没有返回结果，而 handle()系列方法有返回结果。

处理结果的示例代码如下。

```
public void resultTask(){
    CompletableFuture<Integer> future = CompletableFuture.supplyAsync(() -> 3 * 5)
        .thenApply((r) -> r / 0)
        .thenApply((r) -> r * 5)
        .exceptionally((e) -> {
            System.out.println("抛出的异常信息为:" + e.getMessage());
            return 0;
        })
        .handle((t, u) -> {
            System.out.println("结果数据为:" + t);
            return t;
        });
    System.out.println(future.join());
}
```

注意： 13.3.3 节至 13.3.7 节的示例代码位于随书源码的 mykit-concurrent-chapter13 工程下的 io.binghe.concurrent.chapter13.complete.future.CompletableFutureTest 类中。

13.3.8 使用案例

本节使用 CompletableFuture 类模拟实现从淘宝、京东、当当对比某类图书的价格。由于篇幅限制，完整的案例代码在随书源码的 mykit-concurrent-chapter13 工程下的 io.binghe.concurrent.chapter13.example.future 包下给出，这里不再赘述具体的程序实现逻辑，读者可自行下载随书源码后查阅。

13.4 CompletionService

CompletionService 是 JDK 提供的一个批量执行异步任务的高性能接口。

13.4.1 概述

使用 Future 集合 Callable 的方式也能够实现多个任务的并行执行，但是在任务的执行过程中，如果前面的任务执行较慢，则后面的任务需要阻塞等待前面的任务执行完毕后才能执行。

CompletionService 能够提交任务也能获取任务的返回结果，提交任务和获取结果分开执行，多个任务同时执行不会互相阻塞，先获取到的任务是先完成的任务，不再依赖任务提交的顺序。

CompletionService 内部使用阻塞队列和 FutureTask 实现了优先获取先完成的任务，也就是说，从 CompletionService 获取到的结果按照任务完成的先后时间排列。CompletionService 内部的阻塞队列用于保存已经执行完成的 Future 结果。通过调用 Future 的 pool() 方法和 take() 方法，能够获取一个已完成的任务对应的 Future 结果，通过调用 Future 的 get() 方法获取任务最终的执行结果。

使用 CompletionService 批量执行异步任务比使用 Future 结合 Callable 有如下优势。

（1）CompletionService 非常巧妙地将线程池 Executor 和阻塞队列 BlockingQueue 的功能进行了整合，能够让批量异步任务的管理更加简单、便捷，让程序开发人员更加专注业务逻辑的开发，而不用自行编写处理批量异步任务的复杂流程。

（2）CompletionService 能够让异步任务的执行结果有序，也就是说，能够让先执行完的任务先进入阻塞队列。利用这个特性，可以非常方便地实现基于异步任务结果进行的后续有序性处理，同时可以快速实现 Forking Cluster 相关的需求。

（3）CompletionService 能够做到线程池隔离，ExecutorCompletionService 类是 CompletionService 接口的唯一实现类。在使用 ExecutorCompletionService 类的构造方法初始化 CompletionService 对象时，需要传入一个 Executor 线程池。也就是说，每个 CompletionService

接口类型的对象，都可以单独创建一个对应的线程池，这样就能够做到线程池隔离，避免某些非常耗时的任务拖垮整个系统。

> **注意**：Forking Cluster 指系统的一种集群模式，在这种集群模式下，能够以并行的方式查询多个服务，只要有一个查询任务成功返回结果，整个服务就可以返回结果。

13.4.2 接口定义

CompletionService 是 JDK 提供的能够批量执行异步任务的高性能接口，目前，它的实现类为 ExecutorCompletionService。CompletionService 接口的代码如下。

```
public interface CompletionService<V> {
  Future<V> submit(Callable<V> task);
  Future<V> submit(Runnable task, V result);
  Future<V> take() throws InterruptedException;
  Future<V> poll();
  Future<V> poll(long timeout, TimeUnit unit) throws InterruptedException;
}
```

CompletionService 接口的代码定义了 5 个抽象方法，每种方法的具体说明如下。

- Future<V> submit(Callable<V> task)：提交 Callable 任务，返回 Future 结果。
- Future<V> submit(Runnable task, V result)：提交 Runnable 任务和一个接收结果数据的 result，返回 Future 结果。
- Future<V> take()：获取并且移除阻塞队列头部位置的已完成的任务，如果队列不为空则立即返回结果，如果队列为空则阻塞，如果阻塞期间被其他线程中断，则抛出 InterruptedException 异常。
- Future<V> poll()：获取并且移除阻塞队列头部位置的已完成的任务，如果队列不为空则立即返回结果，如果队列为空则返回 null。
- Future<V> poll(long timeout, TimeUnit unit)：在 timeout 时间内获取并且移除阻塞队列头部位置的已完成的任务，如果队列不为空则立即返回结果，如果队列为空则阻塞，直到超过 timeout 时间返回 null。如果阻塞期间被其他线程中断，则抛出 InterruptedException 异常。

13.4.3 ExecutorCompletionService 类的核心实现

在 JDK 1.8 版本中 ExecutorCompletionService 类是 CompletionService 接口的唯一实现类，在 CompletionService 接口中定义了执行异步任务的方法，并且能够实现任务按照执行结果的先后顺序排列，这些功能都是在 ExecutorCompletionService 类中实现的。接下来，就简单剖析下 ExecutorCompletionService 类的实现。

1. 核心成员变量

在 ExecutorCompletionService 类中定义的核心成员变量如下。

```
private final Executor executor;
private final AbstractExecutorService aes;
private final BlockingQueue<Future<V>> completionQueue;
```

各成员变量的具体含义如下。

- executor：执行任务的线程池。
- aes：主要用来创建执行任务。
- completionQueue：存储执行完成的任务。

2. 核心内部类

在 ExecutorCompletionService 类中定义了一个私有内部类 QueueingFuture，代码如下。

```
private class QueueingFuture extends FutureTask<Void> {
    QueueingFuture(RunnableFuture<V> task) {
        super(task, null);
        this.task = task;
    }
    protected void done() { completionQueue.add(task); }
    private final Future<V> task;
}
```

QueueingFuture 类继承了 FutureTask 类，主要是在 FutureTask 类的基础上进行了一些扩展，实现了 FutureTask 类的 done()方法，任务执行完成后就会回调这种方法。在 QueueingFuture 类实现的 done()方法中，直接将任务放入 completionQueue 队列中，这样就能够实现每个任务完成后都立刻被放入 completionQueue 队列，最终 completionQueue 队列中的任务都是按照执行完毕的先后顺序存储的。

3. 构造方法

在 ExecutorCompletionService 类中，主要提供了两种构造方法，代码如下。

```
//构造方法①，只传入一个 Executor 线程池
public ExecutorCompletionService(Executor executor) {
    if (executor == null)
        throw new NullPointerException();
    this.executor = executor;
    this.aes = (executor instanceof AbstractExecutorService) ?
        (AbstractExecutorService) executor : null;
    this.completionQueue = new LinkedBlockingQueue<Future<V>>();
}

//构造方法②，传入一个 Executor 线程池和一个 BlockingQueue 队列
public ExecutorCompletionService(Executor executor,
                                 BlockingQueue<Future<V>> completionQueue) {
    if (executor == null || completionQueue == null)
        throw new NullPointerException();
    this.executor = executor;
    this.aes = (executor instanceof AbstractExecutorService) ?
```

```
        (AbstractExecutorService) executor : null;
    this.completionQueue = completionQueue;
}
```

在 ExecutorCompletionService 类提供的构造方法中，至少需要传入一个 Executor 线程池来完成异步任务，也可以传入一个 BlockingQueue 队列，ExecutorCompletionService 类默认创建的 BlockingQueue 队列是无界的 LinkedBlockingQueue 队列，存在内存溢出的风险。

4．提交任务的方法

ExecutorCompletionService 类实现了 CompletionService 接口定义的提交任务的方法，代码如下。

```
public Future<V> submit(Callable<V> task) {
    if (task == null) throw new NullPointerException();
    RunnableFuture<V> f = newTaskFor(task);
    executor.execute(new QueueingFuture(f));
    return f;
}

public Future<V> submit(Runnable task, V result) {
    if (task == null) throw new NullPointerException();
    RunnableFuture<V> f = newTaskFor(task, result);
    executor.execute(new QueueingFuture(f));
    return f;
}
```

在 ExecutorCompletionService 类提交任务前，会将任务封装成 QueueingFuture 对象，当任务执行完成后，会回调 QueueingFuture 类的 done()方法，将任务放入 completionQueue 队列。

5．获取完成的任务方法

ExecutorCompletionService 类中实现了 CompletionService 接口中定义的获取完成任务的方法，代码如下。

```
public Future<V> take() throws InterruptedException {
    return completionQueue.take();
}
public Future<V> poll() {
    return completionQueue.poll();
}
public Future<V> poll(long timeout, TimeUnit unit)
        throws InterruptedException {
    return completionQueue.poll(timeout, unit);
}
```

获取完成的任务的方法直接从 completionQueue 队列中获取并移除队列头部的元素，具体说明可以参见 13.4.2 节。

13.4.4 使用案例

本节使用 CompletionService 的 Forking Cluster 模式模拟实现从淘宝、京东、当当获取某类商品的价格，只要有一个平台返回价格，服务就返回结果。由于篇幅限制，在随书源码的 mykit-concurrent-chapter13 工程下的 io.binghe.concurrent.chapter13.example. service 包下给出了完整的案例代码，这里不再赘述具体的程序实现逻辑，读者可自行下载随书源码后查阅。

线程池核心技术篇

第 14 章

线程池总体结构

在真实高并发场景下，一般不会直接使用 Thread 类创建线程，而是使用线程池来创建并管理线程。可以这么说，学好线程池对于并发编程是非常重要的。

14.1 线程池简介

线程池的创建和回收是一个非常消耗系统资源的过程，如果在系统中频繁地创建和回收线程，会极大降低程序的执行性能。并且，短时间内创建大量的线程可能造成 CPU 占用 100%、死机或内存溢出等问题。而使用线程池就能非常轻松地解决这些问题。

14.1.1 线程池核心类继承关系

线程池是 Java 从 JDK 1.5 版本开始提供的一种线程使用模式，能够自动创建和回收线程，并管理线程的生命周期。在线程池中能够管理和维护多个线程。

Java 的线程池主要是通过 Executor 框架实现的，涉及 Executor 接口、ExecutorServcie 接口、AbstractExecutorService 抽象类、ScheduledExecutorService 接口、ThreadPoolExecutor 类和 ScheduledThreadPoolExecutor 类。线程池核心类继承关系如图 14-1 所示。

实现线程池最核心的类是 ThreadPoolExecutor，而 ScheduledThreadPoolExecutor 类实现了定时任务功能，能够使提交到线程池中的任务定时、定期执行。为了便于创建线程池，除了图 14-1 所示的接口和类，JDK 还提供了一个 Executors 工具类，Executors 类中封装了创建线程池的各种方法，专门用于创建线程池。不过，在真实的高并发场景下，并不推荐使用 Executors 工具类创建线程池，而是推荐直接使用 ThreadPoolExecutor 类创建线程池。

```
           Executor
              ▲
              │
         ExecutorService
              ▲
         ┌────┴─────────┐
  AbstractExecutorService   ScheduledExecutorService
         ▲                        ▲
         │                        │
   ThreadPoolExecutor             │
         ▲                        │
         └────────┬───────────────┘
                  │
         ScheduledThreadPoolExecutor
```

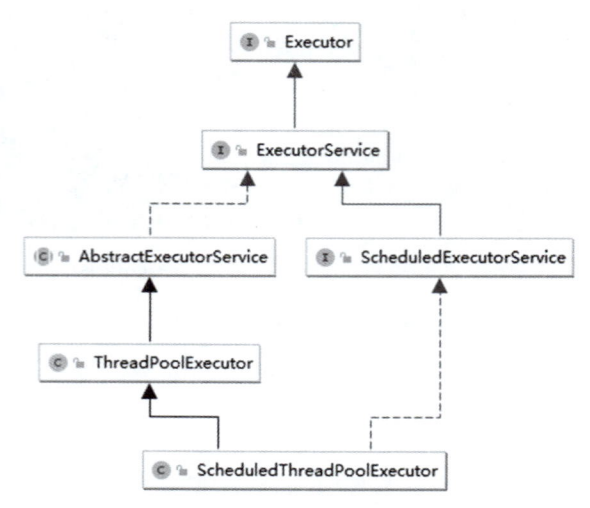

图 14-1　线程池核心类继承关系

14.1.2　线程池的优点

这里，综合对比直接使用 Thread 类创建线程的弊端与使用线程池的优点，来加深读者对线程池的理解。

1. 直接使用 Thread 类创建线程的缺点

直接在程序中使用 Thread 类创建线程的方式是非常不可取的，主要体现在如下几方面。

（1）每次通过 Thread 类创建一个线程对象的性能是非常差的，每次创建 Thread 对象后，调用 Thread 的 start()方法都会在操作系统层面分配一个与之对应的线程，这个过程比较耗时。

（2）直接使用 Thread 类创建线程缺乏有效的统一管理机制，如果在短时间内创建大量线程，线程之间就会竞争系统资源，可能造成 CPU 占用 100%、死机或者内存溢出等问题。

（3）直接使用 Thread 类创建线程提供的线程功能非常有限，例如，无法让线程执行更多的任务、无法定期执行某些任务等。

（4）直接使用 Thread 类创建线程，无法对线程进行有效监控。

2. 使用线程池管理线程的优点

使用线程池能够非常容易地解决直接使用 Thread 创建线程产生的问题，主要体现在如下几方面。

（1）线程池能够复用线程资源，有效减少了线程的创建和回收频率，减少了线程的创建与回收对系统性能造成的影响，比直接使用 Thread 类创建线程的系统性能高。

（2）使用线程池能够有效控制最大并发线程数，提高系统资源的利用率。创建的线程数是可控的，短时间内不会因为创建大量的线程导致线程过多地竞争资源，引起线程阻塞。

（3）在线程池中可以定时或定期执行某个或某些任务，提供了单线程执行任务的机制，也能够控制并发线程数。

（4）线程池提供了监控线程资源的方法，可以对线程池中的线程资源进行实时监控。

14.1.3　Executors 类

为了更加方便地创建线程池，Java 从 JDK 1.5 版本开始专门提供了 Executors 类。Executors 类提供了丰富的创建线程池的方法。

（1）Executors.newCachedThreadPool()方法。创建一个可缓存的线程池，如果线程池中的线程数量超过程序处理的需要，则线程池可以根据具体情况灵活回收空闲的线程。在向线程池提交任务时，如果线程池中没有空闲的线程，则新建线程处理任务。

（2）Executors.newFixedThreadPool()方法。创建一个定长的线程池，可以控制线程池中线程的最大并发数。在向线程池提交任务时，如果线程池中有空闲线程，则分配一个空闲线程执行任务。如果线程池中没有空闲线程，则将提交的任务放入阻塞队列中等待。

（3）Executors.newScheduledThreadPool()方法。创建一个支持定时、周期性执行任务的线程池。

（4）Executors.newSingleThreadExecutor()方法。创建内部只有一个线程的线程池，线程池内部使用唯一的线程执行任务，提交到线程池中的任务都会按照先到先处理的原则串行执行。

（5）Executors.newSingleThreadScheduledExecutor()方法。创建内部只要一个线程，并且支持定时、周期性执行任务的线程池。

（6）Executors.newWorkStealingPool()方法。从 JDK 1.8 版本开始提供的方法，底层使用 ForkJoinPool 实现，创建一个拥有多个任务队列的线程池，可以并行执行任务。

在使用 Executors.newFixedThreadPool()方法和 Executors.newSingleThreadExecutor()方法创建线程池时，内部阻塞队列使用 LinkedBlockingQueue 的默认构造方法进行初始化，默认的 LinkedBlockingQueue 队列的长度为 Integer.MAX_VALUE，使用 Executors.newFixedThreadPool() 方法和 Executors.newSingleThreadExecutor()方法创建的线程池在高并发环境下容易发生内存泄漏。另外，在高并发环境下使用 Executors.newCachedThreadPool()方法创建的线程池容易导致 CPU 占用 100%的问题。

所以，在真正的高并发场景下，不推荐使用 Executors 类创建线程池，推荐直接使用 ThreadPoolExecutor 类创建线程池。

14.1.4　ThreadPoolExecutor 类

ThreadPoolExecutor 是线程池中最核心的类，通过查看 ThreadPoolExecutor 的代码可以得知，在使用 ThreadPoolExecutor 类的构造方法创建线程池时，最终会调用具有 7 个参数的构造方法，代码如下。

```
public ThreadPoolExecutor(int corePoolSize,
                          int maximumPoolSize,
                          long keepAliveTime,
                          TimeUnit unit,
                          BlockingQueue<Runnable> workQueue,
                          ThreadFactory threadFactory,
                          RejectedExecutionHandler rejectHandler)
```

接下来，对 ThreadPoolExecutor 类构造方法中每个参数的具体含义进行简单的介绍。

（1）corePoolSize 参数。表示线程池的核心线程数。

（2）maximumPoolSize 参数。表示线程池中的最大线程数。

（3）keepAliveTime 参数。表示线程没有任务执行状态保持的最长时间。当线程池中的线程数量大于 corePoolSize 时，如果没有新的任务提交，则核心线程外的线程不会立即销毁，需要等待，直到等待的时间超过 keepAliveTime 才会终止。

（4）unit 参数。表示 keepAliveTime 的时间单位。

（5）workQueue 参数。表示线程池中的阻塞队列，存储等待执行的任务。

（6）threadFactory 参数。线程工厂，用来创建线程池中的线程。提供一个默认的线程工厂来创建线程，当使用默认的线程工厂创建线程时，会为线程设置一个名称，使新创建的线程具有相同的优先级，并且是非守护线程。

（7）rejectHandler 参数。表示拒绝处理任务时的策略。当 workQueue 阻塞队列已满、线程池中的线程数已经达到最大，且线程池中没有空闲线程时，如果继续提交任务，就需要采取一种策略来处理这个任务。

其中，在 ThreadPoolExecutor 类的构造方法中，最重要的 3 个参数是 corePoolSize、maximumPoolSize 和 workQueue，这 3 个参数会对线程池的运行过程产生重大的影响。三者的关系如下。

- 如果线程池中运行的线程数小于 corePoolSize，则直接创建新线程处理任务，即使线程池中的其他线程是空闲的。
- 如果运行的线程数大于或等于 corePoolSize 并且小于 maximumPoolSize，则只有当 workQueue 队列满时，才会创建新的线程处理任务。如果 workQueue 队列不满，则将新提交的任务放入 workQueue 队列中。

- 当设置的 corePoolSize 与 maximumPoolSize 相同时，创建的线程池大小是固定的，如果满足有新任务提交、线程池中没有空闲线程，且 workQueue 未满的条件，就把请求放入 workQueue，等待空闲的线程从 workQueue 中取出任务进行处理。
- 如果运行的线程数量大于 maximumPoolSize，同时 workQueue 已满，则通过拒绝策略参数 rejectHandler 来指定处理策略。

线程池提供了 4 种拒绝策略，分别如下。

- 直接抛出异常，这也是默认的策略。实现类为 AbortPolicy。
- 使用调用者所在的线程来执行任务。实现类为 CallerRunsPolicy。
- 丢弃队列中最靠前的任务并执行当前任务。实现类为 DiscardOldestPolicy。
- 直接丢弃当前任务。实现类为 DiscardPolicy。

另外，ThreadPoolExecutor 类提供了启动和停止任务的方法以及可以监控线程池的方法。启动和停止任务的方法如下。

- execute()：将任务提交到线程池执行。
- submit()：将任务提交到线程池执行，并返回结果。
- shutdown()：优雅地关闭线程池，任务都执行完后线程池才退出。
- shutdownNow()：立即关闭线程池，不等待任务执行完线程池就会退出。

可以监控线程池的方法如下。

- getTaskCount()：获取线程池中已经执行和未执行的任务的总数量。
- getCompletedTaskCount()：获取线程池中已完成的任务数量。
- getPoolSize()：获取线程池中当前的线程数量。
- getCorePoolSize()：获取线程池中的核心线程数。
- getActiveCount()：获取当前线程池中正在执行任务的线程数量。

14.2 线程池顶层接口和抽象类

线程池极大地简化了并发编程中线程的创建、复用和管理工作，让开发人员能够更专注于并发编程的业务开发，而不必关注线程管理的细节。理解线程池内部的工作机制和工作原理，有助于开发人员更好地理解并发编程。

14.2.1 接口和抽象类总览

线程池中重要的接口和抽象类包括 Executor 接口、ExecutorServcie 接口、AbstractExecutorService 抽象类和 ScheduledExecutorService 接口，如图 14-2 所示。

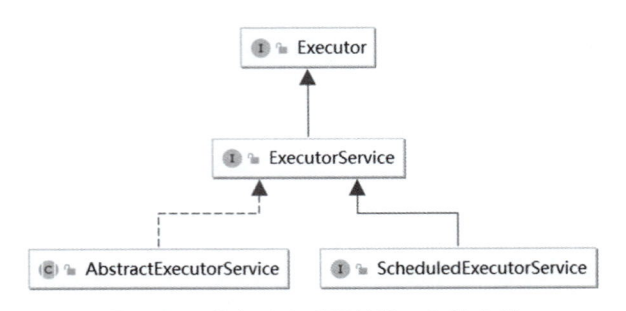

图 14-2　线程池中重要的接口和抽象类

接口和抽象类的作用如下。

- Executor 接口：线程池顶层的接口，提供了一个无返回值的提交任务的方法。
- ExecutorService 接口：派生自 Executor 接口，扩展了很多功能，例如关闭线程池、提交任务并返回结果数据、唤醒线程池中的任务等。
- AbstractExecutorService 抽象类：派生自 ExecutorService 接口，实现了几个非常实用的方法，例如，提交任务并返回结果数据、唤醒线程池中的任务等，供子类调用。
- ScheduledExecutorService 定时任务接口，派生自 ExecutorService 接口，拥有 ExecutorService 接口定义的全部方法，并扩展了定时任务相关的方法。

14.2.2　Executor 接口

Executor 接口是整个线程池设计中顶层的接口，线程池中其他的接口会直接或间接继承 Executor 接口，其他的类会直接或间接实现 Executor 接口。Executor 接口的代码如下。

```
public interface Executor {
    //提交运行任务，参数为Runnable接口对象，无返回值
    void execute(Runnable command);
}
```

Executor 接口的代码比较简单，只提供了一个无返回值的提交任务的 execute(Runnable)方法，使用这个方法不会返回线程池的执行结果。如果使用线程池提交任务后，不需要关心线程池的返回结果，就可以使用 Executor 接口的 execute(Runnable)方法提交任务。但是如果需要获取线程池的执行结果，或者执行完任务需要关闭线程池，就需要使用 Executor 接口的子接口 ExecutorService 来支持。

14.2.3　ExecutorService 接口

ExecutorService 接口是非定时任务类线程池的核心接口，通过 ExecutorService 接口能够向线程池中提交任务（支持有返回结果和无返回结果两种方式）、关闭线程池、唤醒线程池中的任务等。ExecutorService 接口的代码如下。

```java
public interface ExecutorService extends Executor {
    //关闭线程池，线程池中不再接受新提交的任务，但是之前提交的任务可以继续运行，直到完成
    void shutdown();

    //关闭线程池，线程池中不再接受新提交的任务，会停止线程池中正在执行的任务
    List<Runnable> shutdownNow();

    //判断线程池是否关闭
    boolean isShutdown();

    //判断线程池中的所有任务是否结束，只有在调用 shutdown 或者 shutdownNow 方法之后
    //调用此方法才会返回 true
    boolean isTerminated();

    //等待线程池中的所有任务执行结束，并设置超时时间
    boolean awaitTermination(long timeout, TimeUnit unit) throws
InterruptedException;

    //提交一个 Callable 接口类型的任务，返回一个 Future 类型的结果
    <T> Future<T> submit(Callable<T> task);

    //提交一个 Callable 接口类型的任务，并且给定一个泛型类型的接收结果数据参数
    //返回一个 Future 类型的结果
    <T> Future<T> submit(Runnable task, T result);

    //提交一个 Runnable 接口类型的任务，返回一个 Future 类型的结果
    Future<?> submit(Runnable task);

    //批量提交任务并获得 Future 结果列表，Task 列表与 Future 列表一一对应
    <T> List<Future<T>> invokeAll(Collection<? extends Callable<T>> tasks)
        throws InterruptedException;

    //批量提交任务并获得 Future 结果列表，限定处理所有任务的时间
    <T> List<Future<T>> invokeAll(Collection<? extends Callable<T>> tasks,
                        long timeout, TimeUnit unit) throws
InterruptedException;

    //批量提交任务并获得一个已经成功执行的任务的结果
    <T> T invokeAny(Collection<? extends Callable<T>> tasks)
        throws InterruptedException, ExecutionException;

    //批量提交任务并获得一个已经成功执行的任务的结果，限定处理任务的时间
    <T> T invokeAny(Collection<? extends Callable<T>> tasks, long timeout, TimeUnit
unit)
        throws InterruptedException, ExecutionException, TimeoutException;
}
```

这里不再赘述每种方法的具体含义，ExecutorService 接口的方法都是在非定时任务类的线

程池中经常使用的。

14.2.4 AbstractExecutorService 抽象类

AbstractExecutorService 类是线程池的一个抽象类，是核心接口 ExecutorService 的实现类，在 ExecutorService 接口的基础上实现了几个比较实用的方法，提供给子类使用，AbstractExecutorService 类每种方法的实现细节如下。

（1）newTaskFor()方法。AbstractExecutorService 类提供了两个重载的 newTaskFor()方法，代码如下。

```
protected <T> RunnableFuture<T> newTaskFor(Runnable runnable, T value) {
    return new FutureTask<T>(runnable, value);
}

protected <T> RunnableFuture<T> newTaskFor(Callable<T> callable) {
    return new FutureTask<T>(callable);
}
```

从 newTaskFor()方法的代码可以看出，每个 newTaskFor()方法的返回值都是 RunnableFuture，RunnableFuture 是一个实现了 Runnable 和 Future 的接口，主要用于获取线程池中任务的执行结果。在实际使用时，经常使用的是 RunnableFuture 的实现类 FutureTask。newTaskFor 方法的作用是将提交到线程池中的任务封装成 FutureTask 对象，后续将 FutureTask 对象提交到线程池中。

（2）doInvokeAny()方法。doInvokeAny()方法是 AbstractExecutorService 类的一个私有方法，可以批量执行线程池中的任务，最终返回一个结果数据，代码如下。

```
private <T> T doInvokeAny(Collection<? extends Callable<T>> tasks, boolean timed,
long nanos)
    throws InterruptedException, ExecutionException, TimeoutException {
    //提交的任务为空，抛出空指针异常
    if (tasks == null)
        throw new NullPointerException();
    //记录待执行的任务数量
    int ntasks = tasks.size();
    //任务集合中的数据为空，抛出非法参数异常
    if (ntasks == 0)
        throw new IllegalArgumentException();
    ArrayList<Future<T>> futures = new ArrayList<Future<T>>(ntasks);
    //以当前实例对象作为参数构建 ExecutorCompletionService 对象
    // ExecutorCompletionService 负责执行任务，后面调用 poll 返回第 1 个执行结果
    ExecutorCompletionService<T> ecs = new ExecutorCompletionService<T>(this);
    try {
        // 记录可能抛出的异常
        ExecutionException ee = null;
        // 初始化超时时间
        final long deadline = timed ? System.nanoTime() + nanos : 0L;
        Iterator<? extends Callable<T>> it = tasks.iterator();
```

```java
//提交任务,并将返回的结果添加到 futures 集合中
//提交一个任务以确保在进入循环之前开始一个任务
futures.add(ecs.submit(it.next()));
--ntasks;
//记录正在执行的任务数量
int active = 1;
for (;;) {
    //从完成任务的 BlockingQueue 队列中获取并移除下一个将要完成的任务的结果
    //如果 BlockingQueue 队列中的数据为空,则返回 null
    Future<T> f = ecs.poll();
    //获取的结果为空
    if (f == null) {
        //集合中未执行的任务的数量
        if (ntasks > 0) {
            //未执行的任务数量减 1
            --ntasks;
            //提交完成并将结果添加到 futures 集合中
            futures.add(ecs.submit(it.next()));
            //正在执行的任务数量加 1
            ++active;
        }
        //如果所有任务执行完毕,并且返回了结果,则退出循环
        //之所以处理 active 为 0 的情况,是因为 poll()方法是非阻塞方法
        //可能导致未返回结果时 active 为 0
        else if (active == 0)
            break;
        //如果 timed 为 true,则超时获取结果
        else if (timed) {
            f = ecs.poll(nanos, TimeUnit.NANOSECONDS);
            if (f == null)
                throw new TimeoutException();
            nanos = deadline - System.nanoTime();
        }
        //如果没有设置超时时间,并且所有任务都被提交了,则一直阻塞,直到返回一个执行结果
        else
            f = ecs.take();
    }
    //如果获取到执行结果,则将正在执行的任务减 1,从 Future 中获取结果并返回
    if (f != null) {
        --active;
        try {
            return f.get();
        } catch (ExecutionException eex) {
            ee = eex;
        } catch (RuntimeException rex) {
            ee = new ExecutionException(rex);
        }
    }
}
```

```
          if (ee == null)
              ee = new ExecutionException();
          throw ee;
      } finally {
          //如果从所有执行的任务中获取到一个结果，则取消所有执行的任务，不再向下执行
          for (int i = 0, size = futures.size(); i < size; i++)
              futures.get(i).cancel(true);
      }
}
```

doInvokeAny()方法代码的主要逻辑是批量执行线程池中的任务，最终返回一个结果。在doInvokeAny()方法中，不管线程池中执行了多少任务，只要获取到一个任务的返回结果，就会取消线程池中运行的所有任务，并返回结果。

在 doInvokeAny()方法中，向线程池中提交任务使用的是 ExecutorCompletionService 类的submit()方法，代码如下。

```
public Future<V> submit(Callable<V> task) {
    if (task == null) throw new NullPointerException();
    RunnableFuture<V> f = newTaskFor(task);
    executor.execute(new QueueingFuture(f));
    return f;
}

public Future<V> submit(Runnable task, V result) {
    if (task == null) throw new NullPointerException();
    RunnableFuture<V> f = newTaskFor(task, result);
    executor.execute(new QueueingFuture(f));
    return f;
}
```

ExecutorCompletionService 类的 submit()方法本质上还是调用 Executor 接口的 execute()方法向线程池中提交任务。

（3）invokeAny()方法。在 invokeAny()方法中，主要调用 doInvokeAny()方法向线程池中提交多个任务，返回一个结果数据，invokeAny()方法的代码如下。

```
public <T> T invokeAny(Collection<? extends Callable<T>> tasks)
    throws InterruptedException, ExecutionException {
    try {
        return doInvokeAny(tasks, false, 0);
    } catch (TimeoutException cannotHappen) {
        assert false;
        return null;
    }
}

public <T> T invokeAny(Collection<? extends Callable<T>> tasks, long timeout,
TimeUnit unit)
    throws InterruptedException, ExecutionException, TimeoutException {
```

```
        return doInvokeAny(tasks, true, unit.toNanos(timeout));
}
```

可以看出，invokeAny()方法内部通过直接调用 doInvokeAny()方法向线程池中提交多个任务，只要返回一个任务的结果即可。

如果需要线程池在执行多个任务时，只要一个任务返回结果就立即返回，那么可以使用 invokeAny()方法。

（4）invokeAll()方法。invokeAll()方法实现了无超时时间设置和有超时时间设置的逻辑，先来看无超时时间设置 invokeAll()方法，代码如下。

```
public <T> List<Future<T>> invokeAll(Collection<? extends Callable<T>> tasks)
    throws InterruptedException {
    if (tasks == null)
        throw new NullPointerException();
    ArrayList<Future<T>> futures = new ArrayList<Future<T>>(tasks.size());
    //标识所有任务是否完成
    boolean done = false;
    try {
        //遍历所有任务
        for (Callable<T> t : tasks) {
            //将每个任务都封装成 RunnableFuture 对象提交任务
            RunnableFuture<T> f = newTaskFor(t);
            //将结果数据添加到 futures 集合中
            futures.add(f);
            //执行任务
            execute(f);
        }
        //遍历结果数据集合
        for (int i = 0, size = futures.size(); i < size; i++) {
            Future<T> f = futures.get(i);
            //任务没有完成
            if (!f.isDone()) {
                try {
                    //阻塞等待任务完成并返回结果
                    f.get();
                } catch (CancellationException ignore) {
                } catch (ExecutionException ignore) {
                }
            }
        }
        //任务完成（正常完成或异常完成）
        done = true;
        //返回结果数据集合
        return futures;
    } finally {
        //如果发生异常 InterruptedException，则取消已经提交的任务
        if (!done)
            for (int i = 0, size = futures.size(); i < size; i++)
```

```
                    futures.get(i).cancel(true);
        }
    }
}
```

上述代码的总体逻辑为：首先将所有任务封装成 RunnableFuture 对象，调用 execute()方法执行任务，将返回的结果数据添加到 futures 集合后对 futures 集合进行遍历。然后检测任务是否完成，如果没有完成，则调用 get()方法阻塞任务，直到返回结果，此时会忽略异常。最后在 finally 代码块中对所有任务是否完成的标识进行判断，如果存在未完成的任务，则取消已经提交的任务。

有超时时间设置的 invokeAll()方法代码如下。

```
public <T> List<Future<T>> invokeAll(Collection<? extends Callable<T>> tasks,
                                     long timeout, TimeUnit unit)
    throws InterruptedException {
    if (tasks == null)
        throw new NullPointerException();
    long nanos = unit.toNanos(timeout);
    ArrayList<Future<T>> futures = new ArrayList<Future<T>>(tasks.size());
    boolean done = false;
    try {
        for (Callable<T> t : tasks)
            futures.add(newTaskFor(t));
        final long deadline = System.nanoTime() + nanos;
        final int size = futures.size();
        for (int i = 0; i < size; i++) {
            execute((Runnable)futures.get(i));
            // 在添加执行任务时进行超时判断，如果超时则立刻返回 futures 集合
            nanos = deadline - System.nanoTime();
            if (nanos <= 0L)
                return futures;
        }
         // 遍历所有任务
        for (int i = 0; i < size; i++) {
            Future<T> f = futures.get(i);
            if (!f.isDone()) {
                //对结果进行超时判断
                if (nanos <= 0L)
                    return futures;
                try {
                    f.get(nanos, TimeUnit.NANOSECONDS);
                } catch (CancellationException ignore) {
                } catch (ExecutionException ignore) {
                } catch (TimeoutException toe) {
                    return futures;
                }
                //重置任务的超时时间
                nanos = deadline - System.nanoTime();
            }
```

```
            }
            done = true;
            return futures;
        } finally {
            if (!done)
                for (int i = 0, size = futures.size(); i < size; i++)
                    futures.get(i).cancel(true);
        }
}
```

通过代码可以看出，有超时时间设置的 invokeAll()方法与无超时时间设置的 invokeAll()方法主体逻辑相同，只是有超时时间设置的 invokeAll()方法在执行过程中增加了超时判断的逻辑。一个是在添加执行任务时进行超时判断，如果超时，则立刻返回 futures 集合；另一个是在每次对结果进行判断时都添加超时处理逻辑。

从代码中还可以看出，invokeAll()方法内部调用了 Executor 接口的 execute()方法来提交任务。

（5）submit()方法。submit()方法将任务封装成 RunnableFuture 对象提交到线程池，在线程池中执行任务后返回 Future 结果，通过返回的 Future 可以获取具体的执行结果，代码如下。

```
public Future<?> submit(Runnable task) {
    if (task == null) throw new NullPointerException();
    RunnableFuture<Void> ftask = newTaskFor(task, null);
    execute(ftask);
    return ftask;
}

public <T> Future<T> submit(Runnable task, T result) {
    if (task == null) throw new NullPointerException();
    RunnableFuture<T> ftask = newTaskFor(task, result);
    execute(ftask);
    return ftask;
}

public <T> Future<T> submit(Callable<T> task) {
    if (task == null) throw new NullPointerException();
    RunnableFuture<T> ftask = newTaskFor(task);
    execute(ftask);
    return ftask;
}
```

在通过 submit()方法向线程池提交任务时，本质上还是调用了 Executor 接口的 execute()方法。所以，在向非定时任务类的线程池提交任务时，本质上调用的都是 Executor 接口的 execute()方法，只是根据具体实现类的不同，调用的具体实现类的 execute()方法不同。

14.2.5　ScheduledExecutorService 接口

ScheduledExecutorService 接口继承自 ExecutorService 接口，继承了 ExecutorService 接口的所有功能，并且扩展了定时处理任务的能力，ScheduledExecutorService 接口的代码如下。

```
public interface ScheduledExecutorService extends ExecutorService {

    //延迟delay时间来执行command任务，只执行一次
    public ScheduledFuture<?> schedule(Runnable command,
                                       long delay, TimeUnit unit);

    //延迟delay时间来执行callable任务，只执行一次
    public <V> ScheduledFuture<V> schedule(Callable<V> callable,
                                           long delay, TimeUnit unit);

    //延迟initialDelay时间首次执行command任务，之后每隔period时间执行一次
    public ScheduledFuture<?> scheduleAtFixedRate(Runnable command,
                                                  long initialDelay,
                                                  long period,
                                                  TimeUnit unit);

    //延迟initialDelay时间首次执行command任务，之后每延时delay时间执行一次
    public ScheduledFuture<?> scheduleWithFixedDelay(Runnable command,
                                                     long initialDelay,
                                                     long delay,
                                                     TimeUnit unit);
}
```

在 ScheduledExecutorService 接口提供了延时只执行一次任务的方法，也提供了按周期执行任务的方法。

这里需要注意的一个细节是，scheduleAtFixedRate()方法和 scheduleWithFixedDelay()方法对于第二次开始执行任务的时间的计算方法不一样，具体如下。

- scheduleAtFixedRate()方法：按固定的频率执行，从当前任务开始执行的时间计算，一个 period 时间后，会检测当前任务是否执行完毕，如果当前任务执行完毕，则立即执行下一个任务。否则，等待当前任务执行完毕后立即执行下一个任务。
- scheduleWithFixedDelay()方法：按固定的延时执行，从当前任务执行结束时开始计算，在 delay 时间后，开始执行下一个任务。delay 时间指当前任务执行结束至下一个任务开始时的时间。

第 15 章

线程池核心流程

上一章从总体框架层面介绍了线程池实现的功能，本章开始深入线程池实现的细节，探究线程池的底层执行流程。

15.1 线程池正确运行的核心流程

作为线程池中最核心的实现类，ThreadPoolExecutor 类定义的一些常量、方法和内部类对线程池整体逻辑的正确执行起到了至关重要的作用。

注意：有关线程池的创建方式，读者可以参考《深入理解高并发编程：核心原理与案例实战》一书，有关 ThreadPoolExecutor 类构造方法的详细说明，读者可以参见第 14 章的内容。

15.1.1 ThreadPoolExecutor 类的重要属性

ThreadPoolExecutor 类中有几个非常重要的常量和方法。其中，有一个 AtomicInteger 类型的常量 ctl，它是一个原子类对象，贯穿线程池的整个生命周期，主要用来保存线程的数量和线程池的状态。ThreadPoolExecutor 类中与 ctl 常量相关的属性如下。

```
//主要用来保存线程数量和线程池的状态，高3位保存线程状态，低29位保存线程数量
private final AtomicInteger ctl = new AtomicInteger(ctlOf(RUNNING, 0));
//线程池中线程数量的位数（32-3）
private static final int COUNT_BITS = Integer.SIZE - 3;
//表示线程池中的最大线程数量
//将数字1的二进制值向右移29位，再减1
private static final int CAPACITY   = (1 << COUNT_BITS) - 1;
//线程池的运行状态
private static final int RUNNING    = -1 << COUNT_BITS;
private static final int SHUTDOWN   =  0 << COUNT_BITS;
private static final int STOP       =  1 << COUNT_BITS;
private static final int TIDYING    =  2 << COUNT_BITS;
private static final int TERMINATED =  3 << COUNT_BITS;
```

```java
//获取线程状态
private static int runStateOf(int c)     { return c & ~CAPACITY; }
//获取线程数量
private static int workerCountOf(int c)  { return c & CAPACITY; }
private static int ctlOf(int rs, int wc) { return rs | wc; }
private static boolean runStateLessThan(int c, int s) {
    return c < s;
}
private static boolean runStateAtLeast(int c, int s) {
    return c >= s;
}
private static boolean isRunning(int c) {
    return c < SHUTDOWN;
}
private boolean compareAndIncrementWorkerCount(int expect) {
    return ctl.compareAndSet(expect, expect + 1);
}
private boolean compareAndDecrementWorkerCount(int expect) {
    return ctl.compareAndSet(expect, expect - 1);
}
private void decrementWorkerCount() {
    do {} while (! compareAndDecrementWorkerCount(ctl.get()));
}
```

由上述代码可以看出，线程池的核心状态包括 RUNNING、SHUTDOWN、STOP、TIDYING、TERMINATED，这些状态本质上构成了线程池的整个生命周期。

注意：有关线程池中核心状态的说明与状态的流转过程，读者可以参考《深入理解高并发编程：核心原理与案例实战》一书。

除了 ctl 相关的属性，ThreadPoolExecutor 类中还提供了其他重要属性。

```java
//用于存放任务的阻塞队列
private final BlockingQueue<Runnable> workQueue;
//可重入锁
private final ReentrantLock mainLock = new ReentrantLock();
//存放线程池中线程的集合，在访问这个集合时，必须获得 mainLock 锁
private final HashSet<Worker> workers = new HashSet<Worker>();
//在锁内部阻塞等待条件完成
private final Condition termination = mainLock.newCondition();
//线程工厂，以此来创建新线程
private volatile ThreadFactory threadFactory;
//拒绝策略
private volatile RejectedExecutionHandler handler;
//默认的拒绝策略
private static final RejectedExecutionHandler defaultHandler = new AbortPolicy();
```

上述代码对每个属性都给出了详细的注释。

15.1.2　ThreadPoolExecutor 类的重要内部类

ThreadPoolExecutor 类的 Worker 内部类和拒绝策略内部类对线程池的执行至关重要。

1. Worker 内部类

从代码角度来看，Worker 类继承了 AQS 类，实现了 Runnable 接口，本质上是在线程池中执行任务的线程，Worker 类的整体代码如下。

```java
private final class Worker extends AbstractQueuedSynchronizer implements Runnable{
    private static final long serialVersionUID = 6138294804551838833L;
    //真正执行任务的线程
    final Thread thread;
    //第一次执行 Runnable 任务
    Runnable firstTask;
    //用于存放此线程完成的任务数
    volatile long completedTasks;
    //Worker 类唯一的构造方法，传递的 firstTask 可以为 null
    Worker(Runnable firstTask) {
        setState(-1);
        this.firstTask = firstTask;
        this.thread = getThreadFactory().newThread(this);
    }
    //调用外部 ThreadPoolExecutor 类的 runWorker 方法执行任务
    public void run() {
        runWorker(this);
    }

    //是否获取到锁
    //state=0 表示未获取到锁
    //state=1 表示已获取到锁
    protected boolean isHeldExclusively() {
        return getState() != 0;
    }

    //使用 AQS 设置线程状态
    protected boolean tryAcquire(int unused) {
        if (compareAndSetState(0, 1)) {
            setExclusiveOwnerThread(Thread.currentThread());
            return true;
        }
        return false;
    }

    //使用 AQS 尝试释放锁
    protected boolean tryRelease(int unused) {
        setExclusiveOwnerThread(null);
        setState(0);
        return true;
    }
```

```
public void lock()              { acquire(1); }
public boolean tryLock()        { return tryAcquire(1); }
public void unlock()            { release(1); }
public boolean isLocked()       { return isHeldExclusively(); }

void interruptIfStarted() {
    Thread t;
    if (getState() >= 0 && (t = thread) != null && !t.isInterrupted()) {
        try {
            t.interrupt();
        } catch (SecurityException ignore) {
        }
    }
}
}
```

在 Worker 类的构造方法中，首先将同步状态 state 设置为 -1，以防止 runWorker()方法在运行之前被中断。在其他线程调用线程池的 shutdownNow()方法中断线程池时，如果 Worker 类的 state 状态的值大于 0，则会中断线程，如果 state 状态的值为 -1，则不会中断线程。

Worker 类还实现了 Runnable 接口，需要重写 run()方法，而 Worker 的 run()方法本质上调用的是 ThreadPoolExecutor 类的 runWorker()方法，在 runWorker()方法中，会首先调用 Worker 的 unlock()方法，该方法会将 state 设置为 0，所以这时调用 shutdownNow()方法就会中断当前线程，如果此时已经进入了 runWorker()方法，就不会在没有执行 runWorker()方法时中断线程。

2. 拒绝策略内部类

当线程池的 workQueue 阻塞队列已满且没有空闲的线程,同时线程数已经达到了最大值时,如果继续向线程池中提交任务，线程池就会采取拒绝策略来处理这个任务。线程池中默认提供了 4 种拒绝策略，这些拒绝策略都会实现 RejectedExecutionHandler 接口，RejectedExecutionHandler 接口的代码如下。

```
public interface RejectedExecutionHandler {
    void rejectedExecution(Runnable r, ThreadPoolExecutor executor);
}
```

RejectedExecutionHandler 接口提供了一个执行拒绝策略的 rejectedExecution()方法，任何拒绝策略的实现类都要覆写这种方法。

线程池中默认提供的拒绝策略类结构图如图 15-1 所示。

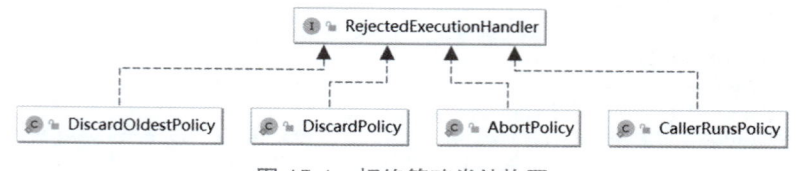

图 15-1 拒绝策略类结构图

由图 15-1 可以看出，线程池中默认提供的 4 种拒绝策略的实现类分别为 AbortPolicy 类、CallerRunsPolicy 类、DiscardOldestPolicy 类和 DiscardPolicy 类。

- AbortPolicy 类：直接抛出异常，这也是默认的策略，如果在调用 ThreadPoolExecutor 的构造方法创建线程池时没有传入指定的拒绝策略，就会默认使用这种策略。
- CallerRunsPolicy 类：用调用者所在的线程执行任务。
- DiscardOldestPolicy 类：丢弃队列中最靠前的任务并执行当前任务。
- DiscardPolicy 类：直接丢弃当前任务。

ThreadPoolExecutor 类提供的 4 个默认拒绝策略的实现类的代码如下。

```java
//用调用者所在的线程执行任务
public static class CallerRunsPolicy implements RejectedExecutionHandler {
    public CallerRunsPolicy() { }
    public void rejectedExecution(Runnable r, ThreadPoolExecutor e) {
        if (!e.isShutdown()) {
            r.run();
        }
    }
}
//直接抛出异常，这也是默认的策略
public static class AbortPolicy implements RejectedExecutionHandler {
    public AbortPolicy() { }
    public void rejectedExecution(Runnable r, ThreadPoolExecutor e) {
        throw new RejectedExecutionException("Task " + r.toString() +
                                   " rejected from " + e.toString());
    }
}
//直接丢弃当前任务
public static class DiscardPolicy implements RejectedExecutionHandler {
    public DiscardPolicy() { }
    public void rejectedExecution(Runnable r, ThreadPoolExecutor e) {
    }
}
//丢弃队列中最靠前的任务并执行当前任务
public static class DiscardOldestPolicy implements RejectedExecutionHandler {
    public DiscardOldestPolicy() { }
    public void rejectedExecution(Runnable r, ThreadPoolExecutor e) {
        if (!e.isShutdown()) {
            e.getQueue().poll();
            e.execute(r);
        }
    }
}
```

可以通过实现 RejectedExecutionHandler 接口，并且重写 RejectedExecutionHandler 接口的 rejectedExecution()方法来自定义拒绝策略。例如下面的代码。

```java
public class BingheRejectedPolicy implements RejectedExecutionHandler {
    public BingheRejectedPolicy() { }
    public void rejectedExecution(Runnable r, ThreadPoolExecutor e) {
        if (!e.isShutdown()) {
            System.out.println("自定义的线程池拒绝策略，
                        主要使用调用者所在的线程来执行任务")
            r.run();
        }
    }
}
```

在调用 ThreadPoolExecutor 的构造方法创建线程池时，可以将自定义的拒绝策略传入构造方法，使用自定义的拒绝策略的代码片段如下。

```java
ThreadPoolExecutor threadPool = new ThreadPoolExecutor(0, 1000,
                    60L, TimeUnit.SECONDS,
                    new LinkedBlockingQueue(1024),
                    Executors.defaultThreadFactory(),
                    new BingheRejectedPolicy());
                    threadPool.submit(task);
```

15.2 线程池执行任务的核心流程

ThreadPoolExecutor 类是线程池中最核心的类之一，能够保证线程池按照正常的业务逻辑执行任务，并且能够保证修改线程池每个阶段状态的原子性。

15.2.1 核心流程概述

ThreadPoolExecutor 类中定义了一个 workers 工作线程集合，代码如下。

```java
private final HashSet<Worker> workers = new HashSet<Worker>();
```

用户可以向线程池中添加需要执行的任务，workers 集合的工作线程可以直接执行任务，或者从任务队列中获取任务后执行。ThreadPoolExecutor 类提供了整个线程池从创建到执行任务，再到消亡的流程方法。

在 ThreadPoolExecutor 类中，线程池的逻辑主要体现在 execute(Runnable)方法、addWorker(Runnable, boolean)方法、addWorkerFailed(Worker)方法和拒绝策略上。

15.2.2 execute()方法解析

execute(Runnable)方法的作用是将 Runnable 类型的任务提交到线程池中执行任务的流程，可以简化成如图 15-2 所示。

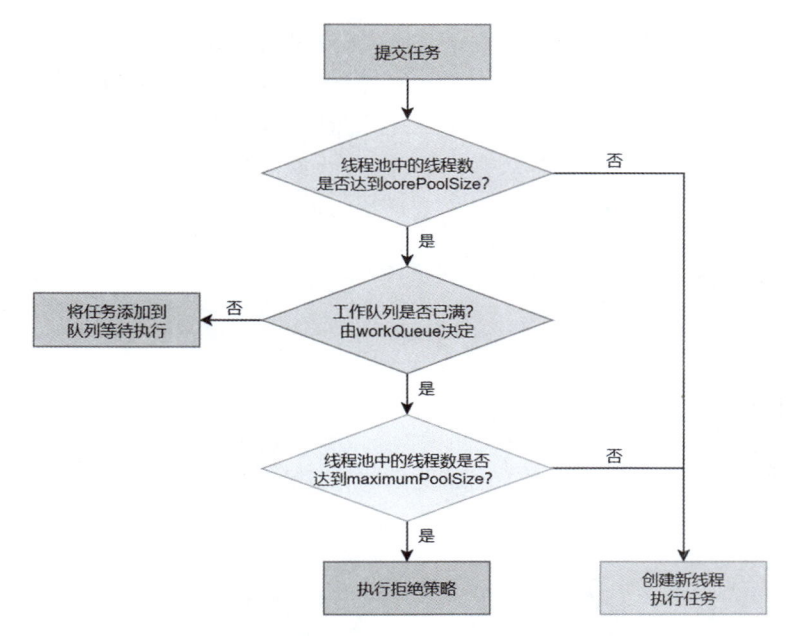

图 15-2　线程池执行任务的流程

注意：关于线程池执行任务流程的详细介绍，读者可以参考《深入理解高并发编程：核心原理与案例实战》一书。

接下来，直接拆解 execute(Runnable)方法的代码，具体分析其执行逻辑，读者可以到 JDK 的 java.util.concurrent.ThreadPoolExecutor 类中查看完整的 execute(Runnable)方法。

（1）如果程池中的线程数小于 corePoolSize 核心线程数，则向 workers 工作线程集合中添加一个核心线程执行任务。代码如下。

```
//线程池中的线程数量小于corePoolSize
if (workerCountOf(c) < corePoolSize) {
   //重新开启线程执行任务
   if (addWorker(command, true))
      return;
   c = ctl.get();
}
```

（2）如果线程池中的线程数量大于 corePoolSize 核心线程数，则判断当前线程池是否处于 RUNNING 状态，如果处于 RUNNING 状态，则添加任务到待执行的任务队列中。代码如下。

```
if (isRunning(c) && workQueue.offer(command))
```

注意：在向任务队列添加任务时，需要判断线程池是否处于 RUNNING 状态，只有线程池处于 RUNNING 状态，才能向任务队列中添加新任务，否则执行拒绝策略。

（3）向任务队列中添加任务成功。由于其他线程可能修改线程池的状态，所以需要对线程池进行二次检查。如果当前线程池的状态不再是 RUNNING 状态，则需要将添加的任务从任务队列中移除，执行后续的拒绝策略。如果当前线程池仍然处于 RUNNING 状态，则判断线程池是否为空。如果线程池中不存在任何线程，则新建一个线程添加到线程池中。代码如下。

```
//再次获取线程池的状态和线程池中线程的数量，用于二次检查
int recheck = ctl.get();
//如果线程池未处于 RUNNING 状态，则从队列中删除任务
if (! isRunning(recheck) && remove(command))
    //执行拒绝策略
    reject(command);
//如果线程池为空，则向线程池中添加一个线程
else if (workerCountOf(recheck) == 0)
    addWorker(null, false);
```

（4）当步骤（3）向任务队列中添加任务失败时，尝试开启新的线程执行任务。如果线程池中的线程数量已经大于线程池中的最大线程数 maximumPoolSize，则不能再启动新线程。此时线程池中的任务队列和线程已满，需要执行拒绝策略，代码如下。

```
//如果任务队列已满，则新增 worker 线程，如果新增线程失败，则执行拒绝策略
else if (!addWorker(command, false))
    reject(command);
```

将 execute(Runnable)方法拆解后，结合线程池执行任务的流程来理解线程池中任务的执行流程就比较简单了。可以说，execute(Runnable)方法的逻辑与一般线程池的执行逻辑大体相同，理解了 execute(Runnable)方法，就理解了线程池的执行逻辑。

15.2.3　addWorker()方法解析

execute(Runnable) 方 法 中 多 次 调 用 了 addWorker(Runnable, boolean) 方 法 。 addWorker(Runnable, boolean)方法总体可以分为 3 部分，第 1 部分使用 CAS 安全地向线程池中添加工作线程；第 2 部分创建新的工作线程；第 3 部分将任务通过安全的并发方式添加到 workers 工作线程集合中，并启动工作线程执行任务。

接下来，直接拆解 addWorker(Runnable, boolean)方法的代码进行深入分析，读者可以到 JDK 的 java.util.concurrent.ThreadPoolExecutor 类中查看完整的 addWorker(Runnable, boolean)方法。

（1）检查任务队列是否在特定条件下为空，代码如下。

```
// 检查队列是否在特定条件下为空
if (rs >= SHUTDOWN &&
    ! (rs == SHUTDOWN &&
        firstTask == null &&
        ! workQueue.isEmpty()))
    return false;
```

（2）通过步骤（1）的校验后进入内层 for 循环，在内层 for 循环中通过 CAS 增加线程池中

的线程数量，如果 CAS 操作成功，则直接退出双重 for 循环。如果 CAS 操作失败，则查看当前线程池的状态是否发生了变化。如果线程池的状态发生了变化，则使用 continue 关键字重新通过外层 for 循环校验任务队列，检验通过后，再次执行内层 for 循环的 CAS 操作。如果线程池的状态没有发生变化，表示上一次 CAS 操作失败了，则继续尝试 CAS 操作。代码如下。

```
for (;;) {
    //获取线程池中的线程数量
    int wc = workerCountOf(c);
    //如果线程池中的线程数量超出限制，则直接返回 false
    if (wc >= CAPACITY ||
        wc >= (core ? corePoolSize : maximumPoolSize))
        return false;
    //通过 CAS 向线程池新增线程数量
    if (compareAndIncrementWorkerCount(c))
        //通过 CAS 保证只有一个线程执行成功，跳出最外层循环
        break retry;
    //重新获取 ctl 的值
    c = ctl.get();
    //如果 CAS 操作失败了，则需要在内循环中重新尝试通过 CAS 新增线程数量
    if (runStateOf(c) != rs)
        continue retry;
}
```

（3）步骤（2）中的 CAS 操作成功表示向线程池中成功添加了工作线程，此时，还没有线程去执行任务。使用全局的独占锁 mainLock 将新增的工作线程 Worker 对象安全地添加到 workers 工作线程集合中，代码如下。

```
//跳出最外层 for 循环，说明通过 CAS 增加线程数量成功
//此时创建新的工作线程
boolean workerStarted = false;
boolean workerAdded = false;
Worker w = null;
try {
    //将执行的任务封装成 worker
    w = new Worker(firstTask);
    final Thread t = w.thread;
    if (t != null) {
        //独占锁，保证操作 workers 集合时同步
        final ReentrantLock mainLock = this.mainLock;
        mainLock.lock();
        try {
            //重新检查线程池状态
            //原因是在获得锁之前可能有线程改变了线程池的状态
            int rs = runStateOf(ctl.get());

            if (rs < SHUTDOWN ||
                (rs == SHUTDOWN && firstTask == null)) {
                if (t.isAlive())
```

```
                throw new IllegalThreadStateException();
            //向 worker 中添加新任务
            workers.add(w);
            int s = workers.size();
            if (s > largestPoolSize)
                largestPoolSize = s;
            //将是否添加了新任务的标识设置为 true
            workerAdded = true;
        }
    } finally {
        //释放独占锁
        mainLock.unlock();
    }
    //如果添加新任务成功，则启动线程执行任务
    if (workerAdded) {
        t.start();
        //将任务是否已经启动的标识设置为 true
        workerStarted = true;
        }
    }
} finally {
    //如果任务未启动或启动失败，则调用 addWorkerFailed(Worker)方法
    if (! workerStarted)
        addWorkerFailed(w);
}
//返回是否启动任务的标识
return workerStarted;
```

以上的主要逻辑是新建 Worker 对象，并获取 Worker 对象的执行线程，如果线程不为空，则获取独占锁。获取锁成功后再次检查线程的状态，以避免在获取独占锁之前有线程修改了线程池的状态，或者关闭了线程池，对后续执行的逻辑产生影响。

如果线程池关闭则释放锁，否则将新增的线程添加到 workers 工作线程集合中，释放锁并启动线程执行任务，将是否启动线程的标识设置为 true。判断线程是否启动，如果没有启动，则调用 addWorkerFailed(Worker)方法。最终返回线程是否启动的标识。

15.2.4 addWorkerFailed()方法解析

在 addWorker(Runnable, boolean)方法中，如果添加工作线程失败或者工作线程启动失败，则会调用 addWorkerFailed(Worker)方法。接下来，分析 addWorkerFailed(Worker)方法的实现，代码如下。

```
private void addWorkerFailed(Worker w) {
    //获取独占锁
    final ReentrantLock mainLock = this.mainLock;
    mainLock.lock();
    try {
        //如果 Worker 任务不为空
```

```
        if (w != null)
            //则将任务从 workers 集合中移除
            workers.remove(w);
        //通过 CAS 将任务数量减 1
        decrementWorkerCount();
        tryTerminate();
    } finally {
        //释放锁
        mainLock.unlock();
    }
}
```

addWorkerFailed(Worker)方法的实现逻辑为获取独占锁，将任务从 workers 工作线程集合中移除，并通过 CAS 将任务的数量减 1，然后释放锁。

15.2.5 拒绝策略执行流程

在 execute(Runnable)方法中，线程池会在适当的时候调用 reject(Runnable)方法执行相应的拒绝策略，reject(Runnable)方法的代码如下。

```
final void reject(Runnable command) {
    handler.rejectedExecution(command, this);
}
```

这里调用的是 handler 的 rejectedExecution 方法，ThreadPoolExecutor 类中对 handler 的定义如下。

```
private volatile RejectedExecutionHandler handler;
```

handler 是 RejectedExecutionHandler 接口类型的一个对象，RejectedExecutionHandler 接口是线程池的拒绝策略接口。也就是说，在 execute(Runnable)方法中，会通过 RejectedExecutionHandler 接口类型的 handler 成员变量在适当的时候调用具体的拒绝策略。

15.3 Worker 线程的核心流程

从代码角度来看，Worker 类继承了 AQS 类，实现了 Runnable 接口，本质上是在线程池中执行任务的线程。在 ThreadPoolExecutor 类的 addWorker(Runnable, boolean)方法中，使用 CAS 安全地更新线程的数量后新建 Worker 线程执行任务。

注意：有关 Worker 类的核心代码解析，读者可以参见 15.1.2 节的内容。

15.3.1 Worker 线程的核心流程概述

在 ThreadPoolExecutor 类中，Worker 线程的核心流程主要涉及 runWorker()方法、getTask()方法、beforeExecute()方法、afterExecute()方法、processWorkerExit()方法、tryTerminate()方法和 terminated()方法。理解了这几种方法的核心流程，也就掌握了 Worker 线程的核心流程。

15.3.2 runWorker()方法解析

Worker 类的 run()方法调用了 ThreadPoolExecutor 类的 runWorker(Worker)方法执行任务。

接下来，直接拆解 runWorker(Worker)方法的代码进行深入分析，读者可以到 JDK 的 java.util.concurrent.ThreadPoolExecutor 类中查看完整的 runWorker(Worker)方法。

（1）获取当前线程的句柄和工作线程中的任务，并将工作线程中的任务设置为空，执行 unlock 方法释放锁，将 state 状态设置为 0，此时可以中断工作线程，代码如下。

```
Thread wt = Thread.currentThread();
Runnable task = w.firstTask;
w.firstTask = null;
//释放锁，将 state 设置为 0，允许中断任务
w.unlock();
```

（2）在 while 循环中进行判断，如果任务不为空，或者从任务队列中获取的任务不为空，则执行 while 循环，否则调用 processWorkerExit(Worker, boolean)方法退出 Worker 工作线程。代码如下。

```
while (task != null || (task = getTask()) != null)
```

（3）如果满足 while 的循环条件，则获取工作线程内部的独占锁，并通过一系列判断来确定是否需要中断当前线程，代码如下。

```
//如果任务不为空，则获取 Worker 线程的独占锁
w.lock();
//如果线程已经停止，或者中断线程后线程终止并且当前线程没有被中断，则中断线程
//大家好好理解一下这个逻辑
if ((runStateAtLeast(ctl.get(), STOP) ||
     (Thread.interrupted() &&
      runStateAtLeast(ctl.get(), STOP))) &&
    !wt.isInterrupted())
    //中断线程
    wt.interrupt();
```

（4）调用执行任务前执行的逻辑，代码如下。

```
//执行任务前执行的逻辑
beforeExecute(wt, task);
```

（5）调用 Runable 接口的 run 方法执行任务，代码如下。

```
//调用 Runable 接口的 run 方法执行任务
task.run();
```

（6）调用执行任务后执行的逻辑，代码如下。

```
//执行任务后执行的逻辑
afterExecute(task, thrown);
```

（7）将完成的任务设置为空，完成的任务数量加 1 并释放工作线程的锁，代码如下。

```
//任务执行完成后，将其设置为空
```

```
task = null;
//完成的任务数量加1
w.completedTasks++;
//释放工作线程获得的锁
w.unlock();
```

（8）退出 Worker 线程，代码如下。

```
//执行退出 Worker 线程的逻辑
processWorkerExit(w, completedAbruptly);
```

当从 Worker 线程中获取的任务为空时，会调用 getTask()方法从任务队列中获取任务。

15.3.3 getTask()方法解析

getTask()方法的主要逻辑是从任务队列中获取任务，代码如下。

```
private Runnable getTask() {
    //轮询是否超时的标识
    boolean timedOut = false;
    //自旋 for 循环
    for (;;) {
        //获取 ctl
        int c = ctl.get();
        //获取线程池的状态
        int rs = runStateOf(c);
        //检测任务队列是否在线程池停止或关闭时为空
        if (rs >= SHUTDOWN && (rs >= STOP || workQueue.isEmpty())) {
            //减少 Worker 线程的数量
            decrementWorkerCount();
            return null;
        }
        //获取线程池中线程的数量
        int wc = workerCountOf(c);
        //检测当前线程池是否正在等待执行任务
        //或者检测当前线程池中的线程数量是否大于 corePoolSize 的值
        boolean timed = allowCoreThreadTimeOut || wc > corePoolSize;
        //如果线程池中的线程数量大于 maximumPoolSize
        //或者线程池在等待执行任务并且轮询超时
        //且当前线程池中的线程数量大于 1 或任务队列为空
        if ((wc > maximumPoolSize || (timed && timedOut))
            && (wc > 1 || workQueue.isEmpty())) {
            //则成功减少线程池中的工作线程数量
            if (compareAndDecrementWorkerCount(c))
                return null;
            continue;
        }
        try {
            //从任务队列中获取任务
            Runnable r = timed ?
                workQueue.poll(keepAliveTime, TimeUnit.NANOSECONDS) :
```

```
            workQueue.take();
    //如果任务不为空则直接返回任务
    if (r != null)
        return r;
    timedOut = true;
} catch (InterruptedException retry) {
    timedOut = false;
    }
  }
}
```

getTask()方法的实现逻辑比较简单，上述代码给出了详细的注释信息。

15.3.4　beforeExecute()方法解析

在 ThreadPoolExecutor 类的 runWorker(Worker)方法真正执行任务前，调用了 beforeExecute(Thread, Runnable)方法，代码如下。

```
protected void beforeExecute(Thread t, Runnable r) { }
```

beforeExecute(Thread, Runnable)方法的方法体为空，可以通过创建 ThreadPoolExecutor 的子类来重写 beforeExecute(Thread, Runnable)方法，使得线程池在正式执行任务前能够执行自定义的业务逻辑。

15.3.5　afterExecute()方法解析

在 ThreadPoolExecutor 类的 runWorker(Worker)方法真正执行任务后，调用了 afterExecute(Runnable, Throwable)方法，代码如下。

```
protected void afterExecute(Runnable r, Throwable t) { }
```

afterExecute(Runnable, Throwable)方法的方法体同样为空，可以通过创建 ThreadPoolExecutor 的子类来重写 afterExecute(Runnable, Throwable)方法，使得线程池在执行任务后执行自定义的业务逻辑。

15.3.6　processWorkerExit()方法解析

processWorkerExit(Worker, boolean)方法的逻辑主要是退出 Worker 线程，并且清理一些系统资源。

接下来，拆解 processWorkerExit(Worker, boolean)方法的代码进行深入分析，读者可以直接到 JDK 的 java.util.concurrent.ThreadPoolExecutor 类中查看完整的 processWorkerExit(Worker, boolean)方法。

（1）如果在执行过程中出现异常，突然中断，则将工作线程数量减 1，代码如下。

```
//执行过程中出现了异常，突然中断
if (completedAbruptly)
```

```
//将工作线程的数量减 1
decrementWorkerCount();
```

（2）获取锁累加完成的任务数量，将完成的任务从 workers 集合中移除并释放锁，代码如下。

```
//获取全局锁
final ReentrantLock mainLock = this.mainLock;
mainLock.lock();
try {
    //累加完成的任务数量
    completedTaskCount += w.completedTasks;
    //将完成的任务从 workers 集合中移除
    workers.remove(w);
} finally {
    //释放锁
    mainLock.unlock();
}
```

（3）尝试终止工作线程，代码如下。

```
//尝试终止工作线程
tryTerminate();
```

（4）判断当前线程池中的线程数量是否小于核心线程数，如果是，则需要在线程池中新建一个线程执行任务；否则直接返回，不再执行后续逻辑。代码如下。

```
//获取 ctl
int c = ctl.get();
//判断当前线程池的状态值是否小于 STOP（RUNNING 或者 SHUTDOWN）
if (runStateLessThan(c, STOP)) {
    if (!completedAbruptly) {
        //如果 allowCoreThreadTimeOut 为 true，则将 min 赋值为 0，否则赋值为 corePoolSize
        int min = allowCoreThreadTimeOut ? 0 : corePoolSize;
        //如果 min 为 0 并且工作队列不为空
        if (min == 0 && ! workQueue.isEmpty())
            //则将 min 的值设置为 1
            min = 1;
        //如果线程池中的线程数量大于 min 的值
        if (workerCountOf(c) >= min)
            //则返回，不再执行程序
            return;
    }
    //调用 addWorker 方法
    addWorker(null, false);
}
```

15.3.7　tryTerminate()方法解析

在 processWorkerExit(Worker, boolean)方法中调用了 tryTerminate()方法来尝试终止工作线程。

接下来，拆解 tryTerminate()方法的代码进行深入分析，读者可以到 JDK 的

java.util.concurrent.ThreadPoolExecutor 类中查看完整的 tryTerminate()方法。

（1）获取 ctl，根据情况设置线程池状态或者中断线程并返回，代码如下。

```
//获取 ctl
int c = ctl.get();
//如果线程池的状态为 RUNNING 或者状态值大于 TIDYING
//或者状态为 SHUTDOWN 并且任务队列为空
//则直接返回程序，不再执行后续逻辑
if (isRunning(c) ||
    runStateAtLeast(c, TIDYING) ||
    (runStateOf(c) == SHUTDOWN && ! workQueue.isEmpty()))
    return;
//如果当前线程池中的线程数量不等于 0
if (workerCountOf(c) != 0) {
    //则中断线程
    interruptIdleWorkers(ONLY_ONE);
    return;
}
```

（2）获取全局锁，通过 CAS 设置线程池的状态，调用 terminated()方法执行逻辑，将线程池的状态设置为 TERMINATED，唤醒所有因为调用线程池的 awaitTermination()方法而被阻塞的线程，最终释放锁，代码如下。

```
//获取线程池的全局锁
final ReentrantLock mainLock = this.mainLock;
mainLock.lock();
try {
    //通过 CAS 将线程池的状态设置为 TIDYING
    if (ctl.compareAndSet(c, ctlOf(TIDYING, 0))) {
        try {
            //调用 terminated()方法
            terminated();
        } finally {
            //将线程池状态设置为 TERMINATED
            ctl.set(ctlOf(TERMINATED, 0));
            //唤醒所有因为调用线程池的 awaitTermination 方法而被阻塞的线程
            termination.signalAll();
        }
        return;
    }
} finally {
    //释放锁
    mainLock.unlock();
}
```

15.3.8　terminated()方法解析

在 tryTerminate()方法中，通过 CAS 将线程池的状态设置为 TIDYING，设置成功后，调用

terminated()方法执行逻辑，代码如下。

```
protected void terminated() { }
```

terminated()方法的方法体为空，可以通过创建 ThreadPoolExecutor 的子类来重写 terminated() 方法，使得 Worker 线程在调用 tryTerminate()方法时，可以执行自定义的 terminated()方法的业务逻辑。

15.4 线程池优雅退出的核心流程

线程池提供了优雅关闭的功能，在关闭线程池时，如果线程池中有正在执行的任务，则将任务执行完毕后，再关闭线程池。

15.4.1 shutdown()方法解析

当使用线程池调用了 shutdown()方法后，线程池就不再接受外界的任务执行了。但是要继续执行调用 shutdown()方法前放入任务队列中的任务。shutdown()方法是非阻塞方法，调用后立即返回，不会等待任务队列中的任务全部执行完毕，代码如下。

```
public void shutdown() {
    //获取线程池的全局锁
    final ReentrantLock mainLock = this.mainLock;
    mainLock.lock();
    try {
        //检查是否有关闭线程池的权限
        checkShutdownAccess();
        //将当前线程池的状态设置为 SHUTDOWN
        advanceRunState(SHUTDOWN);
        //中断 Worker 线程
        interruptIdleWorkers();
        //为 ScheduledThreadPoolExecutor 调用钩子函数
        onShutdown(); // hook for
    } finally {
        //释放线程池的全局锁
        mainLock.unlock();
    }
    //尝试将状态变为 TERMINATED
    tryTerminate();
}
```

总体来说，shutdown()方法的代码比较简单，首先检查是否有权限来关闭线程池，如果有则再次检测是否有中断工作线程的权限；如果没有则抛出 SecurityException 异常。代码如下。

```
//检查是否有关闭线程池的权限
checkShutdownAccess();
//将当前线程池的状态设置为 SHUTDOWN
advanceRunState(SHUTDOWN);
```

```
//中断Worker线程
interruptIdleWorkers();
```

其中，checkShutdownAccess()方法的代码如下。

```
private void checkShutdownAccess() {
    SecurityManager security = System.getSecurityManager();
    if (security != null) {
        security.checkPermission(shutdownPerm);
        final ReentrantLock mainLock = this.mainLock;
        mainLock.lock();
        try {
            for (Worker w : workers)
                security.checkAccess(w.thread);
        } finally {
            mainLock.unlock();
        }
    }
}
```

checkShutdownAccess()方法的代码主要检测是否具有关闭线程池的权限，期间使用了线程池的 mainLock 全局锁。

advanceRunState(int)方法的代码如下。

```
private void advanceRunState(int targetState) {
    for (;;) {
        int c = ctl.get();
        if (runStateAtLeast(c, targetState) ||
            ctl.compareAndSet(c, ctlOf(targetState, workerCountOf(c))))
            break;
    }
}
```

其整体逻辑是判断当前线程池是否处于指定的状态，在 shutdown()方法中传递的状态是 SHUTDOWN，如果是则直接返回；如果不是则将当前线程池的状态设置为 SHUTDOWN。

showdown()方法调用的 interruptIdleWorkers()方法的代码如下。

```
private void interruptIdleWorkers() {
    interruptIdleWorkers(false);
}
```

interruptIdleWorkers()方法调用的是 interruptIdleWorkers(boolean)方法，代码如下。

```
private void interruptIdleWorkers(boolean onlyOne) {
    final ReentrantLock mainLock = this.mainLock;
    mainLock.lock();
    try {
        for (Worker w : workers) {
            Thread t = w.thread;
            if (!t.isInterrupted() && w.tryLock()) {
                try {
```

```
                t.interrupt();
            } catch (SecurityException ignore) {
            } finally {
                w.unlock();
            }
        }
        if (onlyOne)
            break;
    }
} finally {
    mainLock.unlock();
}
}
```

interruptIdleWorkers(boolean)方法的总体逻辑是获取线程池的全局锁，循环所有的工作线程，检测线程是否被中断。如果没有被中断，并且 Worker 线程获得了锁，则执行线程的中断方法并释放线程获取的锁。如果 onlyOne 参数为 true 则退出循环，否则循环所有的工作线程执行相同的操作。最终释放线程池的全局锁。

15.4.2　shutdownNow()方法解析

如果调用了线程池的 shutdownNow()方法，则线程池不仅不会再接受新的任务，还会将任务队列中已有的任务丢弃，正在执行的 Worker 线程也会被立即中断，同时方法会立刻返回。shutdownNow()方法存在一个返回值，也就是当前任务队列中被丢弃的任务列表。shutdownNow()方法的代码如下。

```
public List<Runnable> shutdownNow() {
    List<Runnable> tasks;
    final ReentrantLock mainLock = this.mainLock;
    mainLock.lock();
    try {
        //检查是否有权限关闭
        checkShutdownAccess();
        //设置线程池的状态为 STOP
        advanceRunState(STOP);
        //中断所有的 Worker 线程
        interruptWorkers();
        //将任务队列中的任务移动到 tasks 集合中
        tasks = drainQueue();
    } finally {
        mainLock.unlock();
    }
    //尝试将状态变为 TERMINATED
    tryTerminate();
    //返回 tasks 集合
    return tasks;
}
```

shutdownNow()方法代码的总体逻辑与 shutdown()方法基本相同，只是 shutdownNow()方法将线程池的状态设置为 STOP，中断所有的 Worker 线程，并且将任务队列中的所有任务都移动到 tasks 集合中并返回。

另外，shutdownNow()方法在中断所有的线程时，调用了 interruptWorkers()方法，代码如下。

```
private void interruptWorkers() {
    final ReentrantLock mainLock = this.mainLock;
    mainLock.lock();
    try {
        for (Worker w : workers)
            w.interruptIfStarted();
    } finally {
        mainLock.unlock();
    }
}
```

interruptWorkers()方法的逻辑比较简单，就是获得线程池的全局锁、循环所有的工作线程、依次中断线程，最后释放线程池的全局锁。在 interruptWorkers()方法的内部，实际上调用了 Worker 类的 interruptIfStarted()方法中断线程，代码如下。

```
void interruptIfStarted() {
    Thread t;
    if (getState() >= 0 && (t = thread) != null && !t.isInterrupted()) {
        try {
            t.interrupt();
        } catch (SecurityException ignore) {
        }
    }
}
```

这里本质上还是使用 Thread 类的 interrupt()方法中断线程。

15.4.3 awaitTermination()方法解析

当线程池调用了 ThreadPoolExecutor 类的 awaitTermination(long, TimeUnit)方法后，会阻塞调用者所在的线程，直到线程池的状态被修改为 TERMINATED，或者达到了超时时间才返回。awaitTermination(long, TimeUnit)方法的代码如下。

```
public boolean awaitTermination(long timeout, TimeUnit unit)
    throws InterruptedException {
    //获取剩余时长
    long nanos = unit.toNanos(timeout);
    //获取 Worker 线程的全局锁
    final ReentrantLock mainLock = this.mainLock;
    //加锁
    mainLock.lock();
    try {
        for (;;) {
```

```
            //如果当前线程池状态为 TERMINATED，则返回 true
            if (runStateAtLeast(ctl.get(), TERMINATED))
                return true;
            //如果达到超时时间或超时，则返回 false
            if (nanos <= 0)
                return false;
            //重置剩余时长
            nanos = termination.awaitNanos(nanos);
        }
    } finally {
        //释放锁
        mainLock.unlock();
    }
}
```

awaitTermination(long, TimeUnit)方法的总体逻辑是首先获取全局锁，然后循环判断当前线程池是否是 TERMINATED 状态，如果是则返回 true，否则检测是否超时。如果超时则返回 false，如果未超时则重置剩余时长。

接下来，进入下一轮循环，再次检测当前线程池是否处于 TERMINATED 状态，如果是则返回 true，否则检测是否超时。如果超时则返回 false，如果未超时则重置剩余时长。直到线程池的状态变为 TERMINATED 或者超时。最后释放全局锁。

第 16 章

定时任务线程池

Java 从 JDK 1.5 版本开始提供了定时任务线程池，定时任务线程池的核心类是 ScheduledThreadPoolExecutor，从类结构上看，ScheduledThreadPoolExecutor 类继承自 ThreadPoolExecutor 类，主要提供了定时任务线程池的功能，能够实现周期性的调度任务。

16.1　ScheduledThreadPoolExecutor 类与 Timer 类的区别

ScheduledThreadPoolExecutor 类是从 JDK 1.5 版本开始提供的定时任务线程池，而在 JDK 1.5 版本之前主要使用 Timer 类和 TimerTask 类实现定时任务。

16.1.1　线程实现的区别

在线程的实现方面，Timer 类与 ScheduledThreadPoolExecutor 类是存在区别的。

（1）Timer 类内部的实现是单线程模式，如果某个 TimerTask 任务的执行时间比较久，或者抛出了异常，则会影响其他任务的执行。

（2）ScheduledThreadPoolExecutor 类内部的实现是多线程模式，并且线程池中的线程是可以复用的，某个 ScheduledFutureTask 任务执行时间比较长，不会影响其他任务的调度执行。

16.1.2　系统时间的区别

在获取系统时间层面，Timer 类与 ScheduledThreadPoolExecutor 类的敏感程度不同。

（1）Timer 类内部是基于操作系统的绝对时间实现的，对于操作系统的时间绝对敏感，如果操作系统的时间发生变化，则 Timer 的线程调度不再准确。

（2）ScheduledThreadPoolExecutor 类内部是基于相对时间实现的，操作系统时间的变化不会影响 ScheduledThreadPoolExecutor 对于线程的调度结果。

16.1.3　处理异常的区别

Timer 类与 ScheduledThreadPoolExecutor 类在处理异常方面存在区别。

（1）Timer 类不会捕获异常，其内部是基于单线程实现的，所以某个任务抛出异常，其他任务也会受到影响不再执行。

（2）ScheduledThreadPoolExecutor 类内部基于线程池调度任务，当某个任务抛出异常后，其他任务仍会正常执行。

16.1.4　任务编排的区别

Timer 类与 ScheduledThreadPoolExecutor 类在任务编排方面存在区别。

（1）Timer 类内部不支持对任务进行排序。

（2）ScheduledThreadPoolExecutor 类内部支持对任务进行排序。ScheduledThreadPoolExecutor 类定义了一个静态内部类 DelayedWorkQueue，它本质上是一个有序队列，支持对存储在 DelayedWorkQueue 队列中的任务按照距离下次执行时间的长短进行排序。

16.1.5　任务优先级的区别

Timer 类与 ScheduledThreadPoolExecutor 类在任务执行的优先级上存在区别。

（1）Timer 中执行的 TimerTask 任务没有优先级的概念，按照系统的绝对时间触发。

（2）ScheduledThreadPoolExecutor 类中执行的 ScheduledFutureTask 任务有优先级的概念。ScheduledFutureTask 类实现了 java.lang.Comparable 接口和 java.util.concurrent.Delayed 接口。也就是说，ScheduledFutureTask 类实现了 java.lang.Comparable 接口的 compareTo()方法和 java.util.concurrent.Delayed 接口的 getDelay()方法，可以根据这两种方法实现任务的优先级。

16.1.6　返回结果的区别

Timer 类与 ScheduledThreadPoolExecutor 类在返回结果方面存在区别。

（1）Timer 中执行的 TimerTask 类只是实现了 java.lang.Runnable 接口，无法获取结果。

（2）ScheduledThreadPoolExecutor 类中执行的 ScheduledFutureTask 类继承了 FutureTask 类，FutureTask 类实现了 Future 接口，可以通过 Future 接口的 get()方法获取返回的结果。

注意：有关 Timer 类与 ScheduledThreadPoolExecutor 类实现定时任务的案例，读者可以参考随书源码中的 mykit-concurrent-chapter16 工程。

16.2 定时任务线程池的初始化

从类结构的角度来看，ScheduledThreadPoolExecutor 类是 ThreadPoolExecutor 类的子类，主要在 ThreadPoolExecutor 类实现的线程池基础上实现了定时任务的功能。ScheduledThreadPoolExecutor 类的初始化主要是通过构造方法实现的，ScheduledThreadPoolExecutor 类的构造方法如下。

```java
public ScheduledThreadPoolExecutor(int corePoolSize) {
    super(corePoolSize, Integer.MAX_VALUE, 0, NANOSECONDS, new DelayedWorkQueue());
}
public ScheduledThreadPoolExecutor(int corePoolSize, ThreadFactory threadFactory) {
    super(corePoolSize, Integer.MAX_VALUE, 0, NANOSECONDS,
        new DelayedWorkQueue(), threadFactory);
}
public ScheduledThreadPoolExecutor(int corePoolSize, RejectedExecutionHandler handler) {
    super(corePoolSize, Integer.MAX_VALUE, 0, NANOSECONDS,
        new DelayedWorkQueue(), handler);
}
public ScheduledThreadPoolExecutor(int corePoolSize, ThreadFactory threadFactory, RejectedExecutionHandler handler) {
    super(corePoolSize, Integer.MAX_VALUE, 0, NANOSECONDS,
        new DelayedWorkQueue(), threadFactory, handler);
}
```

ScheduledThreadPoolExecutor 类的构造方法本质上调用了 ThreadPoolExecutor 类的构造方法。

注意：有关 ThreadPoolExecutor 类的构造方法的详细说明，读者可以参见 14.1.4 节的内容。

16.3 定时任务线程池的调度流程

ScheduledThreadPoolExecutor 类中经过一系列方法的调用，最终实现了 ScheduledThreadPoolExecutor 类的调度流程。

16.3.1 schedule()方法解析

schedule()方法实现了延时执行一次任务的功能，在 ScheduledThreadPoolExecutor 类中，提供了两个 schedule()方法，代码如下。

```java
public ScheduledFuture<?> schedule(Runnable command, long delay, TimeUnit unit) {
    //如果传递的 Runnable 对象和 TimeUnit 时间单位为空，则直接抛出空指针异常
    if (command == null || unit == null)
        throw new NullPointerException();
    //封装任务对象，在 decorateTask 方法中直接返回 ScheduledFutureTask 对象
    RunnableScheduledFuture<?> t = decorateTask(command,
        new ScheduledFutureTask<Void>(command, null, triggerTime(delay, unit)));
    //执行延时任务
```

```
        delayedExecute(t);
        //返回任务
        return t;
}
public <V> ScheduledFuture<V> schedule(Callable<V> callable, long delay, TimeUnit
unit)
        //如果传递的Callable对象和TimeUnit时间单位为空，则抛出空指针异常
        if (callable == null || unit == null)
            throw new NullPointerException();
        //封装任务对象，在decorateTask方法中直接返回ScheduledFutureTask对象
        RunnableScheduledFuture<V> t = decorateTask(callable,
            new ScheduledFutureTask<V>(callable, triggerTime(delay, unit)));
        //执行延时任务
        delayedExecute(t);
        //返回任务
        return t;
}
```

通过代码可以看出，ScheduledThreadPoolExecutor 类提供了两种重载的 schedule()方法。两种方法只是第 1 个参数不同，一种传递 Runnable 接口对象，另一种传递 Callable 接口对象。在 schedule()方法的内部，会将传递进来的 Runnable 接口对象和 Callable 接口对象封装成 RunnableScheduledFuture 对象。而 RunnableScheduledFuture 对象本质上是 ScheduledFutureTask 对象。将封装成的 RunnableScheduledFuture 对象传入 delayedExecute()方法中执行定时任务。

16.3.2　decorateTask()方法解析

decorateTask() 方法主要将传入的 Runnable 对象或者 Callable 对象封装成 RunnableScheduledFuture 对象，代码如下。

```
protected <V> RunnableScheduledFuture<V> decorateTask(
    Runnable runnable, RunnableScheduledFuture<V> task) {
    return task;
}

protected <V> RunnableScheduledFuture<V> decorateTask(
    Callable<V> callable, RunnableScheduledFuture<V> task) {
    return task;
}
```

在 decorateTask()方法中，接收一个 Runnable 接口或者 Callable 接口的对象和一个 RunnableScheduledFuture 接口的对象，并且两种 decorateTask()方法都只是将传递进来的 RunnableScheduledFuture 接口的对象返回。在 ScheduledThreadPoolExecutor 类的子类中可以重写这两种方法。

16.3.3　scheduleAtFixedRate()方法解析

scheduleAtFixedRate()方法能够让任务按照固定的频率执行，在当前任务开始执行时计时，

在经过 period 时间后，检测当前任务是否执行完毕，如果当前任务执行完毕，则立即执行下一个任务。否则，需要等待当前任务执行完毕才能执行下一个任务。代码如下。

```
public ScheduledFuture<?> scheduleAtFixedRate(Runnable command, long initialDelay,
long period, TimeUnit unit) {
    //如果传入的 Runnable 对象和 TimeUnit 为空，则抛出空指针异常
    if (command == null || unit == null)
        throw new NullPointerException();
    //如果执行周期 period 传入的数值小于或等于 0，则抛出非法参数异常
    if (period <= 0)
        throw new IllegalArgumentException();
    //将 Runnable 对象封装成 ScheduledFutureTask 任务对象，并设置执行周期
    ScheduledFutureTask<Void> sft = new ScheduledFutureTask<Void>(command,
                        null, triggerTime(initialDelay, unit),
unit.toNanos(period));
    //调用 decorateTask 方法，本质上还是直接返回 ScheduledFutureTask 对象
    RunnableScheduledFuture<Void> t = decorateTask(command, sft);
    //设置执行的任务
    sft.outerTask = t;
    //执行延时任务
    delayedExecute(t);
    //返回执行的任务
    return t;
}
```

scheduleAtFixedRate()方法将传递的 Runnable 对象封装成 ScheduledFutureTask 任务对象，并设置了执行周期，下一次的执行时间相对这一次的执行时间来说多了 period，TimeUnit 决定 period 的单位。采用固定的频率执行定时任务。

16.3.4　scheduleWithFixedDelay()方法解析

scheduleWithFixedDelay ()方法能够让任务按照固定的频率执行。在当前任务执行结束时计时，在经过 delay 时间后，下一个任务开始执行。delay 时间指当前任务执行结束距离下一个任务开始执行之间的时间。代码如下。

```
public ScheduledFuture<?> scheduleWithFixedDelay(Runnable command,
                        long initialDelay, long delay, TimeUnit unit) {
    //如果传入的 Runnable 对象和 TimeUnit 为空，则抛出空指针异常
    if (command == null || unit == null)
        throw new NullPointerException();
    //如果任务延时时长小于或等于 0，则抛出非法参数异常
    if (delay <= 0)
        throw new IllegalArgumentException();
    //将 Runnable 对象封装成 ScheduledFutureTask 任务
    //并设置固定的执行周期
    ScheduledFutureTask<Void> sft = new ScheduledFutureTask<Void>(command,
            null,triggerTime(initialDelay, unit), unit.toNanos(-delay));
    //调用 decorateTask 方法，本质上直接返回 ScheduledFutureTask 任务
```

```
        RunnableScheduledFuture<Void> t = decorateTask(command, sft);
        //设置执行的任务
        sft.outerTask = t;
        //执行延时任务
        delayedExecute(t);
        //返回任务
        return t;
}
```

在将 Runnable 对象封装成 ScheduledFutureTask 对象时，设置了执行周期，但是此时设置的执行周期与 scheduleAtFixedRate 方法设置的执行周期不同。此时设置的执行周期规则为：下一次任务执行的开始时间是这一次任务的完成时间加上 delay 时长，单位由 TimeUnit 决定。采用相对固定的延迟来执行定时任务。

细心的读者会发现在 scheduleWithFixedDelay()方法中设置执行周期时，传递的 delay 值为负，代码如下。

```
ScheduledFutureTask<Void> sft = new ScheduledFutureTask<Void>(command,
                        null, triggerTime(initialDelay, unit),
unit.toNanos(-delay));
```

这里的负数表示的是相对固定的延迟。

ScheduledFutureTask 类的 setNextRunTime()方法会在 run()方法执行完任务后被调用，这种方法更能体现 scheduleAtFixedRate()方法和 scheduleWithFixedDelay()方法的区别，代码如下。

```
private void setNextRunTime() {
    //距离下次执行任务的时间
    long p = period;
    //固定频率执行
    //上次执行任务的时间
    //加上任务的执行周期
    if (p > 0)
        time += p;
    //相对固定的延迟
    //使用系统当前时间
    //加上任务的执行周期
    else
        time = triggerTime(-p);
}
```

在 setNextRunTime()方法中，主要通过下次执行任务的时间来判断是采用固定频率还是采用相对固定的延迟。

16.3.5　riggerTime()方法解析

ScheduledThreadPoolExecutor 类提供了两种 triggerTime()方法，用于获取下一次执行任务的具体时间，代码如下。

```
private long triggerTime(long delay, TimeUnit unit) {
    return triggerTime(unit.toNanos((delay < 0) ? 0 : delay));
}

long triggerTime(long delay) {
    return now() +
        ((delay < (Long.MAX_VALUE >> 1)) ? delay : overflowFree(delay));
}
```

这两种 triggerTime() 方法的代码比较简单，都是获取下一次执行任务的具体时间。需要注意的是：delay < (Long.MAX_VALUE >> 1 判断 delay 的值是否小于 Long.MAX_VALUE 的一半，如果是则直接返回 delay，否则需要处理溢出问题。

16.3.6　overflowFree()方法解析

在 triggerTime() 方法中，处理防止溢出的逻辑使用了 overflowFree() 方法，代码如下。

```
private long overflowFree(long delay) {
    //获取队列中的节点
    Delayed head = (Delayed) super.getQueue().peek();
    //如果获取的节点不为空，则进行后续处理
    if (head != null) {
        //从队列节点中获取延迟时间
        long headDelay = head.getDelay(NANOSECONDS);
        if (headDelay < 0 && (delay - headDelay < 0))
            //将 delay 的值设置为 Long.MAX_VALUE + headDelay
            delay = Long.MAX_VALUE + headDelay;
    }
    //返回延迟时间
    return delay;
}
```

overflowFree 方法主要是将队列中所有节点的延迟时间限制在 Long.MAX_VALUE 内，防止在 ScheduledFutureTask 类的 compareTo() 方法中溢出。

ScheduledFutureTask 类的 compareTo() 方法的代码如下。

```
public int compareTo(Delayed other) {
    if (other == this)
        return 0;
    if (other instanceof ScheduledFutureTask) {
        ScheduledFutureTask<?> x = (ScheduledFutureTask<?>)other;
        long diff = time - x.time;
        if (diff < 0)
            return -1;
        else if (diff > 0)
            return 1;
        else if (sequenceNumber < x.sequenceNumber)
            return -1;
        else
```

```
        return 1;
    }
    long diff = getDelay(NANOSECONDS) - other.getDelay(NANOSECONDS);
    return (diff < 0) ? -1 : (diff > 0) ? 1 : 0;
}
```

ScheduledFutureTask 类的 compareTo()方法的主要作用是对各个延迟任务进行排序，距离下次执行时间短的任务排在前面。

注意：有关 ScheduledFutureTask 类详细的代码解析过程，读者可以关注"冰河技术"微信公众号，阅读相关文章。

16.3.7 delayedExecute()方法解析

delayedExecute()方法主要是执行 ScheduledThreadPoolExecutor 的延迟任务，代码如下。

```
private void delayedExecute(RunnableScheduledFuture<?> task) {
    //如果当前线程池已经关闭，则执行拒绝策略
    if (isShutdown())
        reject(task);
    //如果线程池没有关闭
    else {
        //则将任务添加到阻塞队列中
        super.getQueue().add(task);
        //如果当前线程池处于 SHUTDOWN 状态
        //并且在当前线程池状态下不能执行任务
        //并且成功从阻塞队列中移除任务
        if (isShutdown() &&
            !canRunInCurrentRunState(task.isPeriodic()) &&
            remove(task))
            //则取消任务的执行，并且不会中断执行中的任务
            task.cancel(false);
        else
            //调用 ThreadPoolExecutor 类的 ensurePrestart()方法
            ensurePrestart();
    }
}
```

在 delayedExecute()方法内部调用了 canRunInCurrentRunState()方法，代码如下。

```
boolean canRunInCurrentRunState(boolean periodic) {
    return isRunningOrShutdown(periodic ?
continueExistingPeriodicTasksAfterShutdown :
    executeExistingDelayedTasksAfterShutdown);
}
```

canRunInCurrentRunState 方法会判断线程池在当前状态下能否执行任务。

在 delayedExecute()方法内部还调用了 ThreadPoolExecutor 类的 ensurePrestart()方法，代码如下。

```
void ensurePrestart() {
    int wc = workerCountOf(ctl.get());
    if (wc < corePoolSize)
        addWorker(null, true);
    else if (wc == 0)
        addWorker(null, false);
}
```

首先获取当前线程池中线程的数量，如果线程数量少于 corePoolSize，则调用 addWorker() 方法传递 null 和 true；如果线程数量为 0，则调用 addWorker() 方法传递 null 和 false。

注意：关于 ThreadPoolExecutor 类的 addWorker() 方法的详细解析过程，读者可以参见 15.2.3 节的内容。

16.3.8　reExecutePeriodic()方法解析

reExecutePeriodic()方法主要执行定时任务中已经执行过一次的任务，代码如下。

```
void reExecutePeriodic(RunnableScheduledFuture<?> task) {
    //判断线程池在当前状态下能否执行任务
    if (canRunInCurrentRunState(true)) {
        //将任务放入队列
        super.getQueue().add(task);
        //线程池在当前状态下不能执行任务，并且成功移除任务
        if (!canRunInCurrentRunState(true) && remove(task))
            //取消任务
            task.cancel(false);
        else
            //调用 ThreadPoolExecutor 类的 ensurePrestart() 方法
            ensurePrestart();
    }
}
```

reExecutePeriodic()方法与 delayedExecute()方法存在区别。在调用 reExecutePeriodic()方法时，已经执行过一次任务，所以不会触发线程池的拒绝策略。另外，传入 reExecutePeriodic()方法的任务一定是周期性任务。

16.4　定时任务线程池优雅关闭流程

ScheduledThreadPoolExecutor 类提供了优雅关闭线程池的功能，主要通过 ThreadPoolExecutor 类提供的 onShutdown() 方法完成，onShutdown() 方法主要在 ThreadPoolExecutor 类的 shutdown() 方法中调用，而 ThreadPoolExecutor 类提供的 onShutdown() 方法是一个空方法，代码如下。

```
void onShutdown() {
}
```

ThreadPoolExecutor 类的 onShutdown()方法由子类实现，所以 ScheduledThreadPoolExecutor 类覆写了 onShutdown()方法，实现了具体的逻辑。onShutdown()方法的代码如下。

```java
@Override
void onShutdown() {
    BlockingQueue<Runnable> q = super.getQueue();
    boolean keepDelayed = getExecuteExistingDelayedTasksAfterShutdownPolicy();
    boolean keepPeriodic = getContinueExistingPeriodicTasksAfterShutdownPolicy();
    if (!keepDelayed && !keepPeriodic) {
        for (Object e : q.toArray())
            if (e instanceof RunnableScheduledFuture<?>)
                ((RunnableScheduledFuture<?>) e).cancel(false);
        q.clear();
    }else {
        for (Object e : q.toArray()) {
            if (e instanceof RunnableScheduledFuture) {
                RunnableScheduledFuture<?> t = (RunnableScheduledFuture<?>)e;
                if ((t.isPeriodic() ? !keepPeriodic : !keepDelayed) ||
                    t.isCancelled()) {
                    if (q.remove(t))
                        t.cancel(false);
                }
            }
        }
    }
    tryTerminate();
}
```

先判断线程池调用 shutdown()方法后是否继续执行现有的延迟任务和定时任务，如果不执行，则取消任务并清空队列；如果继续执行，则将队列中的任务强转为 RunnableScheduledFuture 对象后从队列中删除并取消任务。此后调用 ThreadPoolExecutor 类的 tryTerminate()方法。

注意：有关 ThreadPoolExecutor 类的 tryTerminate()方法的详细解析过程，读者可以参见 15.3.7 节的内容。

另外，在随书源码中的 mykit-concurrent-threadpool 工程下，给出了手写线程池的完整案例代码。